Compendium of Plant Genomes

Series Editor

Chittaranjan Kole, Raja Ramanna Fellow, Department of Atomic Energy, Government of India, Kalyani, India

Whole-genome sequencing is at the cutting edge of life sciences in the new millennium. Since the first genome sequencing of the model plant Arabidopsis thaliana in 2000, whole genomes of about 70 plant species have been sequenced and genome sequences of several other plants are in the pipeline. Research publications on these genome initiatives are scattered on dedicated web sites and in journals with all too brief descriptions. The individual volumes elucidate the background history of the national and international genome initiatives; public and private partners involved; strategies and genomic resources and tools utilized; enumeration on the sequences and their assembly; repetitive sequences; gene annotation and genome duplication. In addition, synteny with other sequences, comparison of gene families and most importantly potential of the genome sequence information for gene pool characterization and genetic improvement of crop plants are described.

More information about this series at http://www.springer.com/series/11805

Suguru Tsuchimoto
Editor

The Jatropha Genome

Springer

Editor
Suguru Tsuchimoto
Graduate School of Engineering
Osaka University
Suita, Osaka
Japan

ISSN 2199-4781 ISSN 2199-479X (electronic)
Compendium of Plant Genomes
ISBN 978-3-319-49651-1 ISBN 978-3-319-49653-5 (eBook)
DOI 10.1007/978-3-319-49653-5

Library of Congress Control Number: 2017939319

Printed on acid-free paper

This Springer imprint is published by Springer Nature
The registered company is Springer International Publishing AG
The registered company address is: Gewerbestrasse 11, 6330 Cham, Switzerland

This book series is dedicated to
my wife Phullara, and our children Sourav,
and Devleena

Chittaranjan Kole

Preface to the Series

Genome sequencing has emerged as the leading discipline in the plant sciences coinciding with the start of the new century. For much of the twentieth century, plant geneticists were only successful in delineating putative chromosomal location, function, and changes in genes indirectly through the use of a number of 'markers' physically linked to them. These included visible or morphological, cytological, protein, and molecular or DNA markers. Among them, the first DNA marker, the RFLPs, introduced a revolutionary change in plant genetics and breeding in the mid-1980s, mainly because of their infinite number and thus potential to cover maximum chromosomal regions, phenotypic neutrality, absence of epistasis, and codominant nature. An array of other hybridization-based markers PCR-based markers, and markers based on both facilitated construction of genetic linkage maps, mapping of genes controlling simply inherited traits and even gene clusters (QTLs) controlling polygenic traits in a large number of model and crop plants. During this period a number of new mapping populations beyond F_2 were utilized and a number of computer programs were developed for map construction, mapping of genes, and for mapping of polygenic clusters or QTLs. Molecular markers were also used in studies of evolution and phylogenetic relationship, genetic diversity, DNA-fingerprinting and map-based cloning. Markers tightly linked to the genes were used in crop improvement employing the so-called marker-assisted selection. These strategies of molecular genetic mapping and molecular breeding made a spectacular impact during the last one and a half decades of the twentieth century. But still they remained 'indirect' approaches for elucidation and utilization of plant genomes since much of the chromosomes remained unknown and the complete chemical depiction of them was yet to be unraveled.

Physical mapping of genomes was the obvious consequence that facilitated development of the 'genomic resources' including BAC and YAC libraries to develop physical maps in some plant genomes. Subsequently, integrated genetic-physical maps were also developed in many plants. This led to the concept of structural genomics. Later on, emphasis was laid on EST and transcriptome analysis to decipher the function of the active gene sequences leading to another concept defined as functional genomics. The advent of techniques of bacteriophage gene and DNA sequencing in the 1970s was extended to facilitate sequencing of these genomic resources in the last decade of the twentieth century.

As expected, sequencing of chromosomal regions would have led to too much data to store, characterize, and utilize with the-then available computer software could handle. But development of information technology made the life of biologists easier by leading to a swift and sweet marriage of biology and informatics and a new subject was born—bioinformatics.

Thus, evolution of the concepts, strategies and tools of sequencing and bioinformatics reinforced the subject of genomics—structural and functional. Today, genome sequencing has traveled much beyond biology and involves biophysics, biochemistry and bioinformatics!

Thanks to the efforts of both public and private agencies, genome sequencing strategies are evolving very fast, leading to cheaper, quicker and automated techniques right from clone-by-clone and whole-genome shotgun approaches to a succession of second generation sequencing methods. Development of software of different generations facilitated this genome sequencing. At the same time, newer concepts and strategies were emerging to handle sequencing of the complex genomes, particularly the polyploids.

It became a reality to chemically—and so directly—define plant genomes, popularly called whole-genome sequencing or simply genome sequencing.

The history of plant genome sequencing will always cite the sequencing of the genome of the model plant *Arabidopsis thaliana* in 2000 that was followed by sequencing the genome of the crop and model plant rice in 2002. Since then, the number of sequenced genomes of higher plants has been increasing exponentially, mainly due to the development of cheaper and quicker genomic techniques and, most importantly, development of collaborative platforms such as national and international consortia involving partners from public and/or private agencies.

As I write this preface for the first volume of the new series "Compendium of Plant Genomes", a net search tells me that complete or nearly-complete whole-genome sequencing of 45 crop plants, eight crop and model plants, eight model plants, 15 crop progenitors and relatives, and three basal plants are accomplished, the majority of which are in the public domain. This means that we nowadays know many of our model and crop plants chemically, i.e., directly, and we may depict them and utilize them precisely better than ever. Genome sequencing has covered all groups of crop plants. Hence, information on the precise depiction of plant genomes and the scope of their utilization is growing rapidly every day. However, the information is scattered in research articles and review papers in journals and dedicated web pages of the consortia and databases. There is no compilation of plant genomes and the opportunity of using the information in sequence-assisted breeding or further genomic studies. This is the underlying rationale for starting this book series, with each volume dedicated to a particular plant.

Plant genome science has emerged as an important subject in academia, and the present compendium of plant genomes will be highly useful both to students and teaching faculties. Most importantly, research scientists involved in genomics research will have access to systematic deliberations on the plant genomes of their interest. Elucidation of plant genomes is not onlyof interest for the geneticists and breeders, but also for practitioners of an array of plant science disciplines, such as taxonomy, evolution, cytology, physiology, pathology, entomology, nematology, crop production, biochemistry, and obviously bioinformatics. It must be mentioned that information regarding each plant genome is ever-growing. The contents of the volumes of this compendium are therefore focusing on the basic aspects of the genomes and their utility. They include information on the academic and/ or economic importance of the plants, description of their genomes from a molecular genetic and cytogenetic point of view, and the genomic resources developed. Detailed deliberations focus on the background history of the national and international genome initiatives, public and private partners involved, strategies and genomic resources and tools utilized, enumeration on the sequences and their assembly, repetitive sequences, gene annotation, and genome duplication. In addition, synteny with other sequences, comparison of gene families, and, most importantly, potential of the genome sequence information for gene pool characterization through genotyping by sequencing (GBS) and genetic improvement of crop plants have been described. As expected, there is a lot of variation of these topics in the volumes based on the information available on the crop, model, or reference plants.

I must confess that as the series editor it has been a daunting task for me to work on such a huge and broad knowledge base that spans so many diverse plant species. However, pioneering scientists with life-time experience and expertise on the particular crops did excellent jobs editing the respective volumes. I myself have been a small science worker on plant genomes since the mid-1980s and that provided me the opportunity to personally know several stalwarts of plant genomics from all over the globe. Most, if not all, of the volume editors are my longtime friends and colleagues. It has been highly comfortable and enriching for me to work with them on this book series. To be honest, while working on this series I have been and will remain a student first, a science worker second, and a series editor last. And I must express my gratitude to the volume editors and the chapter authors for providing me the opportunity to work with them on this compendium.

I also wish to mention here my thanks and gratitude to the Springer staff, Dr. Christina Eckey and Dr. Jutta Lindenborn in particular, for all their constant and cordial support right from the inception of the idea.

I always had to set aside additional hours to edit books besides my professional and personal commitments—hours I could and should have given to my wife, Phullara, and our kids, Sourav, and Devleena. I must mention that they not only allowed me the freedom to take away those hours from them but also offered their support in the editing job itself. I am really not sure whether my dedication of this compendium to them will suffice to do justice to their sacrifices for the interest of science and the science community.

Kalyani, India Chittaranjan Kole

Preface to the Volume

Jatropha, *Jatropha curcas* L., a drought-tolerant shrub species of the Euphorbiaceae family, has recently attracted people's attention as a promising biofuel crop. Seeds of jatropha contain non-edible oil that is suitable to produce biodiesel or jet fuel. It is expected to contribute to reduction of CO_2 emission by substituting for fossil fuel, without competing with the food crops. However, because it has been mainly used as a hedge plant for many years, its breeding history as a commercial biofuel crop is not long enough. More extensive studies and breeding efforts would be required for more profitable commercial production. Genome sequence information is an important basis for further genetic studies and breeding. Jatropha is a diploid species, and its estimated genome size is 416 Mb in $n = 11$ chromosomes. Its first draft genome sequence was published in 2011, and since then, more than 300 Mb of the jatropha genome sequence has been determined. Genomic and genetic studies, such as researches on QTL, transcriptional factors, flowering genes, and DNA markers have made progress. On the other hand, metabolomic and physiological studies have also been done, which are essential to improve quality and quantity of the oil, and also to reduce or utilize toxic substances of seeds. To improve agronomic traits, methods to generate genetically modified jatropha have been established, and useful transgenic jatropha plants have been developed based on the genetic and physiological studies. Because jatropha is a commercial crop, translation of the scientific achievements to the practical use is important. Latest results of breeding and studies on practical cultivation would be indispensable for it. The origin of jatropha was deciphered to be Mesoamerica, and only limited genotypes in the Mesoamerican population have been brought to other continents. Tracing dispersal routes of jatropha in the cultural aspect would provide useful information of the genetic resources for breeding. This volume includes achievements in these studies and provides an overview to improve our understanding of jatropha from genomic, metabolomic, genetic, and practical points of view. I thank all the authors for their excellent contributions to this volume and hope that all the chapters will help researchers and breeders to study and improve this promising biofuel plant.

Suita, Japan — Suguru Tsuchimoto

Acknowledgement

I am really grateful for the assistance given by Dr. Chittaranjan Kole, as well as the support by staff members of our lab, in publishing this book.

Contents

Part IV Breeding and Application

Part I

Genome and Molecular Analyses

Genome Analysis

1

Hideki Hirakawa and Shusei Sato

Abstract

In order to accelerate basic and applied researches that involve genetic improvement through molecular breeding, comprehensive analyses of genes and the genome of *Jatropha curcas* have been conducted using both conventional and advanced technologies. The first publicly available draft sequence of the genome of *J. curcas* was reported in 2011, and an updated genome sequence, which is 297 Mb long and covers 99% of the euchromatic regions of the genome, was released in 2012. This genome sequence information has served as a reference for transcriptome analysis and the creation of SSR and SNP markers. The latest genome sequence information with longer scaffold length is now available, and most of the scaffolds have been anchored on the genetic linkage map. The genomic sequence and linkage map provide a valuable resource for basic and applied researches on *J. curcas* as well as comparative genomic analysis.

1.1 Introduction

The tremendous advances in DNA-sequencing technologies and associated bioinformatics and computational processes have allowed us to acquire whole genomes of sequence information on plants of agronomic importance in a relatively short period of time. The publication of the highly accurate sequence of the whole genome of *Arabidopsis thaliana* in 2000 (Arabidopsis Genome Initiative 2000) demonstrated the value of sequence information in plant genetics and genomics for the first time. It also paved the way for sequencing of the whole genomes including draft sequences, which are cost-effective but less accurate; subsequently, genome sequence information was published for a number of plant species, including rice (International Rice Genome Sequencing Project 2005), poplar (Tuskan et al. 2006), grapevine (The French–Italian Public Consortium for Grapevine Genome

H. Hirakawa · S. Sato
Kazusa DNA Research Institute, 2-6-7
Kazusa-Kamatari, Kisarazu, Chiba 292-0818, Japan

S. Sato (✉)
Graduate School of Life Sciences, Tohoku
University, 2-1-1 Katahira, Aoba-Ku, Sendai
980-8577, Japan
e-mail: shuseis@ige.tohoku.ac.jp

S. Tsuchimoto (ed.), *The Jatropha Genome*, Compendium of Plant Genomes,
DOI 10.1007/978-3-319-49653-5_1

Characterization 2007), and *Lotus japonicus* (Sato et al. 2008). The emergence of a massively parallel sequencing system, so-called next-generation sequencing (NGS) technology, has significantly accelerated this trend.

Jatropha curcas is an important oilseed crop with great potential for the production of bio-diesel fuel. *J. curcas* is a diploid species having an estimated genome size of 416 Mb (Carvalho et al. 2008) arranged in $n = 11$ chromosomes (Miller and Webster 1966) with an average GC content of 39% (Carvalho et al. 2008). The size and the base composition of the genome make *J. curcas* an attractive target for functional genomics and molecular breeding.

The sequence information accumulated from the cDNA library from seeds at three stages of fruit maturation was reported in 2010 (Gomes et al. 2010). Since then, a large quantity of information on the gene and genome structures has been published. By using a conventional Sanger sequencing method, large-scale expressed sequence tag (EST) data were generated from the cDNA libraries of developing seeds (Natarajan et al. 2010) and of developing and germinating endosperm (Costa et al. 2010). In addition, transcriptome analyses using NGS have been carried out for leaf and callus (Sato et al. 2011), a mixture of roots, mature leaves, flowers, developing seeds, and embryos (Natarajan and Parani 2011), and three different stages of developing seeds (King et al. 2011). EST-derived simple sequence repeat (SSR) markers have been developed by pyrosequencing the mRNAs (Yadav et al. 2011).

Regarding the genome sequence, a bioenergy crop company, SG Biofuels Inc., announced their completion of the *J. curcas* genome sequencing to 100× coverage using the SOLiD system. However, this information has not been made available to the research community. A draft sequence of the whole genome of *J. curcas* was therefore determined by using a combination of conventional and NGS technologies (Sato et al. 2011), and this sequence was further upgraded by the addition of new data in 2012 (Hirakawa et al. 2012). Because this information was widely available through international databases (DDBJ/GenBank/EMBL) and Web databases (http://www.kazusa.or.jp/jatropha), the draft genome sequence of *J. curcas* served as a resource for acceleration of basic and applied research. Recently, a more complete genome assembly with a longer scaffold length was generated by applying paired-end (PE) and mate-pair (MP) sequence reads of a range of different insert sizes (Wu et al. 2015). Since most of the scaffolds were anchored on a genetic linkage map, the genome sequence and linkage map provide a rich resource of genetic information for breeding and genetic improvement.

In this chapter, the current status of large-scale analyses of the genes and genome of *J. curcas* will be reviewed, and their characteristics and examples of their application will be summarized.

1.2 Genome and Transcriptome Sequence Analyses

1.2.1 Genome Assembly

The first published genome sequence of *J. curcas* was reconstructed by using sequence data obtained by a capillary sequencer (Sanger protocol) and next-generation sequencers. The cultivar name applied to the sequencing was Palawan, which originated in the Philippines. The sequence data obtained using an ABI 3730xl sequencer (Applied Biosystems, USA) consisted of 1,025,000 reads of shotgun sequences and 5300 bacterial artificial chromosome (BAC) end sequences, PE reads with insert size 3 kb and cDNA sequences obtained by using a Genome Sequencer (GS) FLX System (Roche Diagnostics, USA), and PE reads obtained by using an Illumina GAII (Genome Analyzer) sequencer (Illumina Inc., USA) with four kinds of read lengths against shotgun libraries of different insert sizes: 36/36 bases, 50/31 bases, 51/51 bases, and 76/76 bases (read 1/ read 2). The draft genome sequence JAT_r3.0 was determined in 2010 (Sato et al. 2011) and was updated to JAT_r4.5 in 2012 (Hirakawa et al. 2012). Illumina GAII reads with read lengths of 36/36 bases and 50/31 bases were applied in JAT_r3.0, and those with read lengths of 51/51 bases and 76/76 bases were added in JAT_r4.5.

The strategy used to construct the draft genome sequence JAT_r3.0 can be summarized as follows. The PCAP.rep program (Huang et al. 2006) was applied for the assembly of the Sanger reads trimmed by Figaro and Lucy (White et al. 2008) programs, and then, Sanger PCAP contigs and Sanger singlet reads were generated. The MIRA program (Chevreux et al. 1999) was applied for the assembly of the GS FLX reads whose quality was improved by Pyrobayes (Quinlan et al. 2008) and removal of artifacts by Replicate filter (Gomez-Alvarez et al. 2009), and then, 454 MIRA contigs and 454 singlet reads were generated. The contigs and singlets generated by Sanger and GS FLX reads were merged by PCAP.rep. The hybrid contigs were scaffolded by using BAC end sequences using PCAP.rep, and the resultant contigs were further scaffolded by RNA-Seq and cDNA sequences obtained using a GS FLX System and ESTs obtained from NCBI's dbEST (http://www.ncbi.nlm.nih.gov/dbEST/) using GS reference mapper (Roche Diagnostics, USA) and the BLAT program (Kent 2002), followed by generation of the scaffolds, contigs, and singlets. To improve the sequence quality, Illumina GAII PE reads with insert sizes of 36/36 bases and 50/31 bases were assembled by the Velvet program (Zerbino and Birney 2008), and the resulting contigs were mapped onto the scaffolds, contigs, and singlets to correct the InDel errors. Finally, the draft genome sequence (JAT_r3.0) was generated. As a result, the number of contigs and singlets was 150,417, and their total length was 285,858,490 bases. The assembly results are summarized in Table 1.1.

To update the genome information, Illumina GAII PE reads with read lengths of 51/51 bases and 76/76 bases were additionally sequenced. The Illumina GA reads were subjected to quality filtering and adaptor trimming, and the resulting reads were assembled by SOAPdenovo (Li et al. 2010). The scaffolds were merged with the scaffolds of JAT_r3.0 by using PCAP.rep. The merged sequences were named JAT_r4.0. After the assembly, the scaffolds were subjected to the following alterations for analysis: 1) The N regions with sequences corresponding to JAT_r3.0 were replaced; 2) the sequences that largely duplicated (homologous with above 80% length and 100% identity) were removed; 3) the contaminated sequences with hits against bacterial genome sequences in NCBI, mitochondrial genome sequences of *A. thaliana* (accession number: NC_001284.2), chloroplast

Table 1.1 Stastitics of *J. curcas* genome assemblies

Assembly version	JAT_r3.0	JAT_r4.5	JatCur_1.0
Cultivar name	Palawan	Palawan	GZQX0401 (scaffold)
Number of sequences	150,417	39,277	6023
Total length (bases)	286,159,324	297,661,187	318,363,324
Average length (bases)	1902	7579	52,858
Max. length (bases)	29,746	277,264	5,289,327
Min. length (bases)	42	500	200
N50 length (bases)	3832	15,950	746,835
A	17,165,800	97,945,514	87,495,699
T	16,088,187	97,679,966	87,460,404
G	12,723,655	49,793,449	43,659,798
C	10,524,293	49,839,024	43,700,994
N	11,888	2,403,234	56,046,369
Total (ATGC)	56,501,935	295,257,953	262,316,895
G + C% (ATGC)	41.1	33.7	33.3

genome sequences of *J. curcas* (accession number: FJ695500), and sequences in NCBI's UniVec database (http://www.ncbi.nlm.nih.gov/tools/vecscreen/univec/) were removed; and 4) the sequences <500 bp length were removed. The resulting scaffolds were named JAT_r4.5. As a result, the total length of the genome sequence was 221,111,674 bases and consisted of 107,255 scaffolds. The assembly statistics of JAT_r3.0 and JAT_4.5 are compared in Table 1.1.

1.2.2 Gene Finding

In JAT_r3.0, gene prediction and gene modeling were performed by a combination of the methods based on ab initio gene prediction and similarity searches against the plant proteome data sets. For the ab initio method, GeneMark.hmm (Lukashin and Borodovsky 1998) and GENESCAN (Burge and Karlin 1997) with training set of *A. thaliana* were used. For modeling of exon-intron structures, NetGene2 (Hebsgaard et al. 1996) and SplicePredictor (Brendel and Kleffe 1998) programs were applied. In JAT_r4.5, the genes were newly predicted by the Augustus (Stanke et al. 2004) program with the training set of *A. thaliana*. By comparing the genes in JAT_r3.0 and JAT_r4.5, 7124 genes in JAT_r4.5 were appended to the gene set of JAT_r3.0, and 30,203 genes were consequently assigned to JAT_r4.5. The number of genes with a complete structure in JAT_r4.5 was increased about threefold from that in JAT_r3.0. The statistics of the genes predicted in JAT_r3.0 and JAT_r4.5 are summarized in Table 1.2. Similarity searches were performed against the TrEMBL database (http://www.ebi.ac.uk/uniprot) to assign annotation to the predicted genes in JAT_r4.5. A total of 22,088 of the 30,203 (73.1%) genes had significant similarity against the genes in the database. With regard to the gene structure, 25,433 of 30,203 genes (84.2%) had complete structures, and 4770 genes (15.8%) had partial structures. The average length of the 30,203 genes was 1058 bases and that of the 25,433 complete genes was 1109 bases. The number of the genes with GO annotation was 25,954 (85.9%) and that with similarity against the TrEMBL database was 22,088 (73.1%). The domain searches were performed against InterPro database (Mitchell et al. 2015) with the InterProScan program (Jones et al. 2014). To classify the predicted genes into functional categories, GO slim (http://geneontology.org/page/go-slim-and-subset-guide) analysis was performed based on the results of the InterProScan. As a result, the distributions of the GO slim categories in *J. curcas* were similar to the results for *A. thaliana*. The genes were assigned to the metabolic pathways by means of BLASTP (Altschul et al. 1997) searches against the KEGG GENES database (Ogata et al. 1999). By comparing the genes on the pathways among *J. curcas*, *Ricinus communis*, and *A. thaliana*, the genes of *J. curcas* were solely mapped onto 19 pathways, including "galactose metabolism" in carbohydrate metabolism, "biosynthesis of steroids" in lipid metabolism, "glycosylphosphatidylinositol (GPI)–anchor biosynthesis" in glycan biosynthesis and metabolism, and "retinol metabolism" in metabolism of cofactors and vitamins.

Table 1.2 Stastitics of *J. curcas* CDSs

Assembly version	TCs	JAT_r3.0	JAT_r4.5	JatCur_1.0
Cultivar name	–	Palawan	Palawan	GZQX0401
Number of sequences	19,454	58,720	57,437	27,172
Total length (bases)	18,845,949	56,513,823	60,346,622	29,873,889
Average length (bases)	969	962	1051	1099
Max. length (bases)	15,641	8107	16,878	16,764
Min. length (bases)	50	20	22	153
G + C%	41.3	41.1	43.3	43.0

1.2.3 Transcript Sequence Assembly

The transcripts of leaf (534,137 reads; DRX000446) and callus (456,913 reads; DRX000447) in *J. curcas* cv. Palawan were sequenced by using the GS FLX System at the Kazusa DNA Research Institute (KDRI). The SRA database of NCBI (http://www.ncbi.nlm.nih.gov/sra) includes the transcript reads from a mixture of five major tissue libraries (383,937 reads; SRX035761), one seed library (195,692 reads; SRX011411), and one leaf library (2210 reads; SRX020243). In addition, 46,842 EST sequences were registered in dbEST. To construct the tentative consensus sequences (TCs) of *J. curcas*, the transcript reads and EST sequences described above were assembled by GS De Novo Assembler v2.6 software (Roche Diagnostics, USA). As a result, 19,454 TCs were generated, and the average length and GC content were 969 bases and 41.3%, respectively (Table 1.2). A total of 19,435 of 19,454 TCs (99.9%) were assigned to the scaffolds of JAT_r4.5, which means that the coverage of gene space in JAT_r4.5 was increased from that in JAT_r3.0 (95%).

1.2.4 Sequencing of the Organelle Genome

The chloroplast genome sequence has been determined (accession number: NC_012224.1 (Asif et al. 2010)). The total genome size was 163.9 kb, the GC content was 35.4%, and the numbers of genes, proteins, rRNAs, and tRNAs were 130, 84, 8, and 37, respectively (Asif et al. 2010). The total genome size and genome arrangement were similar to those of other plant species. According to the phylogenetic tree of 81 proteins in chloroplasts among 64 taxa, *J. curcas* is close to *Manihot esculanta*, which belongs to the same family as Euphorbiaceae. In the chloroplasts of *J. curcas*, the gene functions of *infA* (translation initiation factor 1) and *rps16* (small subunit ribosomal protein) were lost, and inverted repeats were found in the genic region in the *rpl2* (proteins of large ribosomal subunit) gene. The mitochondrial genome sequence of *J. curcas* has not been determined yet.

1.2.5 Genome Sequencing Projects

The draft genome sequence of *J. curcas* cv. Palawan was determined in 2010 and 2012 at KDRI (Sato et al. 2011; Hirakawa et al. 2012). The genome sequence of JAT_r3.0 had a total length of 297.7 Mb and was comprised of 39,277 scaffolds (BioProject accession number: PRJDA52543). The GC content of the genome sequence was 33.8% (Sato et al. 2011). The genome sequence of JAT_r3.0 has been updated to JAT_r4.5 by adding Illumina reads (Hirakawa et al. 2012). In 2015, the genome sequence of *J. curcas* cv. GZQX0401 was determined at the Chinese Academy of Sciences (BioProject accession number: PRJNA63485). The sequence platforms used were Illumina GAII and HiSeq with seven insert sizes: 200 b, 500 b, 800 b, 2 kb, 5 kb, 10 kb, and 20 kb. The current sequence version is called JatCur_1.0. The genome sequence had a total length of 318.4 Mb and was comprised of 6023 scaffolds. The GC content of the genome sequence was 33.3%, and 27,172 genes were predicted from the genome sequence. In the study, 1208 markers were also developed, and 81.7% of them were mapped onto the 11 pseudomolecules (Wu et al. 2015). The statistics of the assemblies of the studies of genome sequencing of *J. curcas* are compared in Table 1.1. The total length of JAT_r4.5 (39,277 scaffolds, 297,661,187 bases) was close to that of JatCur_1.0 (6023 scaffolds, 318,363,324 bases), while the N50 length of JatCur_1.0 (746,835 bases) was much longer than that of JAT_r4.5 (15,950 bases). The GC contents were not largely different (JAT_r4.5: 33.7%; JatCur_1.0: 33.3%). The scaffolds of JAT_r4.5 and JatCur_1.0 were compared by the LAST program (Kiełbasa et al. 2011) with a score of 281 (corresponding to E value = 1E−100); 36,853 of 39,277 scaffolds of JAT_r4.5 were homologous to scaffolds of JatCur_1.0, while 2424 of 39,277 scaffolds of JAT_r4.5 were non-homologous to scaffolds of JatCur_1.0, whose total length was 2,010,462 bases (0.68%).

To assess the completeness of the gene space, the core eukaryotic genes (CEGs), which are highly conserved among six organisms in eukaryotes (*Homo sapiens*, *Drosophila melanogaster*, *A. thaliana*, *Caenorhabditis elegans*, *Saccharomyces cerevisiae*, and *Schizosaccharomyces pombe*), were employed for the scaffolds of JAT_r3.0, JAT_r4.5, and JatCur_1.0. The 248 CEGs were applied to the analysis by using CEGMA (Parra et al. 2007), and the numbers of the genes having complete and partial structures were 110 (44.4%) and 206 (83.1%) in JAT_r3.0, 218 (87.9%) and 242 (97.6%) in JAT_r4.5, 228 (91.9%), and 244 (98.4%) in JatCur_1.0. Based on these results, the completeness of the core genes was similar between JAT_r4.5 and JatCur_1.0.

In addition, to assess the conservation of the CDSs between Palawan and GZQX0401, the CDSs of JAT_r4.5 and JatCur_1.0 were, respectively, mapped onto each other's scaffolds by using the BLAT program. A total of 52,839 of 57,437 (92.0%) CDSs of JAT_r4.5 were mapped onto the scaffolds of JatCur_1.0, whose total, average, and N50 lengths were 59,366,347, 1124, and 1521 bases. In addition, 4598 of 57,437 (8.0%) CDSs of JAT_r4.5 were not mapped, and the total, average, and N50 lengths of these unmapped CDSs were 980,275, 213, and 201 bases. From these results, most of the CDSs of JAT_r4.5 that were frequently found to be non-homologous to JatCur_1.0 were short, and they might have been specific to cultivar Palawan (JAT_r4.5), or the scaffolds that encoded the corresponding short CDSs have not been sequenced yet in cultivar GZQX0401 (JatCur_1.0). On the other hand, 26,339 of 27,172 (96.9%) CDSs of JatCur_1.0 were mapped onto the scaffolds of JAT_r4.5, whose total, average, and N50 lengths were 29,694,129, 1127, and 1479 bases. In addition, 833 of 27,172 (3.1%) CDSs of JatCur_1.0 were not mapped, and their total, average, and N50 lengths were 179,760, 216, and 195 bases. As a result, the CDSs that were frequently found to be non-homologous to JAT_r4.5 were also short, and they might be specific in cultivar GZQX0401 (JatCur_1.0), or the scaffolds encoded the corresponding short CDSs were not sequenced yet in cultivar Palawan (JAT_r4.5).

1.2.6 miRNA Analysis

According to the similarity searches using BLASTN for 2502 miRNAs, 24 new potential miRNAs were identified from 46,862 ESTs and 1569 GSS sequences (Vishwakarma and Jadeja 2013). The transcription factors for the regulation of cell growth and development, signaling, and metabolism in oil synthesis were found in the 78 potential genes categorized into three miRNA families (Vishwakarma and Jadeja 2013). Transcription sequences (RNA-Seq) were obtained from the three libraries from immature, intermediate, and mature seeds, and 180 conserved miRNAs, 41 precursor miRNAs (pre-miRNAs), and 16 novel pre-miRNAs were identified (Galli et al. 2014). The 426 and 356 sequences were obtained from the two small RNA libraries of *J. curcas* from leaves and seeds ranging from 18 to 26 bases. The small RNA sequences were searched against the genome sequences of *A. thaliana*, rice, grape, poplar, *Euphorbia genistoides*, and *J. curcas*, and 52 miRNAs were identified based on the secondary structure prediction. The target genes of the miRNAs were examined by expression patterns. Among them, 10 miRNAs highly expressed in fruits and seeds were thought to be potentially related to the development or synthesis of fatty acids in seed. One of the identified miRNAs, JcumiR004, was considered to be related to the development and formation of fruit, and this miRNA includes the four target genes for modulating significant oil composition (Wang et al. 2012).

1.2.7 EST Analysis

Currently, (March 2015), 46,865 ESTs are registered in the dbEST. There have been several studies using ESTs to search for SSR markers. In one such study, 43,349 ESTs were used to find

SSR markers, and these ESTs were assembled by the CAP3 program (Huang and Madan 1999). SSRs were extracted by the MIcroSAtellite (MISA) program (Thiel et al. 2003) by setting the repeat length and the number of repeats at 10, 6, 5, 5, 5, and 5 for mono-, di-, tri-, tetra-, penta-, and hexa-nucleotides. As a result, 6108 SSRs were found in the 5175 assembled sequences (Laosatit et al. 2013). The 42,477 ESTs were used for finding SSR markers, and they were assembled to 12,358 contigs and 5730 singlets by the CAP3 program. By searching the microsatellite, 3557 motifs were found from 7.91% of the unigenes. According to the GO analysis, 931 unigenes were related to fatty acid or lipid metabolism pathways, and those over-represented were classified into the GO terms, "Fatty acid metabolic process" and "Fatty acid biosynthetic process" (Grover et al. 2014).

A total of 383,918 reads were obtained using GS FLX Titanium technology (Roche Diagnostics, USA) from the five tissue types: roots, mature leaves, flowers, developing seeds, and embryos of *J. curcas*, and the obtained reads were assembled by GS De Novo Assembler v2.5.2. As a result, 17,457 contigs and 54,002 singlets were generated. The 25,333 ESTs obtained by a capillary sequencer (Sanger protocol) and the assembled contigs from the previous study (Sato et al. 2011) were assembled together, resulting in the generation of 14,327 contigs. Consequently, the sequences of 28,794 unigenes were obtained. Among them, 2320 unigenes were included in the pathways related to oil biosynthesis (Natarajan and Parani 2011).

1.3 Repetitive Sequences of the *J. curcas* Genome

1.3.1 Simple Sequence Repeats (SSRs)

SSRs are tandemly arranged repeats of short DNA motifs (1–6 bp in length) which tend to exhibit length polymorphism due to the variation in the number of repeats. Because of their abundance, codominant nature, and high reproductive nature, SSRs are used as valuable genetic markers applicable for various aspects of molecular genetic studies, such as assessment of genetic diversity and marker-assisted breeding.

In the *J. curcas* genome sequence, more than 40,000 di-, tri-, and tetra-nucleotide SSRs equal to or greater than 15 bp were identified, and thus, the frequency of occurrence of these SSRs can be estimated to be one SSR in every 7.0 kb. The di-, tri-, and tetra-nucleotide SSRs accounted for 46, 34, and 19% of the identified SSRs, respectively. The SSR patterns that appeared frequently were (AT)n, (AAT)n, and (AAAT)n, each representing 71% of di-, 60% of tri-, and 58% of tetra-nucleotide repeat units, respectively. The trinucleotide SSRs, particularly (AAG)n and (AGC)n, were preferentially found in exons. (AT)n, (AG)n, and (AAT)n were enriched in the 5′ and 3′ untranslated regions (UTRs), and (AC)n frequently occurred in introns (Sato et al. 2011; Hirakawa et al. 2012). A similar distribution of SSR proportion has been reported in analysis of the cassava genome (Vásquez and López 2014). Studies are currently underway to create SSR markers and analyze the germplasm diversity, creation of linkage maps, and so on by applying the genome sequence and EST sequences available in the public databases.

1.3.2 Transposable Elements

A search of the *J. curcas* genome sequences using the repeat sequence finding program RECON (Bao and Eddy 2002) unraveled the occurrence of a variety of repeat elements, including the class I and class II transposable element (TE) subfamilies, as well as some elements that were difficult to classify into known subfamilies. The composition of these repeat sequences was analyzed with the RepeatMasker program (http://repeatmasker.org). In total, the identified repetitive sequences accounted for 31.5% of the *J. curcas* genome sequences (Hirakawa et al. 2012). The most abundant repeat category was class I TE (25.2%), in which gypsy-type (16.1%) and copia-type (7.1%) LTR retroelements constituted the major components.

The extensive analysis of gypsy-type retrotransposons in the *J. curcas* genome was reported recently (Alipour et al. 2014). By combining a molecular genetics technique with computer-based mining, this work identified four new gypsy-type retrotransposons, named Jg1-4, which were then grouped into two lineages. Along with the RECON and RepeatMasker analyses, these four retrotransposons bore hit numbers after a BLAST search against *J. curcas* genome sequences. The results of fluorescence in situ hybridization (FISH) revealed that these gypsy-type retrotransposons were accumulated at the pericentromeric region of *Jatropha* metaphase chromosome spreads.

Genome-wide analysis and cytogenetic mapping of copia-type retrotransposons have also been reported (Alipour et al. 2013). The PCR fragments amplified using the degenerated primers for the reverse transcriptase domain of copia-type retroelements were classified into families, which were then grouped into three lineages corresponding to TAR, Angela, and Ale in copia-type families of other plant species. The insertion-site preferences of each family were surveyed by using the *J. curcas* genome sequence, and five of them, which existed in the gene-rich regions, were found to bear potential as appealing candidates for the development of DNA marker systems. The results of FISH analyses on mitotic chromosomes confirmed that the retrotransposons of these families were dispersed throughout all chromosomes with clustering dominantly in the distal part of chromosome arms. These findings indicated that copia-type retrotransposons can be considered a powerful system of molecular markers for elucidating the evolutional and genetic relationships among its various accessions.

With respect to class II transposable elements, significant numbers of miniature inverted-repeat transposable elements (MITEs) were identified in the *J. curcas* genome sequence. MITEs are prevalent in eukaryotic species, including plants, and are believed to be deletion derivatives of DNA transposons. Like autonomous DNA transposons, MITEs usually have terminal inverted repeats (TIRs), flanked by short direct repeats. MITEs are often located in gene-rich euchromatic regions and are associated with genes. The extensive de novo identification of MITEs from 41 plant species using the computer programs MITE Digger, MITE-Hunter, and/or RSPB (Repetitive Sequence with Precise Boundaries) revealed 18,975 elements that were classified into 17 MITE families in the *J. curcas* genome (Chen et al. 2014). Together, these MITE elements covered a total of 4.8 Mb regions of the *J. curcas* genome. The detailed information on these MITE elements is available from the P-MITE database (http://pmite.hzau.edu.cn).

1.4 Protein-Coding Genes

1.4.1 Gene Family Analysis

Based on the obtained *J. curcas* genome sequence information, studies of the gene families with important biological functions have been carried out.

WRKY genes are transcription factors related to development and stress responses, and they have one or two WRKYGQK sequences followed by a zinc finger motif to bind to the W box of target genes. By an intensive search using the WRKY domains of *Arabidopsis* WRKY proteins as query sequences against the *J. curcas* genome (JAT_r3.0) and their own transcriptome sequences, Xiong et al. (2013) identified a total of 58 WRKY genes in the *J. curcas* genome. Comparative analyses of *J. curcas* WRKY genes with *Arabidopsis*, rice, and castor bean WRKY genes revealed that evolutionarily recent WRKY paralogs in the *J. curcas* genome probably arose from early gene duplication events. Among the 58 *J. curcas* WRKY genes, 47 genes were responsible to at least one abiotic stress (Xiong et al. 2013).

In regard to biotic stress, the identification and characterization of disease-resistance genes, including nucleotide-binding site-Leucine-reach repeat (NBS-LRR), is an important approach to accelerate the process of genetic improvement of disease resistance. In the *J. curcas* genome, a total of 91 NBS-LRR genes have been identified by mapping Pfam domains as well as mapping publicly available NBS-LRR mRNA sequences (Sood et al. 2014). By comparing the NBS-LRR genes identified in the *J. curcas* genome sequence with the NBS-LRR genes identified in the castor bean genome, several genes unique to one or common to both genomes were identified (Sood et al. 2014). Along with the 122 defense response-related transcription factors identified in the *J. curcas* genome sequence (Sood et al. 2014), the genome-wide information on NBS-LRR-resistance genes should provide novel insights about the molecular basis of disease-resistance phenotypes.

1.4.2 Tandem Gene Duplication

Tandem gene duplication is one of the major mechanisms of duplication in eukaryotes. In the first and second draft genome sequence reports, the numbers of tandemly arrayed genes, such as NBS-LRR disease-resistance proteins, were described as a characteristic feature of the *J. curcas* genome (Sato et al. 2011; Hirakawa et al. 2012). In the latest genome with longer scaffold length, 3839 tandem gene duplications among 1442 loci in the *J. curcas* genome were confirmed (Wu et al. 2015). The longest tandem gene array consists of 15 cytochrome P450 (CYP) genes. Proteins with kinase active site domains, disease-resistance gene products, UDP-glucuronosyl/-glucosyltransferase domain-containing proteins, and short-chain dehydrogenase/reductases were among the proteins, and domains found most frequently in tandemly repeated genes in the *J. curcas* genome (Wu et al. 2015).

1.5 Comparative Analysis of the *J. curcas* Genome

The obtained *J. curcas* genome sequence information paved the way for proceeding with comparative genome analysis at several levels.

1.5.1 Comparative Analysis Within *J. curcas* Accessions

Understanding the genetic diversity within the *Jatropha* germplasm is critically important in establishing breeding strategies and designing breeding programs. Genetic variations in natural and cultured populations around the world have been studied by using random amplified polymorphic DNA (RAPD), inter-simple sequence repeat (ISSR), and amplified fragment length polymorphism (AFLP) before the genome sequence became available (Sun et al. 2008; Basha and Sujatha 2009; Sudheer-Pamidimarri et al. 2009; Shen et al. 2010). The focus of the study of genetic diversity is shifting to sequence information-based codominant markers, such as SSR and single nucleotide polymorphism (SNP) markers. When the first draft sequence of the *J. curcas* genome was reported, a total of 88 SSR markers were generated from the obtained genome sequence information. These markers were evaluated using the 12 lines of *J. curcas* corrected from meso-America, Africa, and Asia. The number of alleles per locus ranged from one to four with a mean value of 1.31. The markers showed no polymorphisms; those detecting a single allele were most frequent. Polymorphic information content (PIC) values ranged from 0 to 0.45 with a mean value of 0.06 (Sato et al. 2008). The large number of tested markers detecting no polymorphisms and the low mean value of the PIC indicated that the genetic diversity in *Jatropha* lines is generally narrow. Among the tested accessions, the three lines derived from meso-America regions (Guatemala1, Guatemala2, and Mexico2b) are genetically distinct from the other lines derived from Asia and Africa, whereas no significant difference was observed between the Asian and African lines (Sato et al. 2008). A similar trend of slightly higher genetic diversity in varieties from meso-America regions was observed in the analysis of additional SSR markers designed from genome sequence information (Raposo et al. 2014). By using *J. curcas* accessions from Guatemala as a genotyping population, the average number of alleles per locus was increased to 6.9 (among 18 polymorphic loci detected by the analysis of twenty SSR markers). PIC values ranged from 0.114 to 0.886 with a mean value of 0.627. SNP markers have also been used for the genetic diversity analysis. Sandoval et al. (2014) applied both SNP markers and SSR markers to their genetic diversity analysis of 70 accessions from around the world. They found that the variance components based on SNP and SSR marker data were similar, and the genetic diversity in the accessions from meso-America regions was higher than that in accessions from Africa, Asia, and South America, which was consistent with the previous studies (Sandoval et al. 2014). Extensive analysis of SNPs in the *J. curcas* accessions was also carried out based on whole-transcriptome re-sequencing of pooled samples of 12 *J. curcas* accessions (Silva-Junior et al. 2011). By mapping of 66.5 million reads on the unigene set of the draft genome sequence, 60.8 million reads were aligned on the 28,110 unigenes with an average coverage of 152×. Only 18,225 SNPs were detected (with no minor allele frequency or flanking sequence filtering), and this very low number indicated the low level of sequence polymorphism in the transcribed regions of *J. curcas* accessions. Another attempt to create large-scale SNP markers was carried out by using high-throughput sequencing on a GS FLX Systems of the complexity-reduced genomic DNA of 61 accessions. From the 871 assembled contigs contacting 26,940 sequences, a total of 2482 informative SNPs were identified with an average frequency of one SNP per 100 bp. Genotyping of selective SNPs among the 148 global collections of *J. curcas* accessions revealed that a narrow level of genetic diversity existed among the indigenous genotypes as compared to the exotic genotypes of *J. curcas* (Gupta et al. 2012). These SNP markers could be very useful in large-scale marker application in molecular breeding.

1.5.2 Comparative Analysis in Euphorbiaceae

Euphorbiaceae, to which *Jatropha* belongs, is a complex family consisting of 229 genera and 6511 accepted species in "The Plant List" database (http://www.theplantlist.org/browse/A/Euphorbiaceae/). *Euphorbiaceae* includes several economically important species in addition to *Jatropha*, such as an important oilseed crop, castor bean (*R. communis*); an essential food source and bioenergy crop, cassava (*M. esculenta*); and a resource for natural rubber, rubber tree (*Hevea brasiliensis*). The availability of several Euphorbiaceae genomes allows us to take advantage of a comparative genomic approach, which is a powerful tool for gaining insights into genome structure and evolution.

1.5.2.1 Castor Bean Genome Information

Castor bean is a tropical perennial shrub of African origin which is now cultivated in many tropical and subtropical regions around the world. It can be self- and cross-pollinated, and studies performed using castor bean collections from around the world have revealed the relatively low genetic diversity among the castor bean germplasm (Allan et al. 2008; Foster et al. 2010).

Castor bean was the first Euphorbiaceae species whose draft genome sequence information was reported. By assembling ~2.1 million high-quality sequence reads from plasmid and fosmid libraries generated by the Sanger sequence method, a draft genome sequence of castor bean ($2n = 20$) composed of 25,828 scaffolds with an N50 of 496.5 kb was obtained (Chan et al. 2010). When 3500 scaffolds larger than 2 kb were considered, the castor bean genome assembly spanned 325.5 Mb, which was consistent with the genome size estimated by flow cytometry (Arumuganathan and Earle 1991). With the aid of 52,165 EST sequences accumulated from five cDNA libraries, a total of 31,237 potential protein-coding genes were predicted on the obtained genome sequences.

When the first draft genome sequence of *J. curcas* was obtained, the syntenic relation between castor bean and *J. curcas* was analyzed (Sato et al. 2011). A significant level of syntenic relation was detected on 53% of the scaffolds with five or more genes. The ratio of contigs with a syntenic relation between castor bean and *J. curcas* was increased to 88% when the upgraded *J. curcas* genome sequence (JAT_r4.5) was obtained due to the increases in the length of the contigs and in the number of genes with complete prediction (Hirakawa et al. 2012). The syntenic relation was further confirmed by anchoring the castor bean draft genome sequences to the *Jatropha* genetic map (Wu et al. 2015). A total of 410 scaffolds covering 54% of the total scaffold sequences of the castor bean draft genome were anchored to the *Jatropha* genetic map, and 320 well-conserved synteny blocks containing more than 10,000 *J. curcas* genes collinear to castor bean genes on the anchored scaffolds were defined. These syntenic relations could serve as a useful resource for the identification of genes by homology and for information exchange within Euphorbiaceae species.

1.5.2.2 Cassava Genome Information

Cassava is a perennial woody shrub with edible tuberous roots and is grown throughout tropical and subtropical regions of the world, especially Africa, Asia, and the Americas. Its large, starchy roots and edible leaves provide the primary staple food for over 800 million people worldwide (Ceballos et al. 2010). The high starch content (20–40%) makes cassava a desirable energy source both for human consumption and for industrial biofuel applications (Balat and Balat 2009; Schmitz and Kavallari 2009).

The draft genome sequence of cassava was obtained by a whole-genome shotgun strategy using the GS FLX System. A total of 22.4 billion bp of raw sequence data were accumulated, which was ~29 times the estimated size of the genome (770 Mb) (Awoleye et al. 1994), and these reads were assembled into 12,977 scaffolds that spanned a total of 533 Mb, on which 96% of the cassava EST sequences in GenBank were mapped (Prochnik et al. 2012). With the aid of 1.4 million additional EST reads from leaf and root libraries by the GS FLX System, a total of 30,666 genes and 3485 alternative splice forms were predicted on the obtained cassava draft genome sequences. Recently, the draft genome sequences of additional cassava genotypes, W14 (*M. esculenta* ssp. *flabellifolia*), a wild subspecies that shows low storage root yield and low root starch, and KU50, a variety commonly cultivated in Southeast Asia that has sixfold to eightfold higher storage root yield potential, were analyzed by using an integrated assembly strategy combining the sequence reads obtained from Illumina and GS FLX System (Wang et al. 2014). Comparative genomics analysis revealed a considerable amount of genome diversity (SNPs and InDels) in W14 and KU50 when compared with the reference cassava genome (a partially inbred line, AM560). The results of a comparative analysis of the predicted genes from the genomes of W14, KU50, and AM560 revealed that 1584 were unique to W14 or lost in KU50 and AM560, whereas another 1678 genes were specific to the cultivated varieties. The availability of high-quality draft genome sequences for these three genotypes will contribute to the genetic improvement of cassava through a better understanding of its biology.

Detailed information on the reference draft genome sequences of castor bean and cassava can be accessed through the plant comparative genomics portal Phytozome (http://phytozome.jgi.doe.gov). This impressive body of genomic resources will establish the basis of an information exchange for Euphorbiaceae species.

1.5.3 Comparative Mapping

A genetic linkage map is the essential framework for genome-wide identification of associations between DNA markers and traits (Doerge 2002). In addition, a genetic linkage map assists in anchoring the assembled genome sequences to create pseudomolecules and provides a solid basis for comparative mapping. Comparative mapping, in turn, should help to establish the syntenic relationships against the model plant species, such as *A. thaliana*, which would be beneficial for applying the information generated in the model plant species.

A first-generation linkage map was constructed using a mapping population containing two backcross populations consisting of 93 progenies. The F1 populations were produced by an interspecies cross between two accessions of *J. curcas* as female parents and a single *J. integerrima* individual as male parent, and the two BC1F1 populations obtained by backcrossing *J. curcas* parents to each F1 progeny were used as mapping populations (Wang et al. 2011). The mapping populations were genotyped with two types of codominant DNA marker SSRs on the genome sequences and SNPs on the EST sequences. A total of 506 markers were mapped onto 11 linkage groups, out of which 216 were SSR markers and 290 were SNP markers. The total length of the obtained *Jatropha* genetic map was 1440.9 cM, with an average marker spacing of 2.8 cM (range 1.2–4.3 cM in each linkage group). By a similarity search of the 222 ESTs containing SSR and SNP markers mapped on the linkage map against reference genomes, it was revealed that 96.8, 91.0, and 77.5% of *J. curcas* ESTs were homologous to their counterparts in castor bean, poplar, and *Arabidopsis*, respectively. A comparative map between *Jatropha* and *Arabidopsis* using 192 orthologous markers elucidated 38 syntenic blocks and revealed that small linkage blocks were well conserved, but often shuffled (Wang et al. 2011). The linkage map and the data of comparative mapping provide a solid basis for quantitative trait locus (QTL) mapping of agronomic traits, marker-assisted breeding, and cloning genes responsible for phenotypic variations.

Another high-density linkage map was constructed using the BC1 population of an interspecific cross (*J. curcas* × *J. integerrima*) with 1208 SNP, InDel, and SSR markers (Wu et al. 2015). The total genetic distance covered by this linkage map was 1655.8 cM, with an average marker density of 2.1 cM for unique loci. This linkage map was applied for anchoring the scaffolds of the latest genome assembly. A total of 480 scaffolds covering 261.8 Mb (~81.7% of the total scaffold sequences) were anchored to the map to produce eleven pseudochromosomes. This high-density genetic linkage map and anchored genomic sequences provide a valuable resource for fundamental and applied researches as well as for evolutionary and comparative genomics analysis (Wu et al. 2015).

The first intraspecific *J. curcas* map was constructed from four F2 mapping populations created from parental lines displaying differences in a range of traits (King et al. 2013). Genotyping assays were performed using both SNP and SSR markers, and the linkage maps for each mapping population were built individually. Then, the genotype information was merged to create an integrated linkage map containing 502 codominant markers, distributed over 11 linkage groups, with a mean marker density of 1.8 cM per unique locus. By using one of the four mapping populations created from G33 (toxic seed) × G43 (non-toxic seed), linkage analysis for loci controlling phorbol ester biosynthesis was carried out. QTL analysis revealed that a single locus at 41 cM on linkage group 8 was associated with phorbol ester biosynthesis. By anchoring the draft genome sequence contigs of *J. curcas* and castor bean, additional markers were created on the target region to fine map this mutation within 2.3 cM. This first intraspecific *J. curcas* map therefore provides a framework for the dissection of agronomic traits in *J. curcas* and the development of improved varieties by marker-assisted breeding.

1.6 Databases

1.6.1 *J. curcas* Genome Database

The genomic information obtained by the genome assembly has been published, and the data can be downloaded from the database (http://www.kazusa.or.jp/jatropha/, Sato et al. 2011). The current version of the genomic information in the Web database is JAT_r4.5 (Hirakawa et al.

2012). In addition, the genomic information of JAT_r3.0 is still available in the database by clicking the banner for JAT_r3.0. Users can browse the general features of the assembly and search against the CDSs, protein sequences, and keywords in the annotation data for both JAT_r3.0 and JAT_r4.5.

1.6.2 TreeTFDB

TreeTFDB (http://treetfdb.bmep.riken.jp/index.pl, Mochida et al. 2013) is a database of transcription factors for six tree crop species: *J. curcas* (JAT_r3.0), papaya (*Carica papaya*; Phytozome v8.0), cassava (*M. esculenta*; Phytozome v8.0; http://www.phytozome.net), poplar (*Populus trichocarpa*; Phytozome v8.0), castor bean (*R. communis*; Phytozome v8.0), and grapevine (*Vitis vinifera*; Phytozome v8.0). In the current version (1.10), annotated TF models were registered for *J. curcas* (1481 TFs), papaya (1552 TFs), cassava (2638 TFs), poplar (3110 TFs), castor bean (1512 TFs), and grape (1493 TFs). In this database, TFs determined in JAT_r3.0 were registered. The genomic locations of the TFs can be browsed on GBrowse (Stein et al. 2002). In the database, about 60 kinds of TF families have been registered, and the sequences of cDNAs, proteins, 500, 1000, and 3000 bp upstream regions, and domains, and cis-motifs can be obtained from the download site. Blast searches can be conducted for CDSs of the six tree species and cDNAs of *A. thaliana*.

1.6.3 TropiTree

TropiTree (http://ics.hutton.ac.uk/tropiTree/) is a database for the assembled transcripts (unigene) for 24 tropical tree species (*Acacia senegal*, *Acrocarpus fraxinifolius*, *Adansonia digitata*, *Albizia lebbeck*, *Calliandra calothyrsus*, *Diospyros mespiliformis*, *Enterolobium cyclocarpum*, *Faidherbia albida*, *Gliricidia sepium*, *Jacaranda mimosifolia*, *J. curcas*, *Leucaena diversifolia*, *Leucaena leucocephala*, *Moringa stenopetala*, *Prunus Africana*, *Samanea saman*, *Senna siamea*, *Sesbania macrantha*, *Sesbania sesban*, *Tephrosia candida*, *Tipuana tipu*, *Warburgia ugandensis*, *Ziziphus mauritiana*). In addition, EST-SSRs detected from the ESTs and primers for these EST-SSRs can be browsed. Users also can search the genes by homology searches for the query sequences and keyword searches. For *J. curcas*, 13,252 unigenes and 1118 primer sets for EST-SSRs (di-, tri-, and tetra-nucleotide) can be browsed. The unigenes of *J. curcas* are built from the published RNA-Seq data (accession number: ERS399695).

1.6.4 KaPPA-View4

KaPPA-View4 (http://kpv.kazusa.or.jp) is a database for the representation of transcriptome and metabolome data on pathway maps (Sakurai et al. 2011). KaPPA-View4 has two systems, KaPPA-View Classic and KaPPA-View KEGG. The former is based on the traditional KaPPA-View map of *A. thaliana*, which contains information on several genome-sequenced plants, i.e., *A. thaliana*, rice, tomato, *L. japonicus*, soybean, barley, poplar, wheat, grape, and maize (10 plant species). The latter system includes other organisms, e.g., human, mouse, rat, *C. elegans*, *D. melanogaster*, *A. thaliana*, rice, poplar, castor bean, sorghum, grape, maize, *Physcomitrella patens*, *Escherichia coli*, and budding yeast (15 species). The genomic information of *J. curcas* (JAT_r3.0) has been registered on the KaPPA-View4 Jatropha site (http://kpv2.kazusa.or.jp/kpv4-jat/). A total of 8058 of 40,929 genes were mapped onto the KEGG pathway maps.

1.6.5 PGDBj (Plant Genome DataBase Japan)

PGDBj (http://pgdbj.jp) is a portal site integrating the databases related to plant omics studies (Asamizu et al. 2014). The information related to DNA markers, QTL, and plant diseases has been collected from the literature for 80 plant species by manual curation. PGDBj has links to the other databases, such as SABRE2 DB (http://sabre.epd.brc.riken.jp/SABRE2.html, ref_db4) for the resource (clone sequences) of 15 plant species (*A. thaliana*, *Thellungiella halophila*, *Brassica rapa*, *Nicotiana tabacum*, *Solanum lycopersicum*, *L. japonicus*, *Glycine max*, *M. esculenta*, *Striga hermonthica*, *Ipomoea nil*, *Brachypodium distachyon*, *Triticum aestivum*, *Hordeum vulgare*, *Populus nigra*, *Physcomitrella patens*) and KNApSAcK (http://kanaya.naist.jp/KNApSAcK/) for the species–metabolite relationship. Currently, 50,899 metabolites and 111,199 metabolite–species pairs have been registered. Users can search information of metabolites by inputting information such as the organism name, metabolite name, molecular weight, and molecular formula. For oilseed crops, *J. curcas* and *R. communis* have been registered. In *J. curcas*, 563 SSR markers, 3 SCAR markers, and 290 SNP markers have been registered.

1.7 Conclusion

The availability of whole-genome sequences has significantly altered the approach to understanding *J. curcas*. The publicly available draft genome information has led to large-scale genomic analyses, such as transcriptome analyses and analyses for molecular marker creation. Along with the genome sequence information on castor bean and cassava, the data from comparative genome analysis will serve as a basis to transfer knowledge among the Euphorbiaceae species and should ultimately elucidate the genetic systems in Euphorbiaceae and accelerate the breeding process. The new and fine draft genome information on the additional accession will further enhance the DNA marker creation to cover the entire genome and facilitate a higher level of genomic studies, such as genome-wide association study. By putting more effort into the enrichment of accessions with accurate phenotype information, the advanced genome-based strategies will be able to contribute in a significant way to the breeding programs for improving *Jatropha*.

References

Alipour A, Tsuchimoto S, Sakai H, Ohmido N, Fukui K (2013) Structural characterization of copia-type retrotransposons leads to insights into the marker development in a biofuel crop, *Jatropha curcas* L. Biotechnol Biofuels 6:129

Alipour A, Cartagena JA, Tsuchimoto S, Sakai H, Ohmido N, Fukui K (2014) Identification and characterization of novel Gypsy-type retrotransposons in a biodiesel crop, *Jatropha curcas* L. Plant Mol Biol Rep 32:923–930

Allan G, Williams A, Rabinowicz PD, Chan AP, Ravel J, Keim P (2008) Worldwide genotyping of castor bean germplasm (*Ricinus communis* L.) using AFLPs and SSRs. Genet Resour Crop Evol 55:365–378

Altschul SF, Madden TL, Schäffer AA (1997) Gapped BLAST and PSI-BLAST: a new generation of protein database search programs. Nucleic Acids Res 25:3389–3402

Arabidopsis Genome Initiative (2000) Analysis of the genome sequence of the flowering plant *Arabidopsis thaliana*. Nature 408:796–815

Arumuganathan K, Earle ED (1991) Nuclear DNA content of some important plant species. Plant Mol Biol Rep 9:208–218

Asamizu E, Ichihara H, Nakaya A, Nakamura Y, Hirakawa H, Ishii T, Tamura T, Fukami-Kobayashi K, Nakajima Y, Tabata S (2014) Plant Genome DataBase Japan (PGDBj): a portal website for the integration of plant genome-related databases. Plant Cell Physiol 55:e8

Asif MH, Mantri SS, Sharma A, Srivastava A, Trivedi I, Gupta P, Mohanty CS, Sawant SV, Tuli R (2010) Complete sequence and organisation of the *Jatropha curcas* (Euphorbiaceae) chloroplast genome. Tree Genet Genomes 6:941–952

Awoleye F, Duren M, Dolezel J, Novak FJ (1994) Nuclear DNA content and in vitro induced somatic polyploidization cassava (*Manihot esculenta* Crantz) breeding. Euphytica 76:195–202

Balat M, Balat H (2009) Recent trends in global production and utilization of bio-ethanol fuel. Appl Energ 86:2273–2282

Basha SD, Sujatha M (2009) Genetic analysis of *Jatropha* species andinterspecific hybrids of *Jatropha curcas* using nuclear and organelle specific markers. Euphytica 168:197–214

Bao Z, Eddy SR (2002) Automated de novo identification of repeat sequence families in sequenced genomes. Genome Res 12:1269–1276

Brendel V, Kleffe J (1998) Prediction of locally optimal splice sites in plant pre-mRNA with applications to gene identification in *Arabidopsis thaliana* genomic DNA. Nucleic Acids Res 26:4748–4757

Burge C, Karlin S (1997) Prediction of complete gene structures in human genomic. DNA. J Mol Biol 268:78–94

Carvalho CR, Clarindoa WR, Praça MM, Araújoa FS, Carels N (2008) Genome size, base composition and karyotype of *Jatropha curcas* L., an important biofuel plant. Plant Sci 174:613–617
Ceballos H, Okogbenin E, Pérez JC, López-Valle LAB, Debouck D (2010) Cassava. In: Bradshaw JE (ed) Root and tuber crops, handbook of plant breeding, vol 7. Springer, New York, pp 53–96
Chan AP, Crabtree J, Zhao Q, Lorenzi H, Orvis J, Puiu D, Melake-Berhan A, Jones KM, Redman J, Chen G, Cahoon EB, Gedil M, Stanke M, Haas BJ, Wortman JR, Fraser-Liggett CM, Ravel J, Rabinowicz PD (2010) Draft genome sequence of the oilseed species *Ricinus communis*. Nat Biotechnol 28:951–956
Chen J, Hu Q, Zhang Y, Lu C, Kuang H (2014) P-MITE: a database for plant miniature inverted-repeat transposable elements. Nucleic Acids Res 42(Database issue):D1176–D1181
Chevreux B, Wetter T, Suhai S (1999) Genome sequence assembly using trace signals and additional sequence information, In: Computer Science and Biology: Proceedings of the German Conference on Bioinformatics (GCB), 99, pp 45–56
Costa GG, Cardoso KC, Del Bem LE, Lima AC, Cunha MA, de Campos-Leite L, Vicentini R, Papes F, Moreira RC, Yunes JA, Campos FA, Da Silva MJ (2010) Transcriptome analysis of the oil-rich seed of the bioenergy crop *Jatropha curcas* L. BMC Genomics 11:462
Doerge RW (2002) Mapping and analysis of quantitative trait loci in experimental populations. Nat Rev Genet 3:43–52
Foster JT, Allan GJ, Chan AP, Rabinowicz PD, Ravel J, Jackson PJ, Keim P (2010) Single nucleotide polymorphisms for assessing genetic diversity in castor bean (*Ricinus communis*). BMC Plant Biol 10:13
Galli V, Guzman F, de Oliveira LF, Loss-Morais G, Körbes AP, Silva SD, Margis-Pinheiro MM, Margis R (2014) Identifying microRNAs and transcript targets in Jatropha seeds. PLoS One 9:e83727
Gomes KA, Almeida TC, Gesteira AS, Lôbo IP, Guimarães ACR, de Miranda AB, Van Sluys MA, da Cruz RS, Cascardo JCM, Carels N (2010) ESTs from seeds to assist the selective breeding of *Jatropha curcas* L. for oil and active compounds. Genom Insights 3:29–56
Gomez-Alvarez V, Teal TK, Schmidt TM (2009) Systematic artifacts in metagenomes from complex microbial communities. ISME J 3:7–1314
Grover A, Kumari M, Singh S, Rathode SS, Gupta SM, Pandey P, Gilotra S, Kumar D, Arif M, Ahmed Z (2014) Analysis of *Jatropha curcas* transcriptome for oil enhancement and genic markers. Physiol Mol Biol Plants 20:139–142
Gupta P, Idris A, Mantri S, Asif MH, Yadav HK, Roy JK, Tuli R, Mohanty CS, Sawant SV (2012) Discovery and use of single nucleotide polymorphic (SNP) markers in *Jatropha curcas* L. Mol Breed 30:1325–1335
Hebsgaard SM, Korning PG, Tolstrup N, Engelbrecht J, Rouzé P, Brunak S (1996) Splice site prediction in *Arabidopsis thaliana* pre-mRNA by combining local and global sequence information. Nucleic Acids Res 24:3439–3452
Hirakawa H, Tsuchimoto S, Sakai H, Nakayama S, Fujishiro T, Kishida Y, Kohara M, Watanabe A, Yamada M, Aizu T, Toyoda A, Fujiyama A, Tabata S, Fukui K, Sato S (2012) Upgraded genomic information of *Jatropha curcas* L. Plant Biotechnol 29:123–130
Huang X, Madan A (1999) CAP3: A DNA sequence assembly program. Genome Res 9:77–868
Huang X, Yang SP, Chinwalla AT, Hillier LW, Minx P, Mardis ER, Wilson RK (2006) Application of a superword array in genome assembly. Nucleic Acids Res 34:201–205
International Rice Genome Sequencing Project (2005) The map-based sequence of the rice genome. Nature 436:793–800
Jones P, Binns D, Chang HY, Fraser M, Li W, McAnulla C, McWilliam H, Maslen J, Mitchell A, Nuka G, Pesseat S, Quinn AF, Sangrador-Vegas A, Scheremetjew M, Yong SY, Lopez R, Hunter S (2014) InterProScan 5: genome-scale protein function classification. Bioinformatics 30:40–1236
Kent WJ (2002) BLAT—the BLAST-like alignment tool. Genome Res 12:656–664
Kiełbasa SM, Wan R, Sato K, Horton P, Frith MC (2011) Adaptive seeds tame genomic sequence comparison. Genome Res 21:93–487
King AJ, Li Y, Graham IA (2011) Profiling the developing *Jatropha curcas* L. seed transcriptome by pyrosequencing. Bioenerg Res 4:211–221
King AJ, Montes LR, Clarke JG, Affleck J, Li Y, Witsenboer H, van der Vossen E, van der Linde P, Tripathi Y, Tavares E, Shukla P, Rajasekaran T, van Loo EN, Graham IA (2013) Linkage mapping in the oilseed crop *Jatropha curcas* L. reveals a locus controlling the biosynthesis of phorbol esters which cause seed toxicity. Plant Biotechnol J 11:986–996
Laosatit K, Tanya P, Saensuk C, Srinives P (2013) Development and characterization of EST-SSR markers from *Jatropha curcas* EST database and their transferability across jatropha-related species/genus. Biologia 68:41–47
Li R, Zhu H, Ruan J, Qian W, Fang X, Shi Z, Li Y, Li S, Shan G, Kristiansen K, Li S, Yang H, Wang J, Wang J (2010) De novo assembly of human genomes with massively parallel short read sequencing. Genome Res 20:265–272
Lukashin A, Borodovsky M (1998) GeneMark.hmm: new solutions for gene finding. Nucleic Acids Res 26:1107–1115
Miller KI, Webster GL (1966) Chromosome numbers in the *Euphorbiaceae*. Brittonia 18:372–379
Mitchell A, Chang HY, Daugherty L, Fraser M, Hunter S, Lopez R, McAnulla C, McMenamin C, Nuka G, Pesseat S, Sangrador-Vegas A, Scheremetjew M, Rato C, Yong SY, Bateman A, Punta M, Attwood TK, Sigrist CJ, Redaschi N, Rivoire C, Xenarios I, Kahn D, Guyot D, Bork P, Letunic I,

Gough J, Oates M, Haft D, Huang H, Natale DA, Wu CH, Orengo C, Sillitoe I, Mi H, Thomas PD, Finn RD (2015) The InterPro protein families database: the classification resource after 15 years. Nucleic Acids Res 43(Database issue):D213–D221

Mochida K, Yoshida T, Sakurai T, Yamaguchi-Shinozaki K, Shinozaki K, Tran LS (2013) TreeTFDB: an integrative database of the transcription factors from six economically important tree crops for functional predictions and comparative and functional genomics. DNA Res 20:151–162

Natarajan P, Kanagasabapathy D, Gunadayalan G, Panchalingam J, Shree N, Sugantham PA, Singh KK, Madasamy P (2010) Gene discovery from *Jatropha curcas* by sequencing of ESTs from normalized and full-length enriched cDNA library from developing seeds. BMC Genom 11:606. doi:10.1186/1471-2164-11-606

Natarajan P, Parani M (2011) De novo assembly and transcriptome analysis of five major tissues of *Jatropha curcas* L. using GS FLX titanium platform of 454 pyrosequencing. BMC Genom 12:191

Ogata H, Goto S, Sato K, Fujibuchi W, Bono H, Kanehisa M (1999) KEGG: Kyoto encyclopedia of genes and genomes. Nucleic Acids Res 27:29–34

Osorio LRM, Salvador AFT, Jongschaap RE, Perez CAA, Sandoval JEB, Trindate LM, Visser RG, van Loo EN (2014) High level of molecular and phenotypic biodiversity in *Jatropha curcas* from Central America compared to Africa, Asia and South America. BMC Plant Biol 14:77

Parra G, Bradnam K, Korf I (2007) CEGMA: a pipeline to accurately annotate core genes in eukaryotic genomes. Bioinformatics 23:7–1061

Prochnik S, Marri PR, Desany B, Rabinowicz PD, Kodira C, Mohiuddin M, Rodriguez F, Fauquet C, Tohme J, Harkins T, Rokhsar DS, Rounsley S (2012) The cassava genome: current progress, future directions. Trop Plant Biol 5:88–94

Quinlan AR, Stewart DA, Strömberg MP, Marth GT (2008) Pyrobayes: an improved base caller for SNP discovery in pyrosequences. Nat Methods 5:81–179

Raposo RS, Souza IG, Veloso ME, Kobayashi AK, Laviola BG, Diniz FM (2014) Development of novel simple sequence repeat markers from a genomic sequence survey database and their application for diversity assessment in *Jatropha curcas* germplasm from Guatemala. Genet Mol Res 13:106–6099

Sakurai N, Ara T, Ogata Y, Sano R, Ohno T, Sugiyama K, Hiruta A, Yamazaki K, Yano K, Aoki K, Aharoni A, Hamada K, Yokoyama K, Kawamura S, Otsuka H, Tokimatsu T, Kanehisa M, Suzuki H, Saito K, Shibata D (2011) KaPPA-View4: a metabolic pathway database for representation and analysis of correlation networks of gene co-expression and metabolite co-accumulation and omics data. Nucleic Acids Res 39(Database issue):D677–D684

Sato S, Nakamura Y, Kaneko T, Asamizu E, Kato T, Nakao M, Sasamoto S, Watanabe A, Ono A, Kawashima K, Fujishiro T, Katoh M, Kohara M, Kishida Y, Minami C, Nakayama S, Nakazaki N, Shimizu Y, Shinpo S, Takahashi C, Wada T, Yamada M, Ohmido N, Hayashi M, Fukui K, Baba T, Nakamichi T, Mori H, Tabata S (2008) Genome structure of the legume, *Lotus japonicus*. DNA Res 15:227–239

Sato S, Hirakawa H, Isobe S, Fukai E, Watanabe A, Kato M, Kawashima K, Minami C, Muraki A, Nakazaki N, Takahashi C, Nakayama S, Kishida Y, Kohara M, Yamada M, Tsuruoka H, Sasamoto S, Tabata S, Aizu T, Toyoda A, Shin-i T, Minakuchi Y, Kohara Y, Fujiyama A, Tsuchimoto S, Kajiyama S, Makigano E, Ohmido N, Shibagaki N, Cartagena JA, Wada N, Kohinata T, Atefeh A, Yuasa S, Matsunaga S, Fukui K (2011) Sequence analysis of the genome of an oil-bearing tree, *Jatropha curcas* L. DNA Res 18:65–76

Schmitz PM, Kavallari A (2009) Crop plants versus energy plants-on the international food crisis. Bioorg Med Chem 17:4020–4021

Silva-Junior O, Rosado T, Laviola B, Pappas M, Pappas G, Grattapaglia D (2011) Genome-wide SNP discovery from a pooled sample of accessions of the biofuel plant *Jatropha curcas* based on whole-transcriptome Illumina resequencing. BMC Proc 5:P57

Shen J-L, Xiang-nan J, Hui-qun N, Pei-guang S, Shi-hui N, Xiao-yang C (2010) AFLP analysis of genetic diversity of *Jatropha curcas* grown in Hainan, China. Trees 24:455–462

Sood A, Jaiswal V, Chanumolu SK, Malhotra N, Pal T, Chauhan RS (2014) Mining whole genomes and transcriptomes of Jatropha (*Jatropha curcas*) and Castor bean (*Ricinus communis*) for NBS-LRR genes and defense response associated transcription factors. Mol Biol Rep 41:7683–7695

Stanke M, Steinkamp R, Waack S, Morgenstern B (2004) AUGUSTUS: a web server for gene finding in eukaryotes. Nucleic Acids Res 32(Web Server issue):W309–W312

Stein LD, Mungall C, Shu S, Caudy M, Mangone M, Day A, Nickerson E, Stajich JE, Harris TW, Arva A, Lewis S (2002) The generic genome browser: a building block for a model organism system database. Genome Res 12:610–1599

Sudheer-Pamidimarri DV, Singh S, Mastan SG, Patel J, Reddy MP (2009) Molecular characterization and identification of markers for toxic and non-toxic varieties of *Jatropha curcas* L. using RAPD, AFLP and SSR markers. Mol Biol Rep 36:1357–1364

Sun QB, Li LF, Li Y, Wu GJ, Ge XJ (2008) SSR and AFLP markers reveal low genetic diversity in the biofuel plant *Jatropha curcas* in China. Crop Sci 48:1865–1871

The French-Italian Public Consortium for Grapevine Genome Characterization (2007) The grapevine genome sequence suggests ancestral hexaploidization in major angiosperm phyla. Nature 449:463–467

Thiel T, Michalek W, Varshney RK, Graner A (2003) Exploiting EST databases for the development and characterization of gene-derived SSR-markers in barley (*Hordeum vulgare* L.). Theor Appl Genet 106:22–411

Tuskan GA, Difazio S, Jansson S, Bohlmann J, Grigoriev I et al (2006) The genome of black cottonwood, *Populus trichocarpa* (Torr & Gray). Science 15:1596–1604

Vásquez A, López C (2014) *In silico* genome comparison and distribution analysis of simple sequences repeats in cassava. Int J Genom 2014:471461

Vishwakarma NP, Jadeja VJ (2013) Identification of miRNA encoded by *Jatropha curcas* from EST and GSS. Plant Signal Behav 8:e23152

Wang CM, Liu P, Yi C, Gu K, Sun F, Li L, Lo LC, Liu X, Feng F, Lin G, Cao S, Hong Y, Yin Z, Yue GH (2011) A first generation microsatellite- and SNP-based linkage map of Jatropha. PLoS One 6:e23632

Wang CM, Liu P, Sun F, Li L, Liu P, Ye J, Yue GH (2012) Isolation and identification of miRNAs in *Jatropha curcas*. Int J Biol Sci 8:418–429

Wang W, Feng B, Xiao J, Xia Z, Zhou X, Li P et al (2014) Cassava genome from a wild ancestor to cultivated varieties. Nat Commun 5:5110

White JR, Roberts M, Yorke JA, Pop M (2008) Figaro: a novel statistical method for vector sequence removal. Bioinformatics 24:7–462

Wu P, Zhou C, Cheng S, Wu Z, Lu W, Han J, Chen Y, Chen Y, Ni P, Wang Y, Xu X, Huang Y, Song C, Wang Z, Shi N, Zhang X, Fang X, Yang Q, Jiang H, Chen Y, Li M, Wang Y, Chen F, Wang J, Wu G (2015) Integrated genome sequence and linkage map of physic nut (*Jatropha curcas* L.), a biodiesel plant. Plant J. 81:810–821

Xiong W, Xu X, Zhang L, Wu P, Chen Y, Li M, Jiang H, Wu G (2013) Genome-wide analysis of the WRKY gene family in physic nut (*Jatropha curcas* L.). Gene 524:32–124

Yadav HK, Ranjan A, Asif MH, Mantri S, Sawant SV et al (2011) EST-derived SSR markers in *Jatropha curcas* L.: development, characterization, polymorphism, and transferability across the species/genera. Tree Genet Genom 7:207–219

Zerbino DR, Birney E (2008) Velvet: algorithms for de novo short read assembly using de Bruijn graphs. Genome Res 18:821–829

Linkage Mapping and QTL Analysis

2

Jian Ye, Chunming Wang and Genhua Yue

Abstract

Jatropha curcas L. is a potential plant species for biodiesel production. However, its seed yield is too low for profitable production of biodiesel. To improve the productivity, genetic improvement through breeding is essential. Marker-assisted selection (MAS) has the huge potential to accelerate genetic improvement. We mapped 506 markers (216 microsatellites and 290 SNPs from ESTs) onto 11 linkage groups. However, genetic analysis of the yield traits has not been done in jatropha. Quantitative trait loci (QTL) mapping was conducted to identify genetic factors controlling growth and seed yield in jatropha. We identified a total of 28 QTLs for 11 growth and seed traits using a population of 296 backcrossing jatropha trees. QTL and expression QTL analyses were applied to identify genetic factors that are relevant to seed oil traits in jatropha. We screened key genes in auxin pathway including ARF and IAA families and downstream effectors to identify candidate genes controlling seed size in jatropha. *JcARF19* was mapped in the major QTL region and significantly associated with seed length.

2.1 Introduction

The increasing demand for diesel coupled with the continuous rise in price of crude oil has forced us to search for an ecologically sustainable alternative energy source (Fairless 2007). Biodiesel from vegetable oil emerged as a viable alternative, particularly nonedible vegetable oil. *Jatropha curcas* L., also called Physic nut, is a perennial poisonous shrub belonging to the Euphorbiaceous family (Heller 1996). This plant originating from Mexico and central America has been spread to other tropical and subtropical countries and is mainly grown in Asia and Africa. People claimed that *J. curcas* is resistant to a high degree of drought and does not directly compete with food crops (Openshaw 2000; Fairless 2007). The generation interval of *J. curcas* was only 6 months in tropical regions and its genome consists of 11 chromosome pairs (Carvalho et al. 2008). Its seeds contain ca. 30% oil

J. Ye · C. Wang (✉) · G. Yue (✉)
Temasek Life Sciences Laboratory, National, University of Singapore, 1 Research Link, 117604 Singapore, Singapore
e-mail: wangchm@njau.edu.cn

G. Yue
e-mail: genhua@tll.org.sg

S. Tsuchimoto (ed.), *The Jatropha Genome*, Compendium of Plant Genomes,
DOI 10.1007/978-3-319-49653-5_2

that is usable in a standard diesel engine (Shah et al. 2005). Therefore, *J. curcas* is regarded as a promising candidate for producing biodiesel (Fairless 2007; Jain and Sharma 2010).

However, *J. curcas* has been an uncultivated wild-species, and until recently, little (Wang et al. 2011b) is known about its genetics and genome. Although some genes involved in pathways for biosynthesis of fatty acids and lipids have been cloned recently (Zhang et al. 2007; Carvalho et al. 2008; Gu et al. 2011), methods for gene silencing have been established (Ye et al. 2009), and the genome was sequenced (Sato et al. 2011). *J. curcas* has never been domesticated and bred for producing oil in large scale. Detailed selective breeding has not been extensively carried out yet. To make the production of *J. curcas* profitable and sustainable, genetic improvement of oil yield and quality, as well as diseases and pests resistance is demanded.

Molecular breeding, also called marker-assisted selection (MAS), refers to the procedure of the use of DNA markers which are tightly linked to traits to assist phenotypic selection. In comparison with the traditional breeding, molecular breeding possesses several advantages such as selection at seedling stage, no influence of environment, and selection of preferred homozygotes, thus accelerating the genetic improvement. With the rapid development of next-generation sequencing (NGS) technologies, it is now easy to detect and characterize a large number of DNA markers using NGS and polymerase chain reaction (PCR). Genomic resources such as molecular markers, linkage maps, ESTs and genome sequences, as well as mapped quantitative trait loci (QTL) for important traits, are powerful tools to speed up genetic improvement for oil yield and quality through MAS.

2.2 Microsatellite- and SNP-Based Linkage Map of Jatropha

A genetic linkage map is an essential tool in molecular breeding for genetic improvement (Guimarães 2007). Such a map will facilitate genome mapping, genetic dissection of QTL and positional cloning of important genes and provide a scaffold for assembling physical maps and an indispensable tool for functional genomics (Harushima et al. 1998; Meksem and Kahl 2006). Linkage maps have been constructed in a number of economically important species (Harushima et al. 1998; Hayashi et al. 2001; Hwang et al. 2009; Li et al. 2009; Ren et al. 2009; Wang et al. 2011a; Xia et al. 2010) using different markers, such as protein polymorphisms, RAPD, RFLP, AFLP, microsatellites (also termed as single sequence repeats: SSRs), and SNPs in reference families. Although a rather large variety of markers exists, each possessing its own advantages and drawbacks, microsatellites are the most preferred markers for linkage mapping (Goldstein and Schlotterer 1999). Recently, single nucleotide polymorphism (SNP) markers have attracted significant attention in creating dense genetic linkage maps and genome-wide association studies (Wang et al. 2005) because SNPs are the most abundant class of polymorphisms in genomes and can be genotyped cost-effectively (Rafalski 2002).

With the elucidation of genes involved in many biochemical pathways, information generated in the model species, *Arabidopsis thaliana*, holds enormous potential for application in breeding of other crops (Panjabi et al. 2008). Establishment of syntonic relationships between *A. thaliana* and other crops through comparative mapping would be beneficial for the identification of candidate genes contributing to agronomic traits from corresponding regions in *A. thaliana* and also serve as a resource to generate more markers for fine mapping in syntonic regions of other crops (Brown 2007).

2.2.1 Identifying and Genotyping of Microsatellite and SNP Markers

In order to facilitate genetic improvement of jatropha, we constructed a first-generation linkage map comprising 216 microsatellites and 290 SNPs and spanning 1440.9 cM. We generated a comparative map between *J. curcas* and *A. thaliana* containing 192 marker loci derived from expressed sequence tags (ESTs). This linkage map represents the first linkage map of *Jatropha* and could provide an indispensable and powerful tool for QTL analysis, gene mapping, and marker-assisted selection (MAS) in breeding.

A total of 245 microsatellite loci were informative in the mapping panel 296 SNPs were identified which showed polymorphism among the parental lines of the mapping population. To our surprise, in the two *J. curcas* individuals (i.e., PZM16 and ZS-2), all 245 microsatellites were homozygous, while in the *J. integerrima* individual (S001), 65% of these microsatellites were heterozygous, and the remaining microsatellites were homozygous, but the genotypes of the *J. integerrima* individual were different from these of the two *J. curcas* individuals. Therefore, two F_1 hybrid individuals (i.e., CI7041 and CI7018) were all heterozygous at all 245 microsatellite loci. The B_1CF_1 families were highly informative for constructing a linkage map.

2.2.2 Linkage Map

For each of the 245 microsatellite and 296 SNP markers, genotypes were obtained for all 93 offspring. Genotype data of the markers were passed forward into linkage analysis. As a result, 216 microsatellites and 290 SNPs were mapped into 11 linkage groups. The remaining 29 informative microsatellites and 6 informative SNPs were not mapped to the linkage map. The length of the 11 linkage groups ranged from 84.9 to 187.5 cM (Table 2.1; Figs. 2.1 and 2.2). The linkage map covered 1440.9 cM with average marker spacing of 2.8 cM ranging from 1.2 to 4.3 cM. The number of markers on each linkage group ranged from 22 for linkage group 5 (LG 5) to 36 for LG 6 (see Table 2.1; Figs. 2.1, 2.2). LG 11 possessed the highest density of markers with marker spacing shorter than 2 cM, while LGs 1, 4, and 8 had the relatively lower density of markers with marker spacing shorter than 4 cM. The 506 DNA markers were located in 324 discrete positions on 11 linkage groups. Therefore, the average space of discrete positions was 4.4 cM ranging from 2.7 for LG 11 to 5.7 cM for LG 1. The most spaces between two discrete positions of markers were smaller than 20 cM on the linkage map. However, there were still a few spaces where the distances between two discrete positions of markers were larger than 20 cM, such as eSNP0815-Jatr650-Jcuint220 on LG 1, eSNP0081-SNP005-1 on LG 2, eSNP0271-SNP032-1 on LG 3, Jcuint302-eSNP0193 and Jatr600-eSNP0086 on LG 4, and eSNP0114-eSNP0045 and eSNP0195-eSNP0979 on LG 8. Most of the large spaces were located in the end of the linkage groups. Most of the segregation distortion markers were clustered on LGs 1,

Table 2.1 Number of markers and genetic length for each linkage group of the jatropha linkage map (Wang et al. 2011b)

LG	No. of markers	Unique loci	Length (cM)	cM/marker	cM/unique locus
1	44	33	187.5	4.3	5.7
2	44	33	164.8	3.7	5.0
3	48	28	151.7	3.2	5.4
4	40	30	171.9	4.3	5.7
5	49	22	116.7	2.4	5.3
6	64	36	127.7	2.0	3.5
7	36	31	82.9	2.3	2.7
8	38	29	163.5	4.3	5.6
9	36	24	87.9	2.4	3.7
10	34	27	101.4	3.0	3.8
11	73	31	84.9	1.2	2.7
Total	506	324	1440.9	2.8	4.4

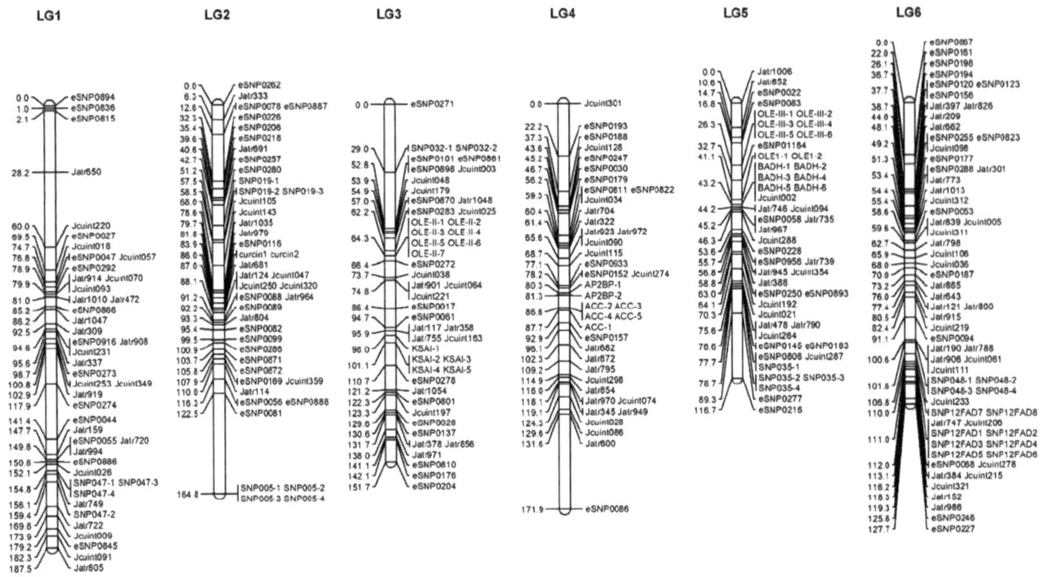

Fig. 2.1 A genetic linkage map of Jatropha. Estimates of map distances between markers are indicated in Kosambi centimorgans (Wang et al. 2011b)

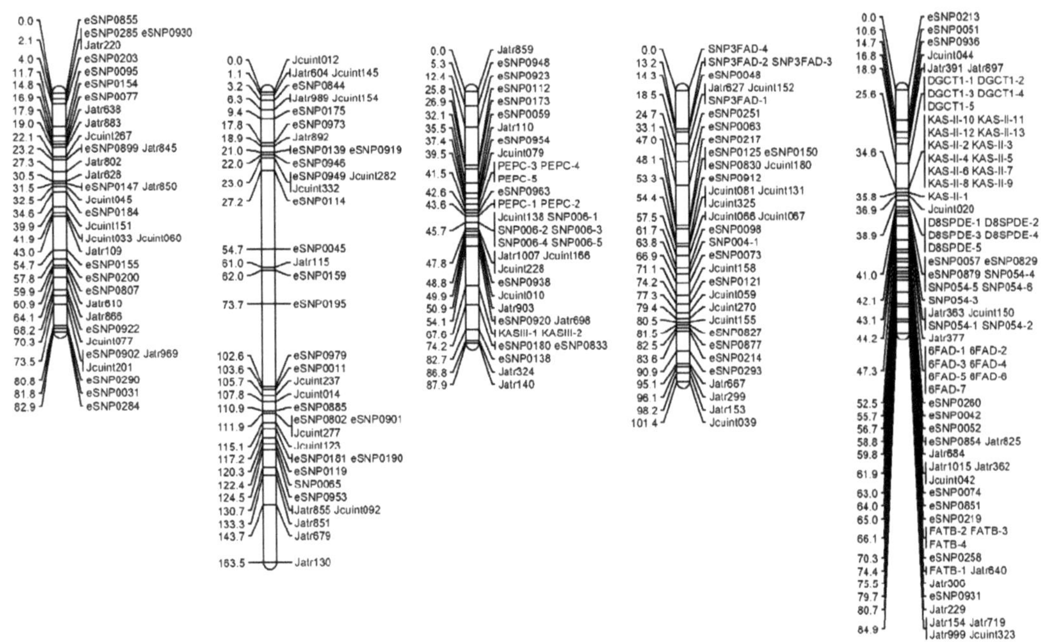

Fig. 2.2 A genetic linkage map of Jatropha. Estimates of map distances between markers are indicated in Kosambi centimorgans (continued) (Wang et al. 2011b)

6, 9, and 10. It can be deduced that the suggested segregation distortion loci at these marker regions might link to deleterious alleles. If these loci in the regions are removed from the analysis, <5% of the remaining markers show significant distortion, as expected by chance. Similar phenomena were reported in the linkage maps we constructed for Asian sea bass (Wang et al. 2011a).

To conduct the comparative genome analysis to identify conserved syntenies, BlastX searches were performed for 222 ESTs containing mapped SSR or SNP markers. Highest percentage of the marker sequences of jatropha could be assigned to 215 ESTs (96.8%) in castor bean, followed by poplar 202 (91.0%) whose genomic sequence has been determined. As searches retrieved 192 (86.5%) sequences from Arabidopsis, we compared the Jatropha linkage maps to sequence maps of the Arabidopsis chromosomes which are available. An uneven distribution of Ath loci originating from each Arabidopsis chromosome was observed in the genome of jatropha. Among the 11 LGs of jatropha, all the LGs except 3, 4, 9, 10, and 11 contained Ath loci from each of the five Arabidopsis chromosomes (Ath Chr1–Ath Chr5). LGs 3 and 10 were devoid of loci from AthChr2 and AthChr4, respectively. LGs 4, 9, and 11 did not contain any locus from AthChr3. The conserved blocks, which were defined as regions that contained at least two Ath loci from the same block region, were drawn on Figs. 2.3 and 2.4. As a result, 176 (79.2%) of the markers were corresponded to sequences mapping to loci in the Arabidopsis genome. A total of 38 genomic blocks from Arabidopsis genome were identified in the genome of jatropha with an average of 2.8 paralogous blocks for each jatropha linkage group. Figures 2.3 and 2.4 show the comparative map of the 11 jatropha linkage groups and the 5 Arabidopsis chromosomes. Conserved synteny blocks were identified in all 11 jatropha linkage groups, each of which contained from 2 to 5 Arabidopsis chromosomal blocks. The largest synteny block conserved between jatropha and Arabidopsis was found in LG 6 with 24 markers spanning 128.7 cM in the Jatropha linkage group and their best matches spanning 3 fragments in chromosomes 1, 3, and 4 of the Arabidopsis genome.

2.3 Mapping QTLs for Oil Traits and EQTLs for Oleosin Genes

Oil containing a high amount of unsaturated fatty acid can find an application as biodiesel feed stock. To make the production of jatropha profitable and sustainable, genetic improvement of oil yield and quality is demanded.

Among the fatty acid present in the jatropha seed oil, linoleic acid (18:2), oleic acid (18:1), palmitic acid (16:0), and stearic acid (18:0) are dominant compositions. Oleic and linoleic acids are the major constituents of jatropha oil (Costa et al. 2010). The breeding goal for jatropha seed oil trait improvement is to increase total oil content and oleic acid and decrease palmitic content (Chhetri et al. 2008).

QTL analysis has been performed to detect the genetic bases of important agronomic or physiological traits, providing valuable information for trait improvement. Genetic markers have made it possible to detect QTLs that are significantly associated with traits and made selection more effective (Wang et al. 2011a). Genetic response can be improved by including the QTLs in marker-assisted selection, which is a method of selection that makes use of phenotypic, genotypic, and pedigree data (Varshney and Tuberosa 2007). This approach treats mRNA expression levels as quantitative traits in a segregating population and maps expression QTL (eQTL) that control expression levels in vivo. For almost any gene analyzed in a segregating population, eQTL analysis can identify the genomic regions influencing its expression level. eQTL that maps to the same genetic location as the gene whose transcript is being measured generally indicates the presence of a *cis*-acting regulatory polymorphism in the gene (*cis*-eQTL). eQTL that maps distant to the location of the gene being assayed most likely identifies the location of *trans*-acting

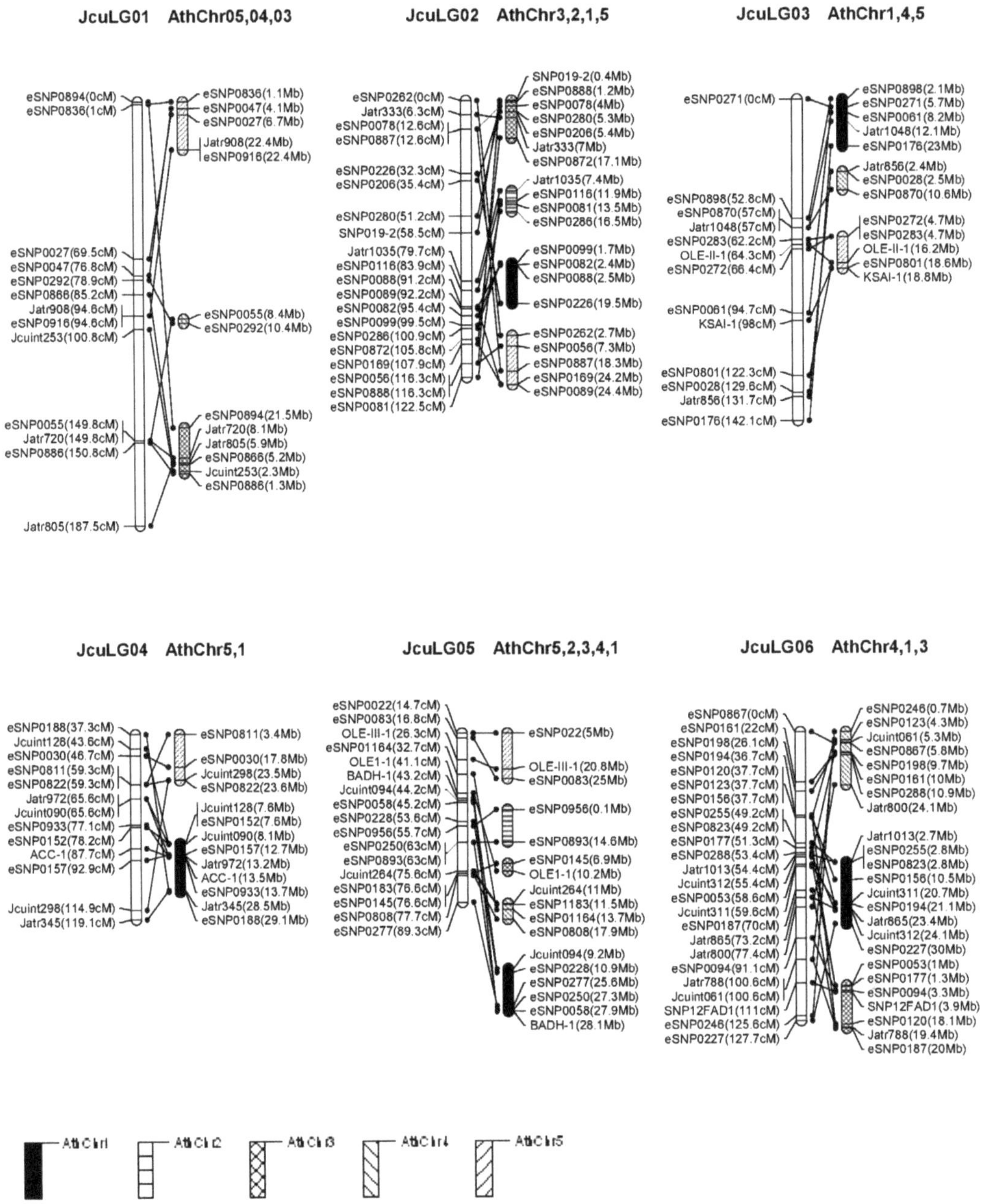

Fig. 2.3 A comparative map between Jatropha (Jcu) and Arabidopsis (Ath). Orthologous Jcu and Ath chromosomes are shown with lines connecting orthologous markers (Wang et al. 2011b)

regulators (*trans*-eQTL) that may control the expression of a number of genes elsewhere in the genome. The genetical genomics approach has been employed for identifying eQTL regulating gene expression (Sladek and Hudson 2006; Yin et al. 2010).

As detailed above, a first-generation genetic linkage map of jatropha has been established, thus providing a necessary tool for a whole-genome scan for QTLs and eQTLs affecting economically important traits including seed oil traits.

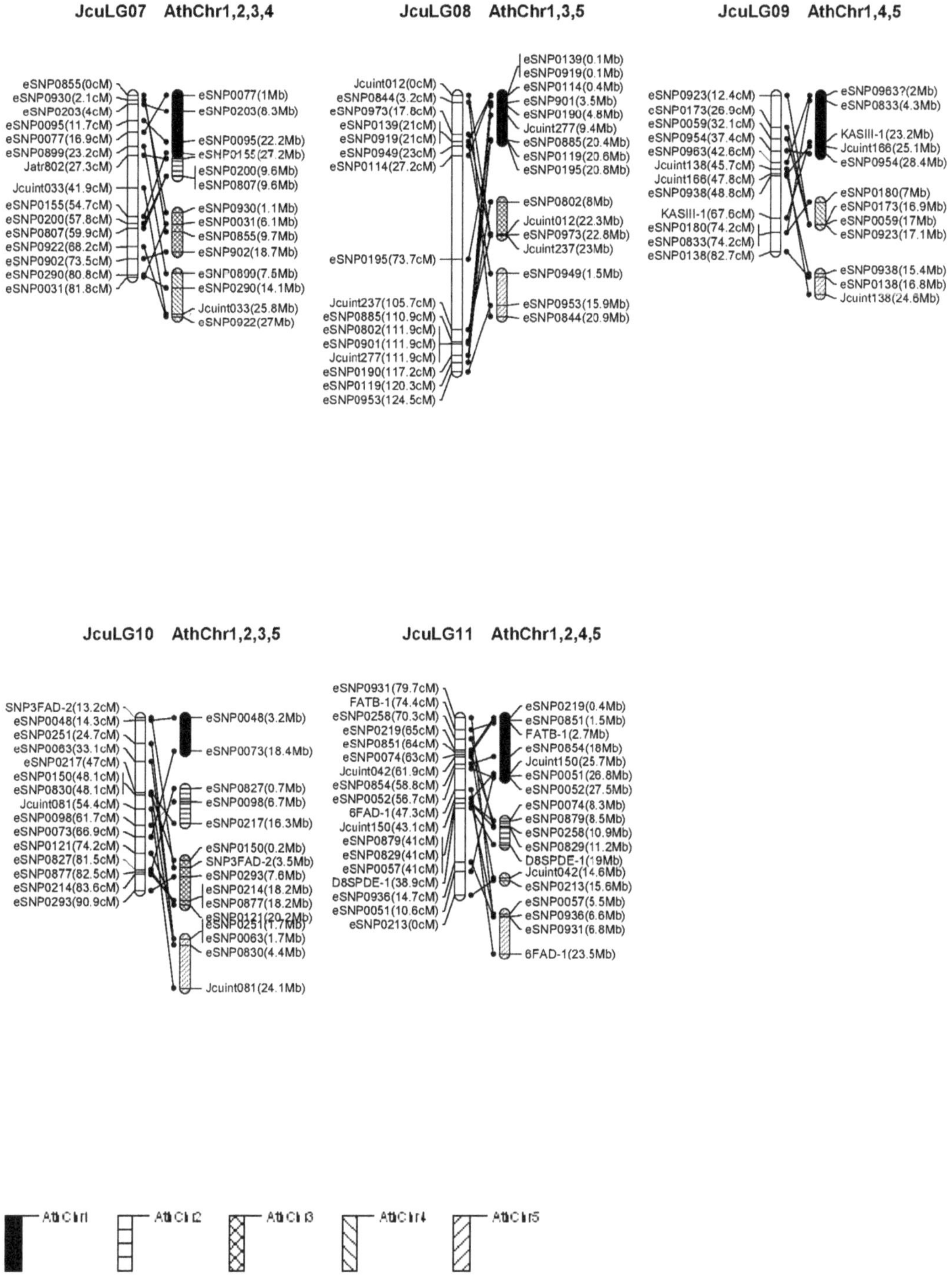

Fig. 2.4 A comparative map between Jatropha (Jcu) and Arabidopsis (Ath). Orthologous Jcu and Ath chromosomes are shown with lines connecting orthologous markers (continued) (Wang et al. 2011b)

Table 2.2 QTLs for seed oil traits and eQTLs for *OleI*, *OleII*, and *OleIII* expressions in Jatropha (Liu et al. 2011)

Trait	QTL[a] (eQTL)	Linkage group	Marker	Position (cM)[b]	LOD peak	R^2 (1%)[c]	Additive effects
C16:0 (%)	*qC16:0-2*	2	Jcuint143	47.4	2.6	0.1	1.36
	qC16:0-7	7	Jatr802	52.1	3.1	7.4	1.42
	qC16:0-9	9	Jatr859	15	2.6	7.2	1.39
C18:0 (%)	*qC18:0-2*	2	curcin2	52.6	2.6	5.3	−0.69
	qC18:0-5	5	Jatr746	37.3	6.9	13	1.15
	qC18:0-6	6	Jcuint036	64	3.9	7.1	−0.84
	qC18:0-7	7	Jatr883	40.3	23	4	0.59
	qC18:0-9	9	Jatr859	0	9.2	17.9	1.26
C18:1 (%)	*qC18::1-1*	1	Jcuint057	0	18.4	36	11.69
	qC18:1-5	5	Jatr739	45.1	2.3	3.4	−3.77
	qC18:1-10	10	Jcuintl80	15.2	4	5.9	4.75
C18:2 (%)	*qC18:2-1*	1	Jcuint057	0	16.5	34.1	−12.07
	qC18:2-6	6	Jatr301	15	2.4	3.8	4.3
	qC18::2-10	10	Jcuintl80	15.2	3	4.6	−4.4
Total oil content (%)	*qOilC-1*	1	Jatr722	55.1	2.3	4.6	−3.72
	qOilC-2	2	Jcuintl43	47.4	2.5	4.9	−3.74
	qOilC-4	4	Jatr872	29.6	5	11.1	−5.56
	qOilC-9	9	Jatr698	18.6	2.5	5.2	3.74
Olel expression ($\Delta\Delta C_T$)	*qOlel-8*	8	Jcuint277	58.2	1.9	5.3	1.71
Olell expression ($\Delta\Delta C_T$)	*qOlell-6*	6	Jatr152	93.4	2.6	6.4	−2.38
Olelll expression ($\Delta\Delta C_T$)	*qOlelll-5*	5	Jatr739	46.2	3.1	11.7	−3.06

[a]QTL (eQTL): starting with "q," followed by an abbreviation of the trait name, the name of the linkage group, and the number of QTLs (eQTLs) affecting the trait on the linkage group. *OleI*, *OleI* expression level; *OleII*, *OleII* expression level; C16:0, C18:0, 18:1, and C18:2, fatty acid compositions of palmitic acid (C16:0), stearic acid (C18:0), oleic acid (C18:1), and linoleic acid (C18:2); OilC: Total oil content
[b]Position from the first marker on each linkage group
[c]Proportion of phenotypic variance (R^2) explained by a QTL (eQTL)
[d]Estimated phenotypic effect of substituting *J. integerrima* alleles with *J. curcas* alleles at QTL (eQTL)

The linkage map, covering 663.0 cM of the genome, converged into 11 linkage groups consisting of 95 DNA markers. The average distance between markers was 7.0 cM. Most of the linkage groups were consistent with those described above. QTL analyses were performed on the means of fatty acid composition, total oil content, and expression levels of *OleI*, *OleII*, and *OleIII* (Table 2.2; Fig. 2.5). We detected 18 QTLs and 3 eQTLs for all traits examined. Individual eQTL or QTL were detected with percentage of variation explained (PVE or r^2) 0.1–36.0%, and 5 of them had PVE exceeding 10%. QTLs or eQTLs with positive and negative allelic effects were identified, with a positive effect implying a higher value for the trait conferred by the allele from PZM16 and vice versa (Fig. 2.6).

Eighteen QTLs were identified dispersed among all the linkage groups except LGs 3 and 11. A QTL of highly significant effect was determined to be located on LG 1 explaining 36% of variation of C18:1 composition and was

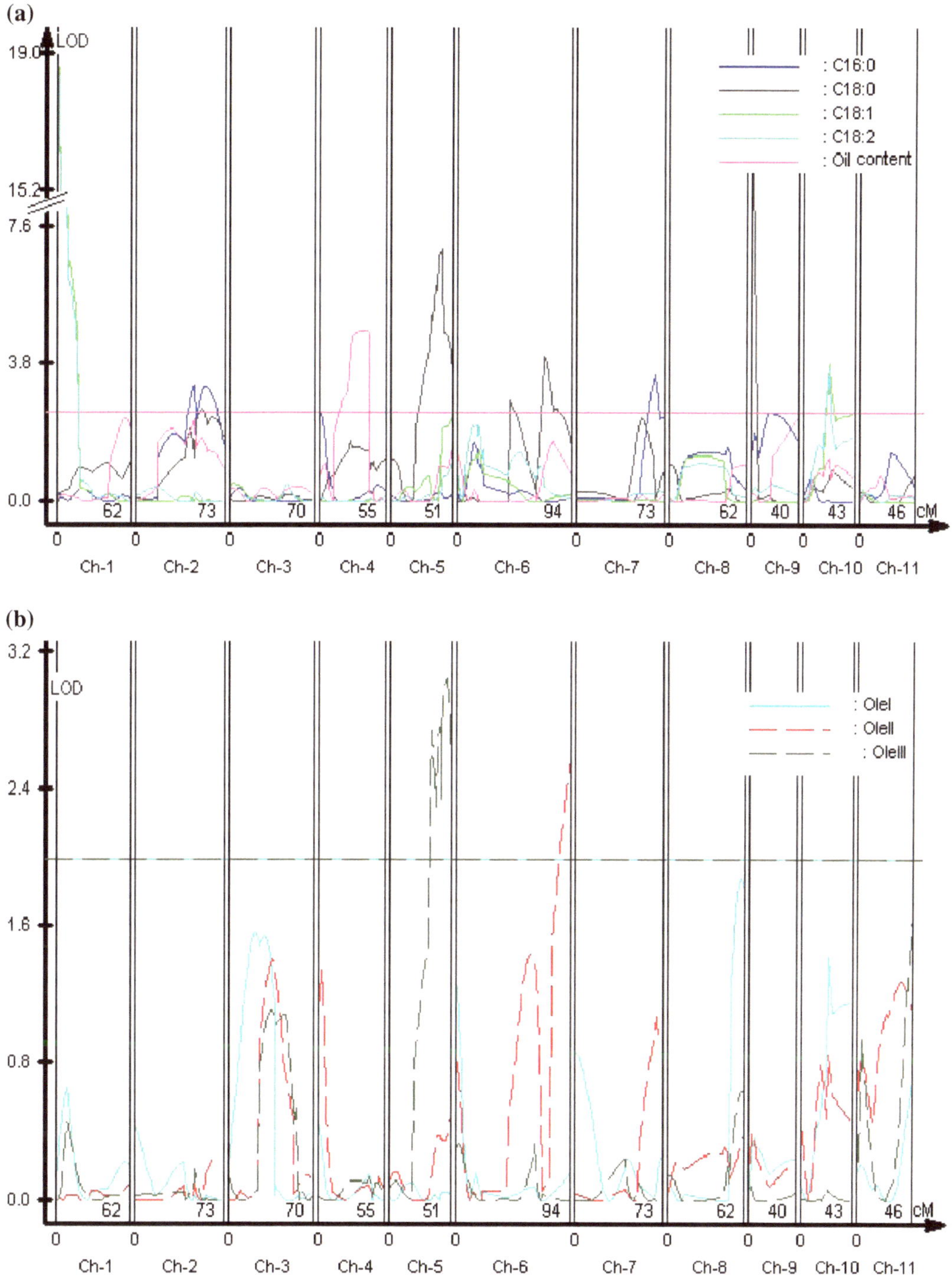

Fig. 2.5 Whole-genome scan for QTL for oil traits and Oleosin gene expression in Jatropha. **a** QTL scans of oil traits on linkage maps. *Horizontal line* indicates 5% LOD significance thresholds (2.5) based on permutation. **b** QTL scans of *OleI*, *OleII* and *OleIII* expressions on linkage maps. *Horizontal line* indicates LOD significance threshold (2.0) (Liu et al. 2011)

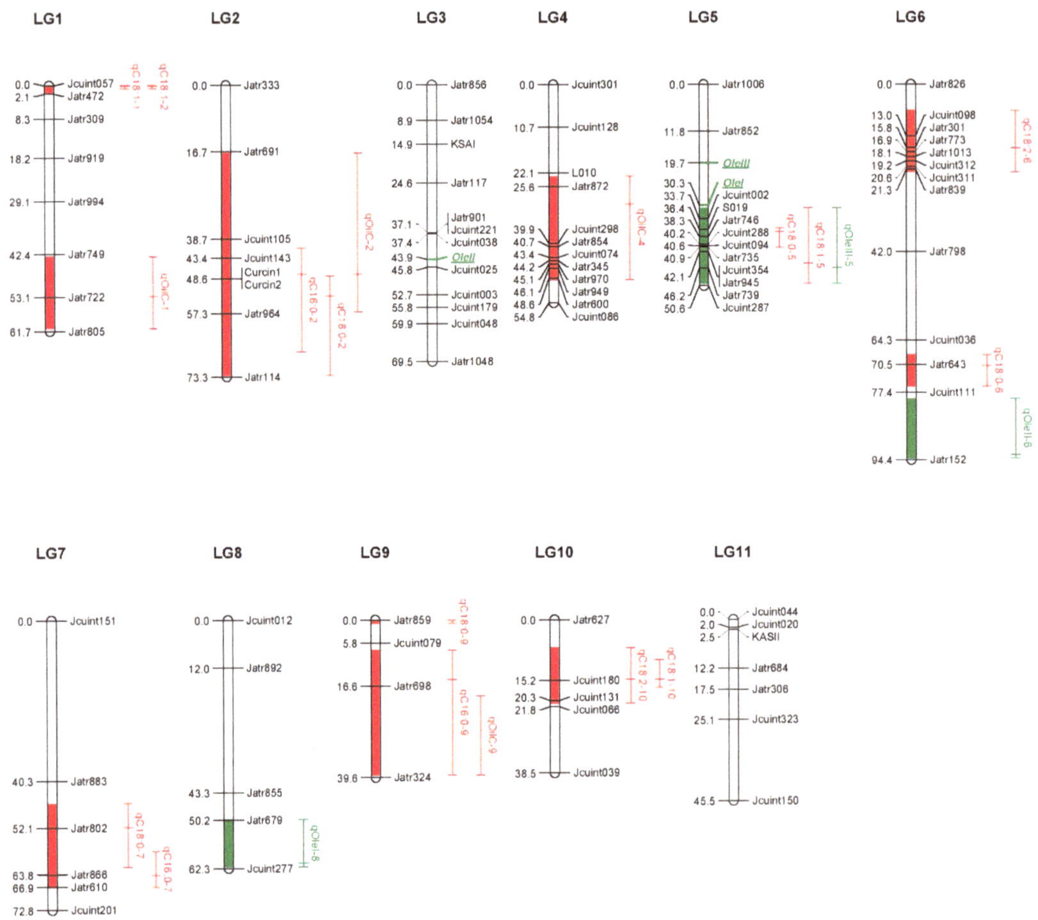

Fig. 2.6 Summary of QTL (eQTL) locations detected on the genome of jatropha. QTLs (eQTLs) represented by *bars* are shown on the left of the linkage groups, close to their corresponding markers. The lengths of the *bars* are proportional to the confidence intervals of the corresponding QTLs (eQTLs) in which the *inner line* indicates position of maximum LOD score (Liu et al. 2011)

found to be associated with C18:2 compositions (Table 2.2; Fig. 2.6). Interestingly, another QTL on LG10 explained 5.9% of variation of C18:1 composition was also associated with C18:2 compositions. Higher values for C18:1 were conferred by the allele from PZM16, while higher values for C18:2 from Hybrid CI7041.

Four QTLs were detected underlying total oil content. At the three QTLs on LGs 1, 2, and 4, respectively, the alleles from hybrid CI7041 contributed high total oil content. The most effective QTL was spotted on LG 4 explaining 11.1% of the variation, whose higher value for total oil content was conferred by the allele from hybrid CI7041.

There were strong QTLs for C18:1 and total oil content detected on LGs 1 and 4, respectively. Mean phenotypic values of each trait were calculated for those progeny with the alternate alleles of the microsatellite markers, inherited from the *J. integerrima* (aa) or *J. curcas* (AA). A two-way ANOVA was performed on the progeny using two allelic combinations (AA, Aa) from markers linked to QTLs in order to investigate associations between phenotypic traits and genotypes of the QTLs. The phenotype values of each allelic combination of the QTLs are listed in Fig. 2.7. Significant differences of phenotype means among different allelic combinations were

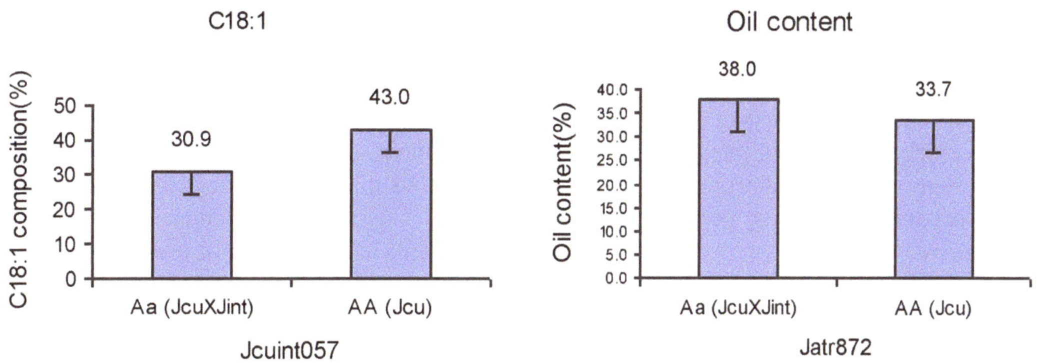

Fig. 2.7 C18:1 composition (*left*) and total oil content (*right*) of plants with different genotypes. Favorite alleles for C18:1 composition are AA from *J. curcas*, and those for total oil content are Aa from hybrid of *J. integerrima* and *J. curcas* (*right*) (Liu et al. 2011)

Table 2.3 SNP markers and real-time PCR primer pairs for *OleI*, *OleII*, and *OleIII* genes (Liu et al. 2011)

Gene	Forward primer (5′–3′) Reverse primer (5′–3′)	PCR product length (bp)	For SNP or real-time PCR use
OleI	CATTGCGCTAGCTGTTGCGACTCC CGCCGCTTTGCCATTTCCATCT	207	SNP and real-time PCR
OleII	GGGGCTATGGGGCTCACAG GTTGAGTTGGTTTATGGGGGATCT	313	SNP and real-time PCR
OleIII	ACAGCCACGATCCCACCAAGTAGT GGACAGAGCTGAGCAGTTTGGACA	443	SNP
OleIII	TGGTGCCGACGGTTATCAC TACATGCTGTCCAAACTGCTCAG	216	Real-time PCR

identified, revealing the effects of alternative alleles inherited from the parents.

Progenies with AA genotype at the marker Jcuint057 located in qC18:1-1 showed the higher C18:1 content (43.0%) than Aa (30.9%). By contrast, progeny with Aa genotype at the marker *Jatr872* located in qOilC-4 showed the higher total oil content (38.0%) than AA (33.7%) (Fig. 2.7). These results suggested the effect of the two QTLs are opposite on these two key oil traits and favorite alleles were differentially from *J. curcas* and *J. integerrima*.

SNP markers were developed in *OleI*, *OleII* and *OleIII* genes (Table 2.3), which were mapped on LGs 5, 3, and 5 respectively (Fig. 2.6). *OleI* and *OleIII* were mapped on LG5 where the QTLs qC18:0-5, qC18:1-5, and qOleIII underlying C18:0, C18:1, and *OleIII* expression clustered. Negative additive effect value of qOleIII-5 indicated that *J. curcas* alleles were positive for *OleIII* expressions, of which LOD score was 3.1. This eQTL of *OleIII* was localized near *OleIII* gene and overlapped with the QTLs controlling C18:0 and C18:1, revealing a *cis*- or *trans*-element for *OleIII* which subsequently controlling the C18:0 and C18:1. One eQTL on LG 8 qOleI-8 was detected underlying *OleI* expression with LOD 1.9 (Table 2.2; Figs. 2.6, 2.7). Additive effect value of qOleI-8 was positive, indicating that *J. integerrima* alleles were positive for *OleI* expressions. To find as many putative QTLs (eQTLs) as possible, and to obtain a clearer understanding of the relationships among examined traits, a threshold eQTL of 1.9 for declaring a suggestive eQTL was employed. Low thresholds may not be useful in plant breeding programs, but they have been shown to help in understanding relationships among traits (Thumma et al. 2001). *OleII* was located on LG 3. One eQTL for *OleII* was detected on LG 6 with LOD 2.6, which closed to qC18:0-6. It is suggested that a *trans*-element for *OleII* could harbor in this region which controlling the C18:0. Additive effect values indicated that *J. curcas* alleles were negative, indicating that the effect of *J. curcas* alleles was positive for *OleII* expressions.

2.4 An Approach for Jatropha Improvement Using Pleiotropic QTLs

Higher seed yield is one of the objectives of jatropha breeding. However, genetic analysis of the yield traits has not been done in jatropha. QTL mapping was conducted to identify genetic factors controlling growth and seed yield in jatropha, a promising biofuel crop.

As in other crops, almost all the economically important traits in jatropha, such as seed yield, biotic or abiotic stress resistance, are quantitative and determined by multiple genes with minor effects, which are described as QTL. In contrast, jatropha had not yet undergone a careful breeding program with systematic selection and improvement of suitable germplasm.

Growth and seed traits were measured in a QTL mapping population, and the frequency distributions of all traits in the progeny showed a continuous distribution. The distribution of phenotypic values showed bidirectional transgressive segregation (Table 2.4), revealing complex genetic bases of these traits. While seed yield in *J. curcas* was higher than that in *J. integerrima*, branch number in *J. integerrima*is significantly higher than that in *J. curcas*. The data implied that *J. integerrima* germplasm could be applied for hybrid breeding to improve agronomic traits, such as branch number in the 4th and 10th months, and the female flower number. Correlation analysis among these traits was performed (Table 2.5), and total seed weight showed a significant correlation with total branch number, female flower number and fruit number, with coefficients 0.364, 0.294, and 0.308, respectively. Therefore, these agronomic traits were suggested to be key factors for seed yields.

QTL analyses were performed on the means of growth traits, branch number, female flower and fruit number, and seed yield (Table 2.6; Fig. 2.8).

Table 2.4 Descriptive statistics on phenotype data of QTL mapping population and parents (*J. curcas* PZMD16, *J. integerrima* S001 and F1 CI7041) (Sun et al. 2012)

N	Trait	Acronym	Mean	SD	Min	Max	PZMD16	CI7041
Growth traits								
1	Height in the 4th month	H4M	70.5	24.6	15.0	134.0	62.16	70
2	Height in the 10th month	H10M	122.9	37 2	33.0	272.0	166	180
3	Diameter in the 4th month	D4M	1.9	0.5	0.6	32	1.45	1.2
4	Diameter in the 10th month	D10M	4.3	1.1	1.1	7.0	62	5
5	Branch number in the 4th month	BN4M	1.8	2.2	0.0	14.0	10	12
6	Branch number in the 10th month	BN10M	6.4	4.0	1.0	19.0	15	17
7	Total branch number	TBN	4.4	2.1	1.0	14.0	35	3.33
8	New branch number per branch	BNPB	2.9	1.1	1.0	8.0	1.75	1.67
Flower, fruit and seed yield								
9	Female flower number	FFN	4.5	2.8	0.0	15.0	4.75	925
10	Fruit number	FRUITNO	9.1	105	0.4	62.0	35	No fruit (hybrid F1)
11	Total seed weight in 2010	TSW	34.1	60.1	0.5	541.4	360.0	90.0

Table 2.5 Correlation coefficients and significance of correlations among growth and yield traits in a QTL mapping population (Sun et al. 2012)

	H4m	H10m	D4m	D10m	BN4m	BN10m	TBN	BNPB	FFN	FruitNo
H10m	0.371***									
D4m	0.652***	0.374***								
D10m	0.171***	0.675***	0.494***							
BN4m	0.170***	0.064	0.229***	0.177***						
BN10m	0.199***	0.189***	0.175***	0.401***	0.415***					
TBN	−0.155*	0.041	−0.092	0.154*	0.067	0.006				
BNPB	−0.027	−0.015	−0.005	0.094	0.141	0.072	0.450***			
FFN	−0.448***	0.138*	−0.207***	0.258***	−0.042	−0.061	0.265***	0.097		
FruitNo	0.026	0.220***	0.077	0.256***	0.198**	0.327***	0.097	0.038	0.215**	
TSW	−0.137*	0.093	0.004	0.262***	0.156*	0.164**	0.364***	0.048	0.294***	0.308***

P values are as follows: $*P < 0.10$, $**P < 0.05$, $***P < 0.01$

Table 2.6 QTLs for growth traits, seed characters (Sun et al. 2012)

Trait	QTL[a]	Linkage group	Marker	Position (cM)[b]	LOD peak	R^2 (%)[c]	Additive effects[d]
Growth							
H4m	*qH4m-3*	3	Jcuint179	55.8	3.50	5	−17.51
	qH4m-7	7	Jatr610	71.9	2.97	4.4	−10.31
H10m	*qH10m-3*	3	Jatr1054	11.9	3.55	6.3	21.53
	qH10m-5	5	Jatr945	42.1	5.31	8.5	21.99
	qH10m-6	6	Jcuint312	25.3	2.56	4	−15.80
	qH10m-9	9	Jatr859	5.0	3.09	4.5	16.30
	qH10m-W	10	Jcuint081	20.6	4.63	6.7	−19.25
D4m	*qD4m-S*	5	Olel	31.3	3.88	5.6	0.26
	qD4m-9	9	Jatr698	24.6	2.70	5.2	0.25
	qD4m-l1	11	Jatr684	14.3	2.98	4.6	0.25
D10m	*qD10m-3a*	3	Jatrl054	11.9	8.11	12.7	0.96
	qD10m-3b	3	Jcuint048	59.9	3.85	4.9	−0.68
	qD10m-5	5	Jatr746	38.3	15.03	21.1	1.11
BN4m	*qBN4m-6*	6	Jcuint036	64.3	4.04	6.9	−1.26
BN10m	*qBN10m-1*	1	Jatr722	54.1	3.44	5.6	2.04
	qBN10m-4	4	Jatr854	41.0	3.51	5.5	−2.19
	qBN10m-6	6	Jcuint111	93.4	3.58	5.9	−2.28
TBN	*qTBN-7*	7	Jatr610	68.9	3.40	8.4	1.30
BNPB	*qBNPB-5*	5	Jatr739	46.2	2.98	7.9	0.62
Flower, fruit, and seed yield							
FFN	*qFFN-2*	2	Jatr691	20.7	2.00	4.5	1.20
	qFFN-5	5	Olel	30.3	3.38	7	1.56
	qFFN-6	6	Jatr301	15.8	6.41	13.6	−2.21
	qFFN-7	7	Jcuint151	2.0	4.16	9.6	1.97
FruitNo	*qFruitNo-1*	1	Jatr749	42.4	2.47	5.5	5.54
	qFruitNo-6	6	Jatr839	28.3	4.97	12.4	−7.61
	qFruitNo-7	7	Jatr866	64.8	3.15	7.3	5.65
TSW	*qTSW-5*	5	Jcuint002	33.7	2.24	5.2	29.33
	qTSW-7	7	Jatr866	66.8	2.70	4.9	26.31

[a]QTL: starting with "q," followed by an abbreviation of the trait name, the name of the linkage group, and the number of QTL affecting the trait on the linkage group

[b]Position from the first marker on each linkage group

[c]Coefficient of determination or the percentage of variance explained (PVE) by the detected QTL

[d]Estimated phenotypic effect of substituting *J. integerrima* alleles with *J. curcas* alleles at QTL

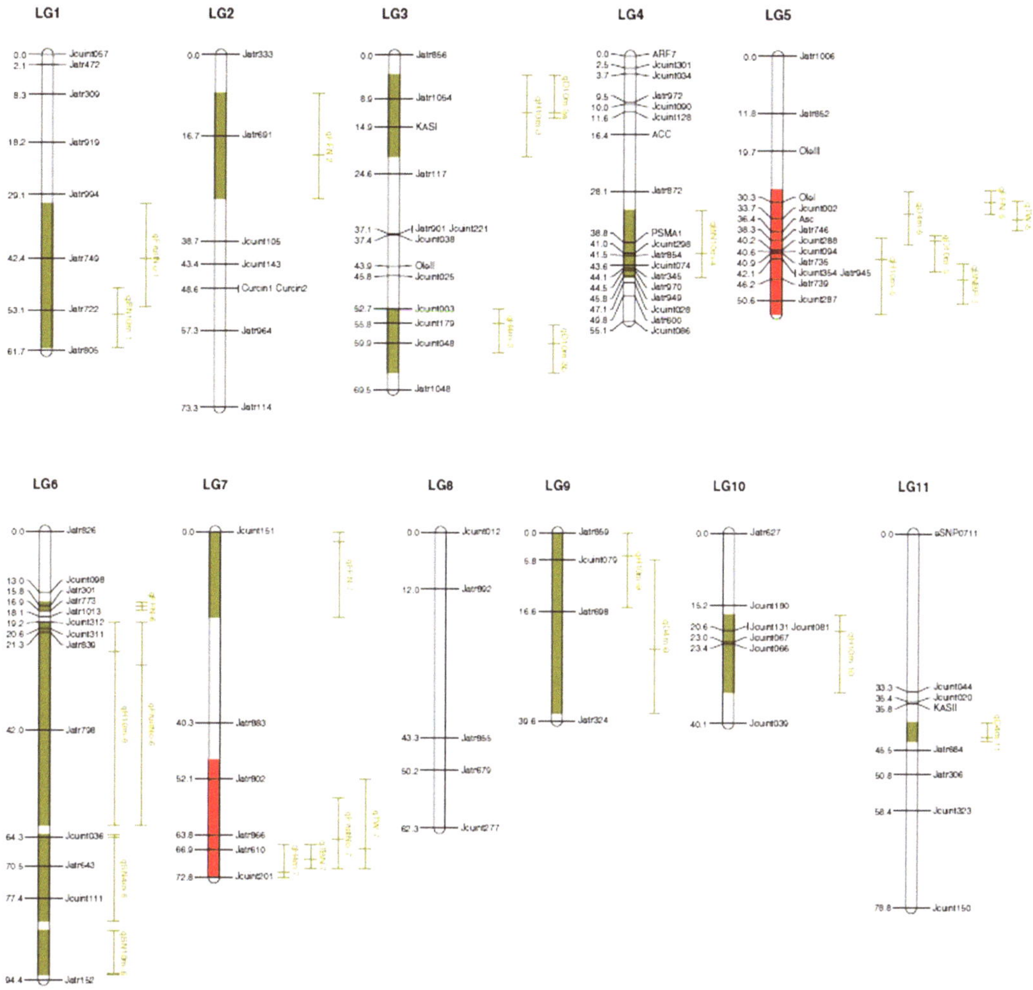

Fig. 2.8 Summary of QTL locations detected. QTL represented by *bars* is shown on the *left* of the linkage groups, close to their corresponding markers. The lengths of the *bars* are proportional to the confidence intervals of the corresponding QTL in which the *inner line* indicates position of maximum LOD score. The confidence intervals of QTL are shown in *green color*, and two QTL clusters are highlighted in *red* (Sun et al. 2012)

We have detected 28 QTLs for all traits examined with LOD threshold 2.0–2.5 determined by permutations. Individual QTLs were detected with percentage of variation explained (PVE or R^2) 3–21.16%, and four of them had PVE exceeding 10%. QTLs with positive and negative additive effects were identified, with a positive effect implying a higher value for the trait conferred by the allele from *J. curcas*, and negative from *J. integerrima* (Table 2.6).

2.4.1 QTLs for Growth Traits

Sixteen QTLs were identified and dispersed among all the linkage groups except LGs 2 and 8. Four QTLs over-lapping on the lower part of LG 5, namely qH10m-5, qD4m-5, qD10m-5, and qTBN-5, were detected underlying plant height in the 10th month, stem diameter in the 4th and 10th months, and total branch number, respectively (Fig. 2.8). Additive effects of these QTLs were positive, indicating that the alleles from *J. curcas* increased these trait values.

Conversely, two QTLs, namely qBN4m-6 and qBN10m-6, were detected on the lower part of LG 6 controlling branch number with negative additive values, indicating *J. integerrima* allele increased branch number.

2.4.2 QTLs for Seed Traits

On LGs 5 and 7, two QTLs of qTSW-5 and qTSW-7 were detected controlling total seed weight, which is one of the most economically important traits. Interestingly, QTLs underlying yield related traits were clustered at these two QTLs. At qWT-5, four QTLs underlying plant height, stem diameter, branch number, and female flower number were detected. Near qTSW-7, three QTLs of qH4m-7, qTBN-7, and qFruitNo-7 were detected, controlling plant height, total branch number, and fruit number, respectively. It was noteworthy that two QTL clusters were detected on LGs 5 and 7, respectively. Five QTLs were detected on the lower part of LG 5 (Fig. 2.9a), and four QTL clusters were detected on lower part of LG 7 (Fig. 2.9b).

2.4.3 Favored Alleles Originated from Two Parents

Two QTL clusters were detected consisting of five and four QTLs, controlling total seed weight, plant height, stem diameter, female flower number, and fruit number. The positive additive effects indicated higher values for the traits conferred by the allele from *J. curcas*. Meanwhile, five QTLs on LG 6, namely qH4m-6, qBN4m-6, qBN10m-6 qFFN-6, and qFruitNo-6, controlling plant height, branch number (in 4th and 10th months post-seed germination), female flower number, and fruit number, respectively, were detected with negative additive effects indicating higher values conferred by *J. integerrima* (Table 2.6).

2.4.4 Major Effects of QTSW-5 and QTSW-7

A two-way analysis of variance (ANOVA) was carried out to assess genetic effects and interactions of the two QTLs of qTSW-5 and qTSW-7 controlling total seed weight. The values of different genotypes are shown in Fig. 2.10. Total seed weight was significantly increased in the presence of these two QTLs. When qTSW-5 presented, total seed weight was improved from 16.66 ± 7.26 to 42.00 ± 5.06 g, and qTSW-7, from 15.97 ± 6.36 to 42.69 ± 6.16 g (Fig. 2.10a).

Interestingly, we found that the two QTLs for seed yield overlapped with other QTLs for other agronomic traits than seed yield itself. ANOVA showed that the QTL qTSW-5 for seed yield affected significantly plant height, stem diameter, new branch number per branch, and female flower number, while qTSW-7 affected plant height, total branch number, and fruit number (Fig. 2.10b).

2.4.5 Effect of Pyramiding QTSW-5 and QTSW-7

The interaction between marker effects for qTSW-5 and qTSW-7 was nonsignificant with a relatively low P value (0.14) (Table 2.7), while the marker effects for qTSW-5 and qTSW-7 were nonadditive (Fig. 2.11). This could be caused by the lack of power in the ANOVA due to an unequal distribution of genotypic classes (Fig. 2.10). Despite the nonsignificance of the interaction of the two QTLs, total seed weight was significantly increased in the presence of the two QTLs. Lines carrying both QTLs produced an average 61.93 ± 7.31 g of seeds, nearly three times as much as any other genotype combinations (Fig. 2.11). Therefore, although total seed weight could be improved by introducing the two QTLs, there would be advantages to be gained by pyramiding the two QTLs.

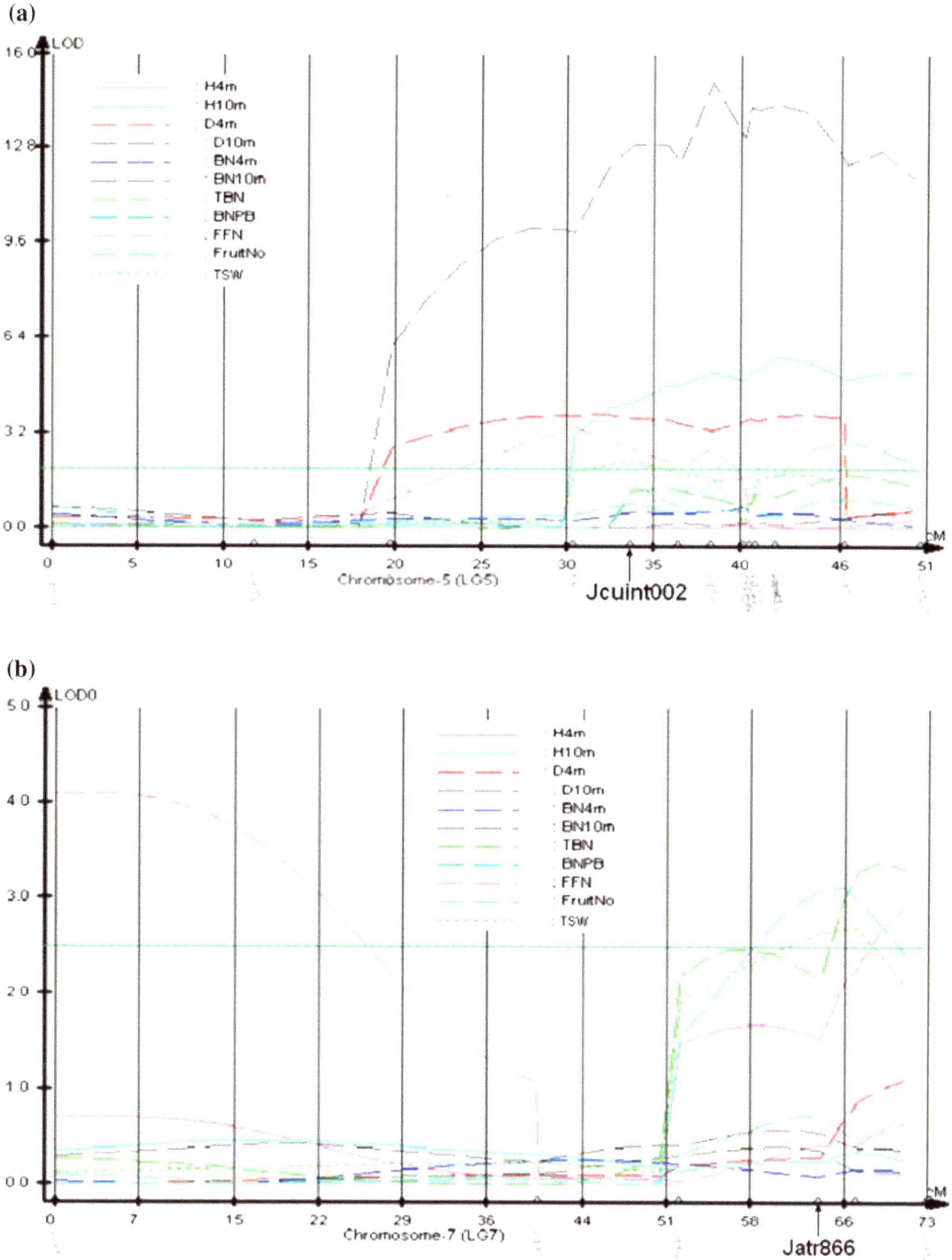

Fig. 2.9 QTL clusters on LGs 5 and 7. QTL scans of growth on linkage maps. *Horizontal line* indicates 5% LOD significance thresholds (2.0) based on permutation. **a** LG 5; **b** LG 7 (Sun et al. 2012)

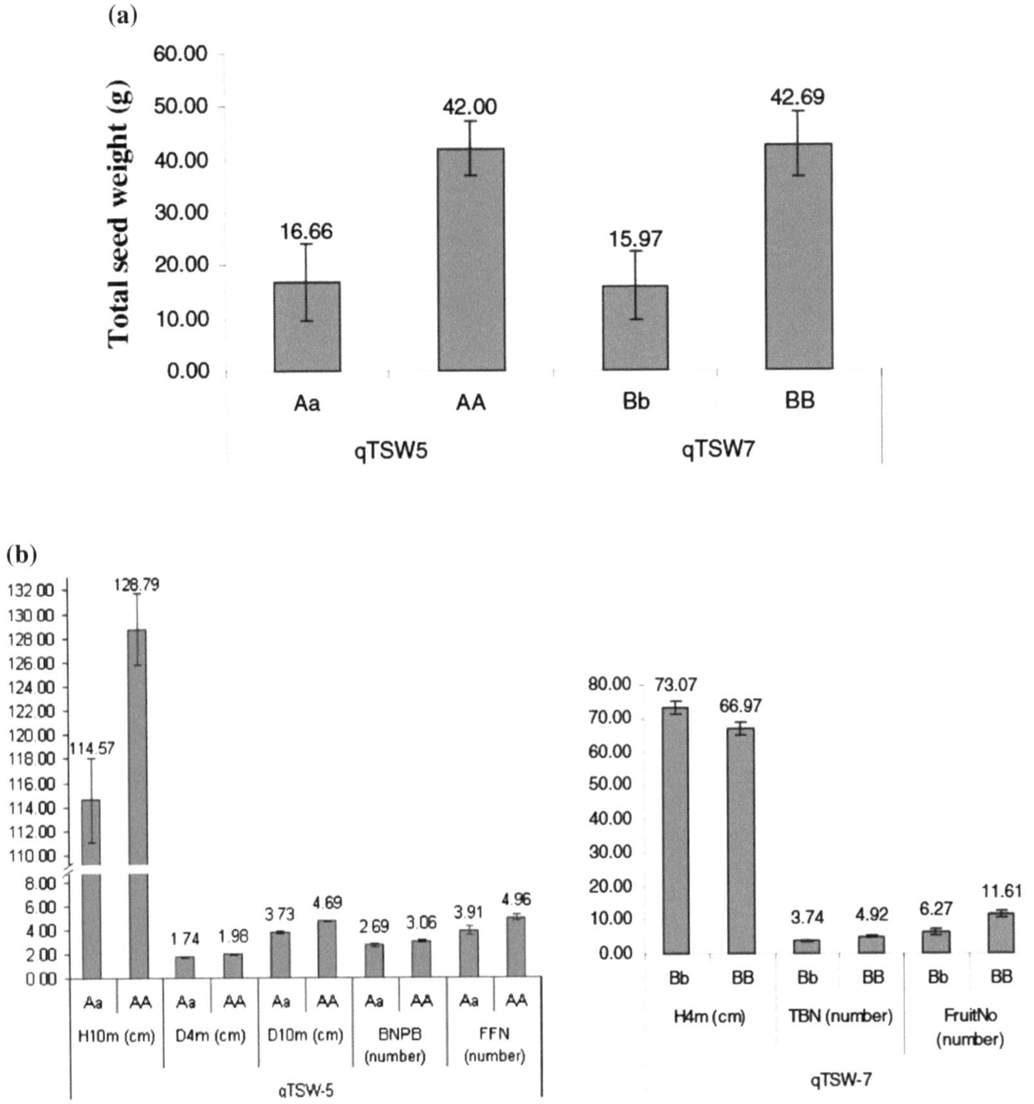

Fig. 2.10 Total seed weight (g) and related traits of plants with different genotypes of qTSW-5 (AA, Aa) and qTSW-7 (BB, Bb); *N* denotes sample number of each genotypic classes; *Error bars* denote standard errors (SEs). **a** Significant major effects of the two QTLs on seed yield; **b** the two QTLs with pleiotropic roles in regulating plant growth and seed yield. Significant at $P < 0.01$ of Bonferroni test (Sun et al. 2012)

Table 2.7 ANOVA of seed yield in the QTL mapping population based on genotypes of the marker loci that are most closely linked to the QTLs (Sun et al. 2012)

Effect	*d.f.*	MS	*F*	*P*
1 (Jcuint002, *qTSW-5*)	1	24943.135	8.190	0.005*
2 (Jatr866, *qTSW-7*)	1	27739.644	9.110	0.003*
1 × 2	1	6703.480	2.200	0.140

* Significant at $P = 0.01$

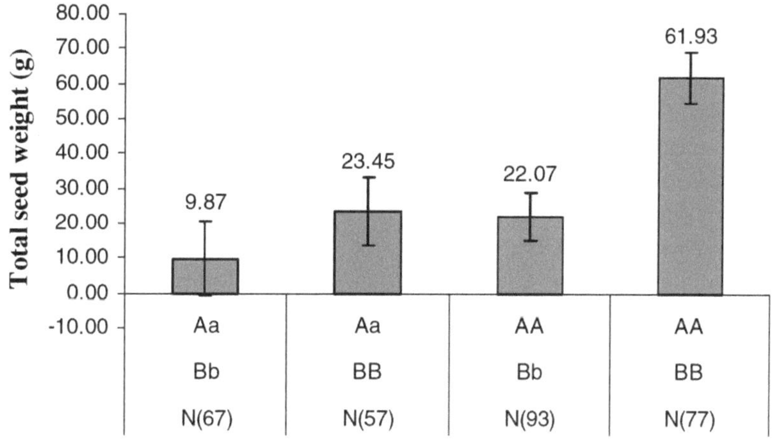

Fig. 2.11 Effects of pyramiding the two QTLs of qTSW-5 (genotypes of AA, Aa) and qTSW-7 (BB, Bb) on seed yield. *Error bars* denote SEs (Sun et al. 2012)

2.5 Identification of Candidate Genes *JcARF19* and *JcIAA9* Associated with Seed Size Traits

Larger seeds provide the germinating seedling a larger supply of nutrients; thus, it increases its competitiveness during seedling establishment and tolerance toward adverse environmental condition (Harper et al. 1970).

The big size of jatropha seeds can be utilized to produce high-quality biodiesel fuel. Interest in using jatropha to help alleviate the energy crisis is increasing and, in fact, jatropha is becoming one of the world's key crops for biodiesel production (Ye et al. 2012) because of its high content of oil in its seeds, tolerance to drought, and ability to thrive in arid soil. Woody bioenergy plants such as jatropha have competitive economic advantages as many years' economic returns after one-time planting investment. Since its short life cycle time, jatropha has been regarded as a potential model woody plant. Therefore, it has been widely used to rehabilitate wastelands, improve the environment, and enhance the quality of rural life by providing economic value for marginal farmlands (Wu et al. 2011). Furthermore, the reduction of greenhouse gas emission for generating 1 GJ of energy by using biodiesel can be at least 40 and up to 107% with respect to fossil diesel according to three independent investigations in South America, Europe, and Asia (Almeida et al. 2011; Bailis and Baka 2010; Kumar et al. 2012). It is, however, a nondomesticated plant with poor and unpredictable seed productivity in large-scale plantation, and little is known about seed size- and yield-determining genes (Sanderson 2009).

Phytohormones play pivotal roles in the developmental processes of diverse yield traits. Plant growth and development are sustained by continuous cell division, which is mainly regulated by auxin (Durbak et al. 2012). Interaction between auxin and other hormones, such as cytokinin, and how they determine a specific developmental output are still poorly understood, especially in the reproductive organs that are important in seed yield traits (Ioio et al. 2008; Durbak et al. 2012; Ruan et al. 2012). Auxin response factors (ARFs) are transcription factors that play a crucial role in auxin signaling pathway. AUX/IAA (indole-3-acetic acid) proteins control the transcription of auxin-inducible genes by binding to the paired ARFs (Berleth et al. 2004). *ARF2* is a key regulator of seed size in Arabidopsis (Okushima et al. 2005a). In *arf2* mutants, the ovule and seed cavity are enlarged compared with the wild type due to extra cell divisions in the integument. The reduction of seed numbers prevents its application in agriculture (Okushima et al. 2005a). In rice, another ARF family gene *ARF8* appears to be associated with seed development, indicating that different ARF family members may control seed development in various plants (Yang et al. 2006; Xue et al. 2009). Arabidopsis mutations of *ARF2*, *ARF9*, *ARF13*, and *ARF14* had weak effects on auxin-inducible gene expression, whereas mutations in *ARF3*, *ARF4*, *ARF6*, *ARF7*, *ARF10*, and *ARF19* had strong effects. This suggested that some ARFs play a major role in auxin-inducible gene expression, while others may play an auxiliary or redundant role at the young seedling stage (Lee et al. 2009). On the other hand, ten gain-of-function AUX/IAA mutants were identified, namely *IAA1*/AXR5, *IAA3*/SHY2, *IAA6*/SHY1, *IAA7*/AXR2, *IAA12*/BDL, *IAA14*/SLR, *IAA17*/AXR3, *IAA18*, *IAA19*/MSG2, and *IAA28*, with each exhibiting reduced auxin response in various aspects of development and growth (Berleth et al. 2004). *ARF7* and *ARF19* activate the transcription of several lateral organ boundaries-domain (LBD) genes, including *LBD16* and *LBD29*. As a transcriptional regulator, *LBD16* may also activate the downstream transcriptional network for lateral root (LR) initiation (Okushima et al. 2007). The lists of auxin-regulated genes of which expression is inhibited in the mutants contain putative downstream targets of *ARF7* and *ARF19* (Okushima et al. 2005b).

2.5.1 QTL and Candidate Gene Mapping

We identified seven ARFs, four Aux/IAA, and four downstream genes from a cDNA library of jatropha seed (Gu et al. 2012). We identified the genomic sequence of these genes from databases of public (Sato et al. 2011) and Temasek Life Sciences Laboratory (unpublished data, Dr. Yan Hong, Temasek Life Sciences Laboratory, Singapore). Alignment and phylogenetic analyses were conducted with references to ARF and Aux/IAA proteins from Arabidopsis and related known proteins from other plants. We named the seven ARF genes as *ARF1*, *ARF2*, *ARF4*, *ARF5*, *ARF7*, *ARF9*, and *ARF19*, the four IAA genes as *IAA3*, *IAA9*, *IAA14*, and *IAA19*, and four downstream genes as *LBD16*, *LBD18*, *LBD19*, and *ARGOS*, individually, with the support of alignment and phylogenetic analysis data. SNP markers were further developed for these genes. All these genes were mapped onto the linkage map (Fig. 2.12a). Seed traits were measured in the QTL mapping population. Seed size in *J. curcas* was much bigger than in *J. integerrima*. Correlation analysis among these traits was performed; we observed significant correlations among the seed traits, including single seed weight, seed length, seed width, and seed height (Ye et al. 2014). We performed QTL analyses on seed traits (Fig. 2.12) and detected 21 QTLs for all the traits. Most importantly, we detected a major QTL qSL11-a controlling seed length with a high LOD score of 16.69 and PVE 29.6% on LG 11, where harboring *ARF19* gene. The peak of the major QTL qSL11-a was located at the position of *ARF19*, controlling seed length with the highest contribution rate in this study. The result implied that *ARF19* is a strong candidate for the major factors affecting seed length (Fig. 2.12a, b).

2.5.2 Candidate Gene LD Mapping

We carried out a LD mapping in this QTL region. Fifteen amplicons were genotyped in the vicinity of the *ARF19* gene. The *ARF19* gene was mapped between two SNP markers, eSNP0536 and KASII [Ketoacylacyl carrier protein (ACP) synthase II gene] (Fig. 2.12b). KASII was found next to the marker Jcuint020. A striking level of linkage disequilibrium was observed among amplicons across the QTL region (Fig. 2.12c), with most of r^2 and D' values higher than 0.8. Association tests were performed in the 15 informative markers (Fig. 2.12c). The pattern of statistical association in the simple regression model indicated generally high P values in the 15 amplicons, with the highest [$-\log10(P) = 13.67$ and contribution rate $R^2 = 56.32$] in the *ARF19* gene (Ye et al. 2014). After controlling for population structure, a single nucleotide polymorphism (SNP) within the *ARF19* gene still showed significant signals (Fig. 2.12c).

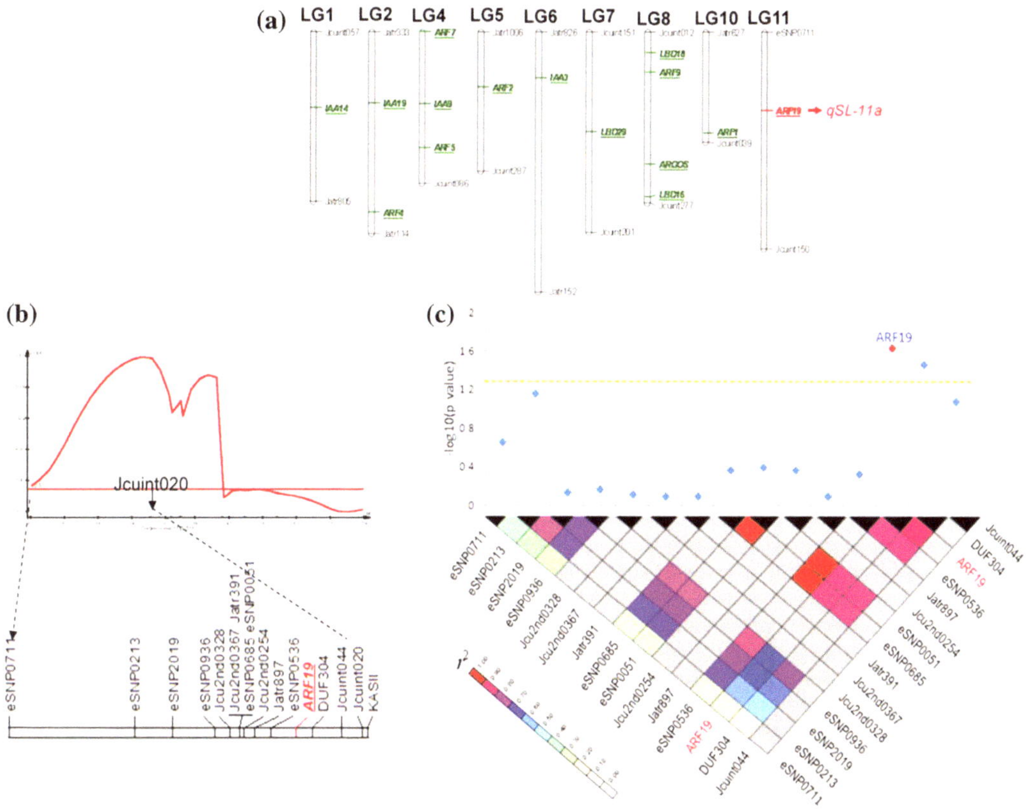

Fig. 2.12 Candidate gene mapping. **a** Jatropha ARF and Aux/IAA genes on the linkage map. **b** qSL11-a with LOD 17.1 on LG11, the QTL with strong effect on seed length, was detected in the region harboring ARF19. *Horizontal line* indicates 5% LOD significance thresholds (2.5) based on permutation. **c** LD mapping of nucleotide polymorphisms in qSL11-a and seed length across a diverse set of Jatropha accessions. An SNP within the ARF19 gene shows significant signal (Ye et al. 2014)

2.5.3 IAA9, Involved in the Auxin Signal Transduction, Defines Seed Length

Expressions of these genes were observed in the mapping population. We found that *IAA9* was mapped near a QTL qSL-4 of seed length affecting seed length. To further investigate the effects of *ARF19* and *IAA9* on seed size, we searched eQTLs for the two genes by investigating their expressions in the QTL mapping population. Expressions of *ARF19* and *IAA9* genes in leaf were higher in *J. integerrima* than in *J. curcas* but lower in *J. integerrima* seed. The expression levels of genes were correlated with each other in both of leaf and seed. The gene expressions were highly correlated with seed length and plant height. A total of four eQTLs were found to control *ARF19* and *IAA9* expression (Table 2.8 and Fig. 2.13). We detected an eQTL cluster controlling ARF19 and IAA9 gene expressions on LG 9, which harbors a QTL controlling plant growth and seed size (Sun et al. 2012). The results indicated that ARF19 and IAA9 together with their QTLs/eQTLs significantly affected plant growth and seed size.

2.5.4 ARF19 and IAA9 Interaction

The SNP was found in the domain IV of *ARF19* where a highly conserved amino acid serine (Ser1079) in *J. curcas* was mutated to glycine in *J. integerrima* due to a G/A mutation. Average seed length of plants with serine in the *ARF19*

Table 2.8 eQTLs for *ARF* and *IAA* genes (Ye et al. 2014)

Trait	eQTL[a]	Linkage group	Marker	Position (cM)[b]	LOD peak	R^2 (%)[c]	Additive effects[d]
Gene expression in leaf							
ARF19-L	*qARF19-9*	9	Jcuint079	5.81	1.80	4.2	−1.13
IAA9-L	*qIAA9-9*	9	Jcuint079	6.81	2.50	6.5	−1.34
Gene expression in seed							
IAA9-S	*qIAA9-2*	2	Jatr964	58.31	3.79	13.4	3.50
IAA9-S	*qIAA9-4*	4	Jatr872	31.11	3.31	11.6	1.82

[a]eQTL: starting with "q," followed by an abbreviation of the gene name, the name of the linkage group, and the number of eQTLs affecting the trait on the linkage group
[b]Position from the first marker on each linkage group
[c]Coefficient of determination or the percentage of variance explained (PVE) by the eQTL
[d]Estimated phenotypic effect of substituting *J. integerrima* alleles with *J. curcas* alleles at eQTL

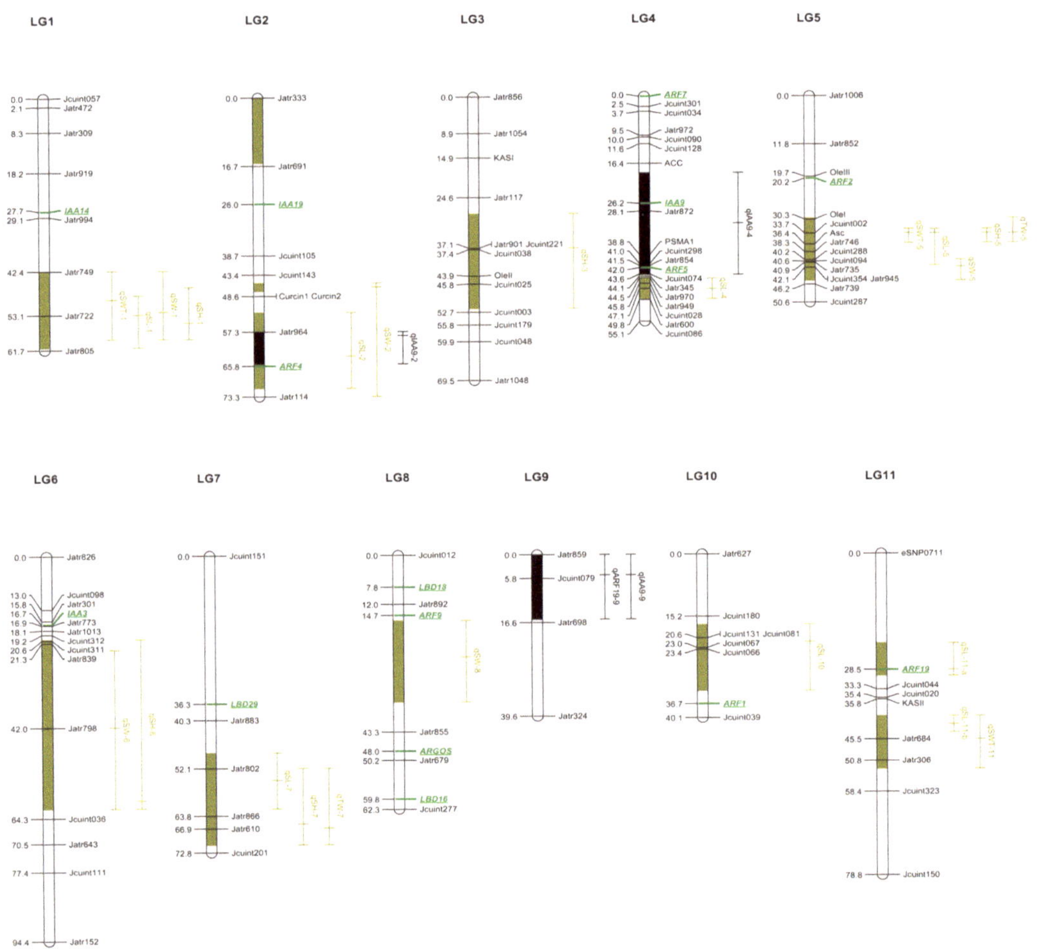

Fig. 2.13 Summary of QTL (eQTL) locations detected. QTL (eQTL) represented by bars are shown on the *left* of the linkage groups, close to their corresponding markers. The lengths of the bars are proportional to the confidence intervals of the corresponding QTLs (eQTLs) in which the *inner line* indicates position of maximum LOD score. ARF and Aux/IAA genes were highlighted in *bold, italic, underline*, and *green*. eQTL regions are *dark* on LGs (Ye et al. 2014)

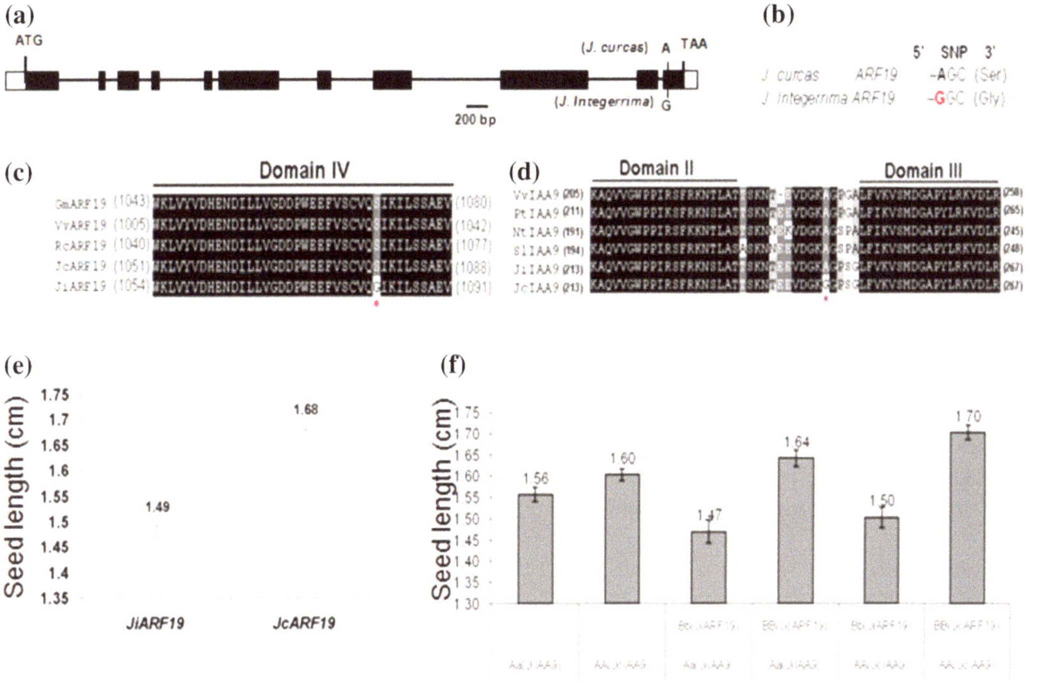

Fig. 2.14 Variants of ARF19 and IAA9 in Jatropha species differing in their seed length. **a** ARF19 gene structure and one key natural variation between alleles from *J. curcas* and *J. Integerrima*. *Unfilled box*: UTR; *filled box*: coding exon; lines connecting the boxes: intron. **b** Occurrence of SNP was identified in ARF19 in Jatropha with *longer* and *shorter* seed length. SNP was identified in ARF19 of *J. curcas* and *J. integerrima*. Amino acid alignment of ARF19 (**c**) and IAA9 (**d**). The red star in **c** and **d** showed the mutation of JiARF19 and JcIAA9 on conserved domains. Jc*Jatrophacurcas*, Ji*Jatrophaintegerrima*, Rc*Ricinuscommunis*, Vv*Vitisvinifera*, Pt*Populustrichocarpa*, Sl*Solanumlycopersicum*, Nt*Nicotianatabacum*. **e** Seed lengths of Jatropha plants with the serine or glycine in ARF19. **f** Seed lengths in plants with JcIAA9 and JiIAA9, and interaction between ARF19 and IAA9 (Ye et al. 2014)

gene (1.68 ± 0.01 cm) was significantly longer than that with glycine (1.49 ± 0.02 cm) (Fig. 2.14e). Because of the positive seed length/SNP correlation and the critical position of SNP in domain IV, we deduced that the SNP mutation was crucial to seed size of jatropha. We further investigated the interaction of *ARF19* and *IAA9* on seed length through ANOVA (Fig. 2.14f). Both genes affected seed size significantly (Ye et al. 2014). Seed length in plants with *JcIAA9* (1.60 ± 0.01 cm) was significantly longer than that with *JiIAA9* (1.56 ± 0.02 cm), suggesting a mild effect of *IAA9* on seed length. Further analysis on the difference of seed length among plants with a combination of *IAA9* and *ARF19* showed that a plant with *JcIAA9* and *JcARF19* (from *J. curcas*) was even longer (1.70 ± 0.02 cm), whereas plant with *JiIAA9* and *JiARF19* (from *J. integerrima*) was even shorter (1.47 ± 0.03 cm). This evidence suggested that *ARF19* and *IAA9* interacted with each other, affecting seed length phenotype, consistent with the results in eQTL analysis.

2.5.5 Gene Structure of ARF19 and IAA9

The intron–exon structure of this genomic region of *ARF19* in jatropha was conserved in fabids plants (Fig. 2.14a). Sequencing of the two jatropha full-length *ARF19*cDNAs (*JcARF19* and *JiARF19*) revealed the occurrence of SNPs in ARF19 domain IV, which was the protein–protein interaction site to mediate auxin signal transduction (Fig. 2.14b). A phylogenic tree was generated and showed that *ARF19* genes identified from Jatropha are homologs of *Arabidopsis thaliana*

ARF19 (Ye et al. 2014). The SNP present in the *J. curcas* and *J. integerrima* (Fig. 2.14b) and amino acid affected in key domains were indicated (Fig. 2.14c). Meanwhile, one SNP in *IAA9* was found in the linker region between domains II and III of *IAA9* (Fig. 2.14d). The amino acid involved in the SNP of *IAA9* conserved in many plants including *J. integerrima* as Alane, while encoded Glyine in *J. curcas* (Fig. 2.14d).

In conclusion, we provided evidences that *ARF19* and *IAA9* contributed to seed length phenotype. *ARF19* and *IAA9*, involved in auxin pathway signal transduction, are conserved in higher plants. It will be feasible to increase crop yield by integrating favored alleles of *ARF19* and *IAA9* into elite varieties of seed crops including jatropha.

Acknowledgements The work is part of the project "Genetic Improvement of Jatropha" initiated and coordinated by Professor Nam-Hai Chua, Rockefeller University, USA. We thank Weijie Tang, State Key Laboratory of Crop Genetics and Germplasm Enhancement, Nanjing Agricultural University, 210095, Nanjing, China, for his editing work on this chapter. The funders had no role in study design, data collection and analysis, decision to publish, or preparation of this chapter.

References

Almeida J, Achten WMJ, Duarte MP, Mendes B, Muys B (2011) Benchmarking the environmental performance of the Jatropha biodieselsystem through a generic life cycle assessment. Environ Sci Technol 45:5447–5453

Bailis RE, Baka JE (2010) Greenhouse gas emissions and land use change from Jatropha curcas-based jet fuel in Brazil. Environ Sci Technol 44:8684–8691

Berleth T, Krogan NT, Scarpella E (2004) Auxin signals—turning genes on and turning cells around. Curr Opin Plant Biol 7(5):553–563

Brown JR (2007) Comparative genomics: basic and applied research. CRC Press, Boca Raton

Carvalho CR, Clarindo WR, Praça MM, Araújo FS, Carels N (2008) Genome size, base composition and karyotype of *Jatropha curcas* L., an important biofuel plant. Plant Sci 174(6):613–617. doi:10.1016/j.plantsci.2008.03.010

Chhetri AB, Tango MS, Budge SM, Watts KC, Islam MR (2008) Non-edible plant oils as new sources for biodiesel production. Int J Mol Sci 9(2):169–180

Costa GG, Cardoso KC, Del Bem LE, Lima AC, Cunha MA, de Campos-Leite L, Vicentini R, Papes F, Moreira RC, Yunes JA (2010) Transcriptome analysis of the oil-rich seed of the bioenergy crop *Jatropha curcas* L. BMC Genom 11(1):462

Durbak A, Yao H, McSteen P (2012) Hormone signaling in plant development. Curr Opin Plant Biol 15(1):92–96

Fairless D (2007) Biofuel: the little shrub that could–maybe. Nature 449(7163):652–655

Goldstein DB, Schlotterer C (eds) (1999) Microsatellites: evolution and applications

Gu K, Chiam H, Tian D, Yin Z (2011) Molecular cloning and expression of heteromeric ACCase subunit genes from *Jatropha curcas*. Plant Sci 180(4):642–649

Gu K, Yi C, Tian D, Sangha JS, Hong Y, Yin Z (2012) Expression of fatty acid and lipid biosynthetic genes in developing endosperm of *Jatropha curcas*. Biotechnol Biofuels 5(1):47

Guimarães EP (2007) Marker-assisted selection: current status and future perspectives in crops, livestock, forestry and fish. Food and Agriculture Organization, Rome (incomplete)

Harper JL, Lovell P, Moore K (1970) The shapes and sizes of seeds. Annu Rev Ecol Syst 1:327–356

Harushima Y, Yano M, Shomura A, Sato M, Shimano T, Kuboki Y, Yamamoto T, Lin SY, Antonio BA, Parco A (1998) A high-density rice genetic linkage map with 2275 markers using a single F2 population. Genetics 148(1):479–494

Hayashi M, Miyahara A, Sato S, Kato T, Yoshikawa M, Taketa M, Hayashi M, Pedrosa A, Onda R, Imaizumi-Anraku H (2001) Construction of a genetic linkage map of the model legume *Lotus japonicus* using an intraspecific F2 population. DNA Res 8 (6):301–310

Heller J (1996) Physic nut. *Jatropha curcas* L. promoting the conservation and use of underutilized and neglected crops, 1. IBPGR, Roma (in full)

Hwang T-Y, Sayama T, Takahashi M, Takada Y, Nakamoto Y, Funatsuki H, Hisano H, Sasamoto S, Sato S, Tabata S (2009) High-density integrated linkage map based on SSR markers in soybean. DNA Res:dsp010

Ioio RD, Nakamura K, Moubayidin L, Perilli S, Taniguchi M, Morita MT, Aoyama T, Costantino P, Sabatini S (2008) A genetic framework for the control of cell division and differentiation in the root meristem. Science 322(5906):1380–1384

Jain S, Sharma MP (2010) Biodiesel production from *Jatropha curcas* oil. Renew Sustain Energy Rev 14 (9):3140–3147. doi:10.1016/j.rser.2010.07.047

Kumar S, Singh J, Nanoti SM, Garg MO (2012) A comprehensive life cycle assessment (LCA) of Jatropha biodiesel production in India. Bioresource Technol 110:723–729

Lee DJ, Park JW, Lee HW, Kim J (2009) Genome-wide analysis of the auxin-responsive transcriptome downstream of iaa1 and its expression analysis reveal the diversity and complexity of auxin-regulated gene expression. J Exp Bot 60(13):3935–3957

Li F, Kitashiba H, Inaba K, Nishio T (2009) A *Brassica rapa* linkage map of EST-based SNP markers for identification of candidate genes controlling flowering time and leaf morphological traits. DNA Res 16 (6):311–323

Liu P, Wang CM, Li L, Sun F, Liu P, Yue GH (2011) Mapping QTLs for oil traits and eQTLs for oleosin genes in jatropha. BMC Plant Biol 11:132
Meksem K, Kahl G (eds) (2006) The handbook of plant genome mapping: genetic and physical mapping. Wiley, Oxford
Okushima Y, Fukaki H, Onoda M, Theologis A, Tasaka M (2007) ARF7 and ARF19 regulate lateral root formation via direct activation of LBD/ASL genes in Arabidopsis. Plant Cell Online 19(1):118–130
Okushima Y, Mitina I, Quach HL, Theologis A (2005a) AUXIN RESPONSE FACTOR 2 (ARF2): a pleiotropic developmental regulator. Plant J 43(1):29–46
Okushima Y, Overvoorde PJ, Arima K, Alonso JM, Chan A, Chang C, Ecker JR, Hughes B, Lui A, Nguyen D (2005b) Functional genomic analysis of the AUXIN RESPONSE FACTOR gene family members in *Arabidopsis thaliana*: unique and overlapping functions of ARF7 and ARF19. Plant Cell Online 17(2):444–463
Openshaw K (2000) A review of *Jatropha curcas*: an oil plant of unfulfilled promise. Biomass Bioenergy 19(1):1–15
Panjabi P, Jagannath A, Bisht NC, Padmaja KL, Sharma S, Gupta V, Pradhan AK, Pental D (2008) Comparative mapping of *Brassica juncea* and *Arabidopsis thaliana* using intron polymorphism (IP) markers: homoeologous relationships, diversification and evolution of the A, B and C Brassica genomes. BMC Genom 9(1):113
Rafalski A (2002) Applications of single nucleotide polymorphisms in crop genetics. Curr Opin Plant Biol 5(2):94–100
Ren Y, Zhang Z, Liu J, Staub JE, Han Y, Cheng Z, Li X, Lu J, Miao H, Kang H (2009) An integrated genetic and cytogenetic map of the cucumber genome. PLoS One 4(6):e5795
Ruan Y-L, Patrick JW, Bouzayen M, Osorio S, Fernie AR (2012) Molecular regulation of seed and fruit set. Trends Plant Sci 17(11):656–665
Sanderson K (2009) Wonder weed plans fail to flourish. Nature 461:328–329
Sato S, Hirakawa H, Isobe S, Fukai E, Watanabe A, Kato M, Kawashima K, Minami C, Muraki A, Nakazaki N, Takahashi C, Nakayama S, Kishida Y, Kohara M, Yamada M, Tsuruoka H, Sasamoto S, Tabata S, Aizu T, Toyoda A, Shin-i T, Minakuchi Y, Kohara Y, Fujiyama A, Tsuchimoto S, Kajiyama S, Makigano E, Ohmido N, Shibagaki N, Cartagena JA, Wada N, Kohinata T, Atefeh A, Yuasa S, Matsunaga S, Fukui K (2011) Sequence analysis of the genome of an oil-bearing tree, *Jatropha curcas* L. DNA Res 18(1):65–76
Shah S, Sharma A, Gupta MN (2005) Extraction of oil from *Jatropha curcas* L. seed kernels by combination of ultrasonication and aqueous enzymatic oil extraction. Bioresour Technol 96(1):121–123. doi:10.1016/j.biortech.2004.02.026
Sladek R, Hudson TJ (2006) Elucidating *cis*-and *trans*-regulatory variation using genetical genomics. Trends Genet 22(5):245–250
Sun F, Liu P, Ye J, Lo LC, Cao S, Li L, Yue GH, Wang CM (2012) An approach for jatropha improvement using pleiotropic QTLs regulating plant growth and seed yield. Biotechnol Biofuels 5:1–10
Thumma BR, Naidu BP, Chandra A, Cameron DF, Bahnisch LM, Liu C (2001) Identification of causal relationships among traits related to drought resistance in *Stylosanthes scabra* using QTL analysis. J Exp Bot 52(355):203–214
Varshney RK, Tuberosa R (2007) Genomics-assisted crop improvement, vol 1. Springer, Berlin
Wang CM, Bai ZY, He XP, Lin G, Xia JH, Sun F, Lo LC, Feng F, Zhu ZY, Yue GH (2011a) A high-resolution linkage map for comparative genome analysis and QTL fine mapping in Asian seabass, *Lates calcarifer*. BMC Genom 12(1):174
Wang CM, Liu P, Yi C, Gu K, Sun F, Li L, Lo LC, Liu X, Feng F, Lin G (2011b) A first generation microsatellite-and SNP-based linkage map of Jatropha. PLoS One 6(8):e23632
Wang WY, Barratt BJ, Clayton DG, Todd JA (2005) Genome-wide association studies: theoretical and practical concerns. Nat Rev Genet 6(2):109–118
Wu J, Liu Y, Tang L, Zhang F, Chen F (2011) A study on structural features in early flower development of *Jatropha curcas* L. and the classification of its inflorescences. Afr J Agr Res 6:275–284
Xia JH, Liu F, Zhu ZY, Fu J, Feng J, Li J, Yue GH (2010) A consensus linkage map of the grass carp (*Ctenopharyngodon idella*) based on microsatellites and SNPs. BMC Genom 11(1):135
Xue L-J, Zhang J-J, Xue H-W (2009) Characterization and expression profiles of miRNAs in rice seeds. Nucleic Acids Res 37(3):916–930
Yang JH, Han SJ, Yoon EK, Lee WS (2006) Evidence of an auxin signal pathway, microRNA167-ARF8-GH3, and its response to exogenous auxin in cultured rice cells. Nucleic Acids Res 34(6):1892–1899
Ye J, Qu J, Bui HTN, Chua NH (2009) Rapid analysis of *Jatropha curcas* gene functions by virus-induced gene silencing. Plant Biotechnol J 7:964–976
Ye J, Hong Y, Qu J, Wang C (2012) Improvement of Jatropha oil bygenetic transformation. Springer Science Publishers, New York
Ye J, Liu P, Zhu CS, Qu J, Wang XH, Sun YW, Sun F, Jiang YL, Yue GH, Wang CM (2014) Identification of candidate genes JcARF19 and JcIAA9 associated with seed size traits in Jatropha. Funct Integr Genom 14(4):66–757
Yin Z, Meng F, Song H, Wang X, Xu X, Yu D (2010) Expression quantitative trait loci analysis of two genes encoding rubisco activase in soybean. Plant Physiol 152(3):1625–1637
Zhang Y, Wang Y, Jiang L, Xu Y, Wang Y, Lu D, Chen F (2007) Aquaporin JcPIP2 is involved in drought responses in *Jatropha curcas*. Acta Biochim Biophys Sin 39(10):787–794

Transcription Factors in Jatropha

3

Keiichi Mochida and Lam-Son Phan Tran

Abstract

Sequence-specific transcription factors (TFs) assume the role of molecular switches that regulate various biological processes in transcriptional networks. Therefore, genome-wide identification and functional characterization of TF-encoding genes is a practical approach to elucidate cellular systems as well as to discover the master switch that regulates cellular functions. In this chapter, we addressed genome-wide TF repertory in jatropha (*Jatropha curcas*) plant. In the jatropha genome, we identified 1481 putative TF-encoding gene models that are classified into 61 TF families by using a computational pipeline that we designed to identify plant TFs. Bioinformatics-based analyses used to characterize the putative TF-encoding genes, such as gene ontology annotation, structural analysis of promoter regions, and comparative analysis of putative orthologs, will aid us in inferring functions of each TF. Finally, we

K. Mochida (✉)
Cellulose Production Research Team, Biomass Engineering Research Division, RIKEN Center for Sustainable Resource Science, 1-7-22 Suehiro-Cho, Tsurumi-Ku, Yokohama, Kanagawa 230-0045, Japan
e-mail: keiichi.mochida@riken.jp

K. Mochida
Kihara Institute for Biological Research, Yokohama City University, 641-12 Maioka-Cho, Totsuka-Ku, Yokohama, Kanagawa 244-0813, Japan

L.-S.P. Tran (✉)
Signaling Pathway Research Unit, RIKEN Center for Sustainable Resource Science, 1-7-22 Suehiro-Cho, Tsurumi, Yokohama 230-0045, Japan
e-mail: son.tran@riken.jp

K. Mochida
Institute of Plant Science and Resources, Okayama University, Chuo 2-20-1, Kurashiki, Okayama 710-0046, Japan

S. Tsuchimoto (ed.), *The Jatropha Genome*, Compendium of Plant Genomes,
DOI 10.1007/978-3-319-49653-5_3

introduce a Web-accessible database TreeTFDB (http://treetfdb.bmep.riken.jp/index.pl), housing information on the TF sets from jatropha genome and five other tree species, cassava (*Manihot esculenta*), poplar (*Populus trichocarpa*), castor bean (*Ricinus communis*), and grapevine (*Vitis vinifera*).

3.1 Transcription Factors in Plants

Sequence-specific transcription factors (TFs) act as the molecular switch that regulates various biological processes such as development, morphogenesis, responses to environmental stimuli, and metabolic processes at transcriptional level (Tran and Mochida 2010; Tran et al. 2010). Transcription of target genes is activated or repressed through recognizing and binding of TFs to specific *cis*-regulatory DNA sequences at their promoter regions. Because of the hierarchical regulatory structure between a TF and multiple target genes in a transcriptional regulatory network, the TF-encoding gene is often a key gene or a master gene of certain biological functions, such as physiological response and metabolism. Molecular interactions of TFs with other transcriptional regulators or with noncoding RNAs are also crucial in the transcriptional regulatory network. Binding of TFs to specific *cis*-regulatory DNA sequences (*cis*-motifs) mediates transcription of the target genes; therefore, variety of combinations of *cis*-motifs in the promoter region enables spatial and temporal gene expression. Identification of the TFs and their target genes is primary importance when deciphering the transcriptional regulatory network involved in a biological phenomenon as well as in obtaining key regulatory genes to engineer biological function of interest.

In the year 2000, the whole genome sequence of a model plant, *Arabidopsis thaliana*, was the first reference genome sequence to be determined in higher plants. Nearly, 1500 genes encoding TFs were identified from ca. 26,000 genes annotated in its genome (Riechmann et al. 2000). The genome-wide identification of TF-encoding genes in *Arabidopsis* determined the distribution of gene families with DNA-binding domains (DBDs) including the plant-specific ones. In the past decade, the availability of datasets of whole genome sequences of many plant species has allowed us to identify the TF-encoding genes across the genome and to elucidate complex transcriptional regulatory networks in various plants using high-throughput experimental techniques and computational methods (Mochida and Shinozaki 2010). Recent revolutionary improvements in DNA sequencing technologies have accelerated the process of the whole genome sequencing in various plant species, including those with larger genomes (Mochida and Shinozaki 2011, 2013). Such technological innovations in life science have provided opportunity to discover TF-encoding genes and to explore *cis*-regulatory motifs at the genome-wide level. Identification, characterization and functional annotation of TF repertoires, and annotation of *cis*-motifs in promoter regions of genes at the genome-wide level provide an essential information resource to promote functional analysis of TFs by combinatorial use of high-throughput experimental techniques and mutant plant resources. Furthermore, integration of information of TF repertories from various plant species will provide an insight into the evolution of TFs and their regulatory networks.

To date, various functions of a number of TFs involved in biological processes such as development, response to environmental stimuli, and metabolic processes have been elucidated based on molecular genetics and omic analyses in model plants such as *Arabidopsis* and rice (*Oryza sativa*). In plant development, the spatio-temporal expression of TF-encoding genes in particular cells enables well-ordered expression of downstream genes involved in processes of organogenesis. One of the typical examples is the regulatory transcriptional network in flower development (Kaufmann et al. 2010b; O'Maoileidigh et al. 2014). A number of TFs have been characterized in developmental regulations such as phytohormone signaling pathways (Ohashi-Ito et al. 2013; Aya et al. 2014). In crop species, particular mutation of TFs involved in plant morphology was selected during the domestication and breeding (Doebley et al. 2006; Sakuma et al. 2011). In response to environmental and biological stimuli, a number of TFs are up-regulated in cells with the perception of environmental changes (Yamaguchi-Shinozaki and Shinozaki 2006; Tran et al. 2007b; Umezawa et al. 2007; Hirayama and Shinozaki 2010; Yan et al. 2014). Some of stress-inducible TFs have been known to activate downstream genes that are involved in cellular responses to stress tolerance. TFs that regulate biosynthesis of particular metabolites could be the key genes that will help elucidate the diversified and complex plant metabolome as well as the gene resources that will be used to artificially synthesize and engineer metabolic processes of useful compounds (Hirai et al. 2007; Li et al. 2013; Ji et al. 2014; Lee et al. 2014; Zhang et al. 2014; Zhu et al. 2014).

Jatropha (*Jatropha curcas*) is a valuable, nonedible energy crop used for fuel production such as biodiesel (Abdulla et al. 2011). Therefore, identification, characterization, and functional annotation of the TF repertoires in jatropha genome will provide a significant information basis not only for understanding biological properties but also for accelerating molecular breeding of this crop species. Genome-wide comparison of the TF repertories with other plant species will impart the evolution of the TFs and transcriptional regulatory networks. Furthermore, information resources on TF annotations will facilitate the study of master regulators involved in plant productivity, fruit quality, and sensing/response and adaptation to environmental changes in jatropha.

In this chapter, we focused on the genome-wide TF repertoire in jatropha plant. We introduced a computational procedure that we established to identify putative TF-encoding genes in plants and its application to the jatropha genome. Then, we described the genome-wide properties of jatropha TFs including the results of the comparative analyses with TFs of other plant species, and functional annotations in various contexts such as *cis*-motif, gene ontology (GO), and gene expression patterns. Finally, we introduced a Web-accessible database of TFs, TreeTFDB (http://treetfdb.bmep.riken.jp/index.pl), to accelerate the research pertaining to tree biotechnology such as biomass engineering for development of bio-based energy and material resources through genetic engineering of TFs. TreeTFDB is the first database housing information on TF sets from jatropha, and it also comprises five other tree species, including cassava (*Manihot esculenta*), poplar (*Populus trichocarpa*), castor bean (*Ricinus communis*), and grapevine (*Vitis vinifera*).

3.2 Genome-wide Identification of TF-Encoding Genes

In 2000, Riechmann et al. compared the genome-wide TF repertories among *Arabidopsis*, fly, worm, and budding yeast and revealed a differential composition of genes encoding proteins in TF families between higher plants, animals, and yeast. In *Arabidopsis*, they represented TF families with plant-specific DNA-binding domains such as EREBP/AP2, NAC, WRKY, ABI3/VP1, ARF, and some of C2C2-Zn (CO-like, Dof, and YABBY). The protein families of Myb, bHLH, MADS, and HSP, which are commonly found in plants and animals, showed an increased number of members of genes in *Arabidopsis* compared with those in animals (Riechmann et al. 2000). Such genome-wide identification and comparative analysis of TF repertories among organisms provided an insight into the evolution of TFs and biological systems (Feller et al. 2011; Yamasaki et al. 2013) in species lineages. Furthermore, generation of a genome-wide catalogue of TF-encoding genes in species provides the primary information resources for functional elucidation based on molecular genetics and omics in combination with various bio-resources and genomic tools (Mochida and Shinozaki 2010).

With recent acceleration of the whole genome sequencing, a framework to rapidly explore genome-wide TF repertories is required for the construction of a comprehensive TF catalogue in the target species as well as for the comparison of TF compositions among species. We established computational pipeline procedure for genome-wide identification of putatively TF-encoding genes in plants (Mochida et al. 2009b). To identify such genes in genome-sequenced plant species, we prepared 62 hidden Markov models (HMMs) corresponding to DNA-binding domains of 61 TF families found in plants. Of those, 51 HMMs were retrieved from Pfam database (Finn et al. 2014), and other 11 models were originally created using HMMbuild of the HMMER2 package (http://hmmer.janelia.org/) based on multiple alignments of homologous genes. With the dataset of HMMs, the initial computational search of TFs in each family is performed using Pfam-HMM search program with threshold for E value set to $E < 1e-5$. To further classify genes with MYB domain, the MYB super family was divided into three subgroups, (R1)R2R3_MYB, MYB_related, and atypical_MYB. The putative MYB proteins were searched against previously classified *Arabidopsis* MYB genes using BLAST search ($E < 1e-5$), and each top hit combination was applied to the subgroup classification. To avoid possible contamination with a pseudo-response regulator or histidine kinase in the GARP_ARRB family, genes containing domains of CCT, CHASE, HATPase_c, and HisKA together with Response_reg of Pfam domains were searched by InterProScan (Mulder and Apweiler 2008). Genes with these domains were subsequently removed from the GARP_ARRB family. Finally, our computational procedure provided a list of putative TF-encoding genes in 61 families (including three MYB subfamilies) of TF repertories.

Initially, we applied this computational procedure to identify TF-encoding genes in soybean (Mochida et al. 2009b; Le et al. 2011; Ha et al. 2013). In soybean, the putative TF-encoding genes predicted by DBD prediction using Pfam-HMM were classified into four categories based on their certainty of prediction as follows: The first group of TFs (Category A) consists of TF-encoding genes showing sequence similarity (BLASTN search with identity thresholds $\geq 95\%$ and $E \leq 1e-100$) with GenBank entries with a functional description as TFs in soybean. Category A genes were classified with the highest confidence level based on the GenBank entries with associated citations of the PubMed database. The second group of TFs (Category B) comprises TFs with putative homologs in well-annotated plants species *Arabidopsis* and/or rice with significant sequence similarities (blastp $E \leq 1e-30$). The third group of TFs (Category C) is comprised of possible TFs that show a significant hit with each of the HMMs used for DBD prediction (Pfam-HMM $E \leq 1e-20$). The last group contains TFs that have promiscuous hit with each of the HMMs used for DBD prediction (Pfam-HMM $E \leq 1e-5$). Genes showing ambiguous DBD hits were manually qualified by referencing to sequence similarity with TFs annotated in *Arabidopsis* and/or rice. We have already applied this computational procedure to 13 species in legumes, grasses, and tree crops. The catalogues of TF-encoding genes in various species created by using a consolidated prediction method have allowed us to compare genome-wide compositions of TF genes and integrated information resources on TFs in these plants.

3.3 The TF Repertory of Jatropha

To identify the TF-encoding genes in jatropha genome, we applied the computational procedure described in Sect. 3.2 (Mochida et al. 2013b) (Fig. 3.1). The whole genome sequence and gene structural annotations were retrieved from the jatropha genome database (JAT_r3.0) maintained by the Kazusa DNA Research Institute (http://www.kazusa.or.jp/jatropha/) (Sato et al. 2011) and subjected to HMM-based search in each of the TF families. In total, 1481 putative TF-encoding gene models showed a significant match with DBDs in each TF family (Table 3.1), representing 2.52% of

Table 3.1 Predicted TF models in the jatropha genome

TF families	No. of TF-encoding gene models	TF families	No. of TF-encoding gene models
(R1)R2R3_Myb	110	HRT	1
ABI3VP1	54	HSF	21
AP2_EREBP	111	JUMONJI	16
ARF	14	LFY	1
ARID	5	LIM	7
Alfin-like	5	LUG	1
Aux_IAA	26	MADS	54
BBR-BPC	4	MBF1	2
BES1	7	Myb_related	34
C2C2_Zn-CO-like	22	NAC	86
C2C2_Zn-Dof	22	Nin-like	8
C2C2_Zn-GATA	19	PHD	83
C2C2_Zn-YABBY	7	PLATZ	9
C2H2_Zn	86	PcG	29
C3H-TypeI	59	S1Fa-like	1
CAMTA	3	SAP	1
CCAAT_Dr1	1	SBP	14
CCAAT_HAP2	6	SRS	6
CCAAT_HAP3	9	TCP	17
CCAAT_HAP5	7	TUB	6
CPP	6	Trihelix	12
E2F_DP	7	ULT	2
EIL	3	VOZ	1
GARP_ARRB	5	WRKY_Zn	59
GARP_G2-like	23	Whirly	0
GRAS	41	ZIM	11
GRF	5	atypical_MYB	19
GeBP	6	bHLH	87
HB	56	bZIP	136
HMG-box	12	zf-HD	9
		zf-TAZ	7
Total (nonredundant)			1481

58,720 gene models that have been annotated in JAT_r3.0, which includes partial and/or pseudo-genes (Sato et al. 2011). This percentage of TFs to the total number of annotated genes was slightly smaller than that in *Arabidopsis*. Generally, ~7% of the total number of genes are classified into TFs (Riechmann et al. 2000; Guo et al. 2005; Mitsuda and Ohme-Takagi 2009; Mochida et al. 2011). In the *Arabidopsis* genome, for example, there are 1968 TFs, which account for 7.23% of the total number of genes in reference to plant TF database (PlnTFDB, http://plntfdb.bio.uni-potsdam.de/v3.0/). As another example, in the rice genome, 1597 TFs [3.06% of proteome annotations in the Rice Annotations Project Database (RAP DB ver2), http://rapdb.

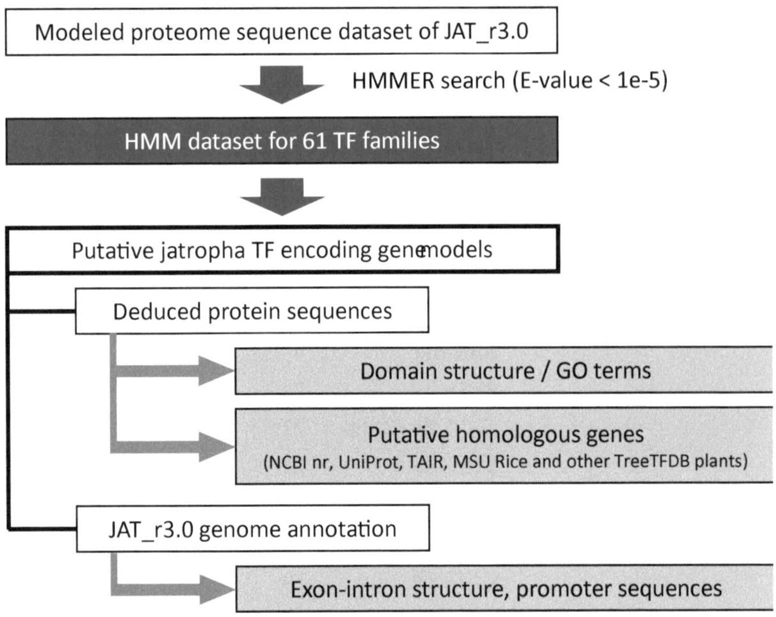

Fig. 3.1 Schematic workflow of the computational pipeline used to discover and annotate genes encoding putative TFs in jatropha

dna.affrc.go.jp/] and 2095 TFs (3.69% of annotations reported in Rice Genome Pseudomolecules Release 6, http://rice.plantbiology.msu.edu/annotation_pseudo_current.shtml) are predicted as TF-encoding genes in GramineaeTFDB (http://gramineaetfdb.psc.riken.jp) and PlnTFDB, respectively. Among the putative TFs found in the jatropha genome, those of 912 genes showed significant similarities with putative homologs of TFs predicted in *Arabidopsis* and/or rice. Further comparative analyses will facilitate inferring the functions of these putatively homologous counterparts of TF-encoding genes in model plants. Phylogenetic analysis of each TF family will determine lineage specific members of the TFs in jatropha. In queried HMMs to the jatropha genome, the domain search for the TF Whirly family did not result in any genes with corresponding DBD. The number of predicted TFs in each family might be modified along with future updating of genome sequences and of the version of gene structural annotation. Currently available assembly of jatropha genome sequence is based on Illumina short reads and longer reads from Roche 454 sequencers (Sato et al. 2011). Computational methodology for de novo genome assembly and newly added available sequence resources of jatropha genome from various types of sequencing platforms will further decipher jatropha genome. Additionally, comprehensive collection of transcriptome sequences based on RNA sequencing (RNA-seq) analysis will improve gene structural annotation and confirm their expression (Eksi et al. 2013).

3.4 Functional Annotations of Jatropha TFs

The majority of TFs in emerging crop species such as jatropha remains experimentally uncharacterized and with unknown function. Therefore, comparative genomics analysis between jatropha TFs and those in model plant species, such as *Arabidopsis* and rice, may provide significant clues of their functions. The GO provides a common context for classifying gene functions, and it might be useful for inferring functions of TFs (Mochida et al. 2009b, 2011, 2013b). We assigned GO terms to the predicted jatropha TFs based on sequence similarity with their respective *Arabidopsis* TFs. Sequence similarity search (blastp $E \leq 1e{-}10$) found 1361 TFs hits to *Arabidopsis* counterparts with at least one GO annotation. For example, under the ontology of biological process, TFs were

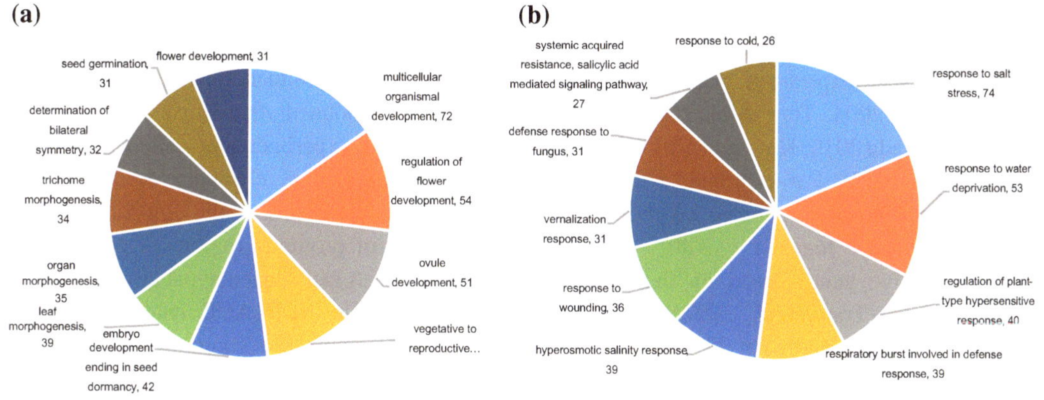

Fig. 3.2 Representative distributions of the Gene Ontology (GO) terms for biological processes associated with jatropha transcription factor encoding genes. The top 11 abundant GO terms detected in "developmental process" (**a**), and the top 10 GO terms detected in "response to stress" (**b**) category were assigned based on sequence similarity search against annotated *Arabidopsis* genes. *Gene numbers* are displayed next to the terms

classified into GO slim categories of "response to stress" and "developmental processes." Figure 3.2a, b shows the top 21 and 20 for "response to stress" and "developmental processes," respectively, of the lower class GO categories of "response to stress" and "developmental processes," respectively. TFs classified in "response to stress" were further classified into putative subprocesses of response to abiotic and biotic stress factors. Those in "developmental processes" were further classified into putative subprocesses, such as organogenesis and morphogenesis. Although molecular genetics studies are needed to elucidate molecular functions of TFs, comparative analysis with model species may facilitate creating a hypothesis regarding molecular functions of TFs prior to experiments.

Phylogenetic relationship between jatropha TFs and putative homologs of model plants will also provide useful information for inferring and comparing their functions in the evolutionary contexts. In various species, a combinatorial approach of gene expression analysis and phylogenetic analysis revealed functional divergence within TF families (Mangelsen et al. 2008; Stevens et al. 2008). For example, the dehydration responsive members of the soybean NAC family of at least 152 TF-encoding genes were first predicted based on phylogenetic analysis with NACs from model plants, and then verified using quantitative real-time PCR analysis (Le et al. 2011). Recent advances in sequencing technologies have enabled us to rapidly access transcriptome datasets of tissues of interest in a target species. Until now, 59 entries of the RNA sequencing of *J. curcas* are publicly available from NCBI SRA (as of Nov. 26, 2014). Following the genome-wide identification of TFs, survey of the resources of available transcriptome datasets will provide transcriptional features of uncharacterized TFs. Expression profiles of TFs in various tissues are essential to reveal those function in transcriptional networks involved in diverse biological processes in jatropha. Furthermore, integrated functional annotations of TFs provide a significant knowledge foundation to narrow down useful candidate TFs that can be applied to improve productivity of jatropha by biotechnological approaches.

3.5 *Cis*-Motif Analysis of Jatropha Genome

Specific interactions between *cis*-motif sequences at promoter regions of target genes and DBDs of TFs represent a central molecular machinery that regulates special and temporal expression of target genes (Tran et al. 2007c; Thao and Tran

2012). It has been reported that various *cis*-motifs have been essential for tissue specific and stress-responsive expression patterns of genes (Kasuga et al. 2004; Tran et al. 2007b; Tsuwamoto and Harada 2010; Kim et al. 2011; Figueroa and Browse 2012). For example, a number of environmental stress-responsive *cis*-motifs have been reported together with identified stress-responsive genes based on the transcriptome analyses (Tran et al. 2007a). Elucidated interaction between particular TFs and *cis*-motifs has considered regulatory networks in response to abiotic stress conditions such as cold, salt, heat, and drought in plants (Hirayama and Shinozaki 2010). Such information on the *cis*-motif architecture at the promoter regions has provided essential knowledge for us to artificially manipulate transcriptional networks. Transgenic plants with stress-responsive promoters and TF-encoding genes that execute particular regulatory networks in response to environmental stimuli have demonstrated enhanced tolerance to adverse conditions (Kasuga et al. 1999, 2004; Nakashima et al. 2007). Moreover, recent computational analysis of the promoter regions and transcriptome analysis suggested positive correlation between multi-stimuli response genes and density of *cis*-elements in the promoter regions (Vandepoele et al. 2009). Therefore, in addition to the genome-wide identification of TF repertories, comprehensive analysis of promoter architecture with *cis*-motif distribution will effectively aid in functional prediction of genes and elucidating the transcriptional regulatory networks.

Within abiotic stress response, 13 major stress-responsive *cis*-motifs, of which 12 are stress-inducible motifs and one is drought-repressible motif, were reported previously (Yamaguchi-Shinozaki and Shinozaki 2005). In addition, a number of *cis*-motifs that are essential for regulation of various biological processes such as development and hormonal response have been identified in a number of studies on the promoter function in plants. The PLACE database (http://www.dna.affrc.go.jp/PLACE/) is an information resource that consolidates published *cis*-motifs (Higo et al. 1999). The AtcisDB (Arabidopsis *cis*-regulatory element database) in AGRIS Web site (http://arabidopsis.med.ohio-state.edu/) is another comprehensive information resource that consists of the upstream regions of annotated *Arabidopsis* genes with a description of experimentally validated and predicted *cis*-regulatory elements (Yilmaz et al. 2011). To characterize the global *cis*-motif architecture of promoter regions of genes in jatropha, we searched sequence patterns of 467 *cis*-motifs housed in the PLACE database, 12 major abiotic stress-inducible *cis*-motifs, and 99 *cis*-motifs in AtcisDB in promoter regions of genes in the jatropha genome. The upstream 500-, 1000-, and 3000-bp sequences from the putative transcription start site for each gene were retrieved from the jatropha genome sequence (JAT_r3.0) based on the gene structural annotation. The sequence pattern of each *cis*-motif was searched, and a compile of the distributions in the promoter regions was created. The comprehensive dataset on the *cis*-motifs in jatropha will be a useful resource for eliciting promoter functions of genes as well as in designing further experiments such as yeast one-hybrid analysis and ChIP-seq analysis to elucidate molecular interactions in particular regulatory network modules (Kaufmann et al. 2010a; Zhu et al. 2012). In *Arabidopsis*, integrated analysis of transcriptome datasets and promoter structure information represented *cis*-motifs commonly found at gene promoters in response to various stress environments as well as those specifically associated with genes in response to particular stress environments as secondary motifs (Ma and Bohnert 2007).

In model organisms, ChIP-seq analyses of TFs have been carried out to identify TF-occupied sequences, which could provide evidence of molecular interactions between TFs and promoters of candidate genes. In addition to *cis*-motifs, epigenetic modifications of promoter regions such as hypermethylation and chromatin modifications are another important paradigm that affects the interaction between promoters and TFs, altering the expression patterns of genes (Fojtova et al. 2003; Hemmes et al. 2012). Combining epigenome profiling datasets together with *cis*-motif and gene expression in model organisms will provide new insights into the transcriptional regulatory networks. In crops such as jatropha, comparative genomics with model organisms emphasize the importance of rapid knowledge integration from model organisms and application in breeding process.

3.6 TreeTFDB

With the increase of the available data on completed genome sequences in plants, the genome-wide identification of TF repertories and compilation of the related information in the Web-accessible databases is possible for various plant species. For examples, RARTF (Iida et al. 2005) and DRTF (Gao et al. 2006) databases for *Arabidopsis* and rice, respectively, and SoybeanTFDB (Mochida et al. 2009b) and SoyDB (Wang et al. 2010) for soybean have been available as species-specific databases for catalogued TFs in these particular plant species. Databases of TFs from multiple species, such as Grassius (Yilmaz et al. 2009) and GramineaeTFDB (Mochida et al. 2011) for grass species, LegumeTFDB (Mochida et al. 2010) for legume species, and PlnTFDB (Perez-Rodriguez et al. 2010) and PlantTFDB (Zhang et al. 2011; Jin et al. 2014) covering a wider range of plant species, have also been established on the Web. The SoybeanTFDB (http://soybeantfdb.psc.riken.jp/), LegumeTFDB (http://legumetfdb.psc.riken.jp/), and GramineaeTFDB (http://gramineaetfdb.psc.riken.jp) have been established by the authors. As the fourth lineup of our TF databases, TreeTFDB (http://treetfdb.bmep.riken.jp/index.pl) comprises the TF sets from six tree crops: jatropha, papaya, cassava, poplar, castor bean, and grapevine. TreeTFDB was the first database that included the TFs from jatropha.

All the data generated through computational analysis of jatropha TFs together with TFs from the other five tree species were integrated to develop TreeTFDB and establish a convenient information resource of TFs from genome-sequenced tree species. The database implements MySQL (Oracle Corporation, Redwood City, CA, USA) as the relational database management system. The Web interface of the database is designed by Perl CGI and Java script running on the Apache Web server. The definition strings of hit subjects resulted from the sequence similarity searches of protein domains that are queried by InterProScan, the *cis*-motif names from PLACE and AtcisDB databases, stress-responsive *cis*-motifs, and the assigned GO terms were assembled into a keyword database to enable users to query TFs by various associated annotations. A BLAST server was implemented to provide a similarity search interface for queried sequences using NCBI BLAST together with sequences of the six tree species as well as those of *Arabidopsis*. Generic Genome Browser (GBrowse) (Donlin 2009) was also implemented for genome sequence and gene structural annotation of each of the six tree species to visualize the gene structure together with *cis*-motifs found in the promoter regions. Multiple sequence alignments of TFs sharing conserved protein sequences were provided, which were generated from clustered TF proteins at different identity levels (30, 60, and 90%) in the six tree species. These functionalities of the user interface have been well implemented not only for the TF databases but also for other Web-accessible information resources established by the authors (Mochida et al. 2009a, 2013a). Cross-references of the corresponding identifiers in other resources such as putative homologs and cDNAs were also implemented into the database with the URLs to the original referenced data in order to provide hyperlinks on the Web interface with seamless navigations. All of the data in the TreeTFDB are accessible not only through a Web interface but also as downloadable files from the Web site.

The Web interface of TreeTFDB is represented in Fig. 3.3, and detailed functionalities of the database are also introduced in the Help page. The detailed page of each entry in TreeTFDB consists of the following 10 properties: Summary, HMM search results, sequence, BLAST results, clustered protein family, corresponding to other identifiers, Gene Ontology, InterProScan result, and *cis*-motif prediction. The Summary part includes the accession identifier, family of TFs, lengths of cDNA and predicted proteins, link to the genome browser, and embedded genome browser image (Fig. 3.3a). The HMM search part provides raw data of HMM search results, which includes the HMM data name for TF prediction, hit score and *E* value, and hit region of the queried protein sequence and its peptide sequence (Fig. 3.3b). Sequences of the cDNA and protein are shown in the FASTA format in the sequence part. The two "similarity search" buttons allow users to navigate to the similarity search page using respective sequences as queries (Fig. 3.3c). The section for BLAST results provides the results of a BLAST-based sequence similarity search of the respective entry used as a query against various databases, including NCBI nr, UniProt, *Arabidopsis* proteins of TAIR, rice proteins of TIGR/MSU RICE, poplar proteins of Phytozome, as well as proteins of the six tree species used in TreeTFDB. Each listed dataset consists of the name of the database used, blast program used, and description of the hit entries with scores and *E* values. Each description includes internal hyperlinks to the raw data of each blast result. Columns for the *Arabidopsis* and rice databases show the cross-references and links to representative databases of TFs established in each organism, including DATF, RARTF, PlnTFDB, and AtTFDB in *Arabidopsis* and DRTF, Grassius, and PlnTFDB in rice. In columns for the database of *C. papaya, J. curcas, M. esculenta, P. trichocarpa, R. communis* and *V. vinifera*, internal hyperlinks lead the users to a detailed page of hit entries in TreeTFDB itself (Fig. 3.3d). By using the CD-HIT package (Li and Godzik 2006), the entries of the six tree species are hierarchically clustered into protein clusters using global amino acid sequence identity thresholds of 100–30% in 10% decrements. The hierarchically clustered data are imported into the Web-based hierarchical-structure-viewing interface, showing the results in thresholds of 30, 60 and 90%. On the interface, users also can view multiple alignments for each clustered TFs (Fig. 3.3e). All cDNA sequences of entries in TreeTFDB are assigned to possible identical sequences found in the following cDNA-related resources of corresponding species: Affymetrix GeneChip target sequence, TIGR Gene Index (TIGR GI), NCBI UniGene, and PlantGDB. Although currently there are no cDNA-related sequence resources for jatropha, these information will be integrated to the database in the future when jatropha transcript resources are available. The Gene Ontology part consists of GO terms associated with the TF entry by protein domain search using InterProScan or sequence similarity search to find the *Arabidopsis* counterpart with GO annotations using BLASTP (Fig. 3.3f). The "InterProScan result" part shows an embedded figure of the results of InterProScan (Fig. 3.3g). The *cis*-motif prediction part provides *cis*-motif prediction results by patterns matched with *cis*-motifs. By clicking on either "500 bp," "1000 bp," or "3000 bp," the users can view the single-base resolution architecture of the 500-, 1000-, or 3000-bp promoter regions and the genomic sequence of the TF-encoding gene, together with the identified *cis*-motifs (Fig. 3.3h).

Integration of various available annotations, such as *cis*-motifs, GO annotations and putative homologs in model organisms, assists in inferring the function of TFs. Therefore, TreeTFDB is a useful information resource that aids in functional prediction of jatropha TFs as well as in building hypotheses prior to experiments to confirm jatropha TF functions. Further implementation of transcriptome datasets and information about available mutants will make the database more attractive. For example, in the GramineaeTFDB, our TF database for grass species, we supplied hyperlinks to expression profiles of the TF genes of maize, rice and barley, and thanks to the available public GeneChip datasets. Information about the availability of FOX and Ds mutant lines in rice and maize TFs, respectively, are also provided in GramineaeTFDB (Mochida et al. 2011). For crop species such as jatropha, proactive integration of

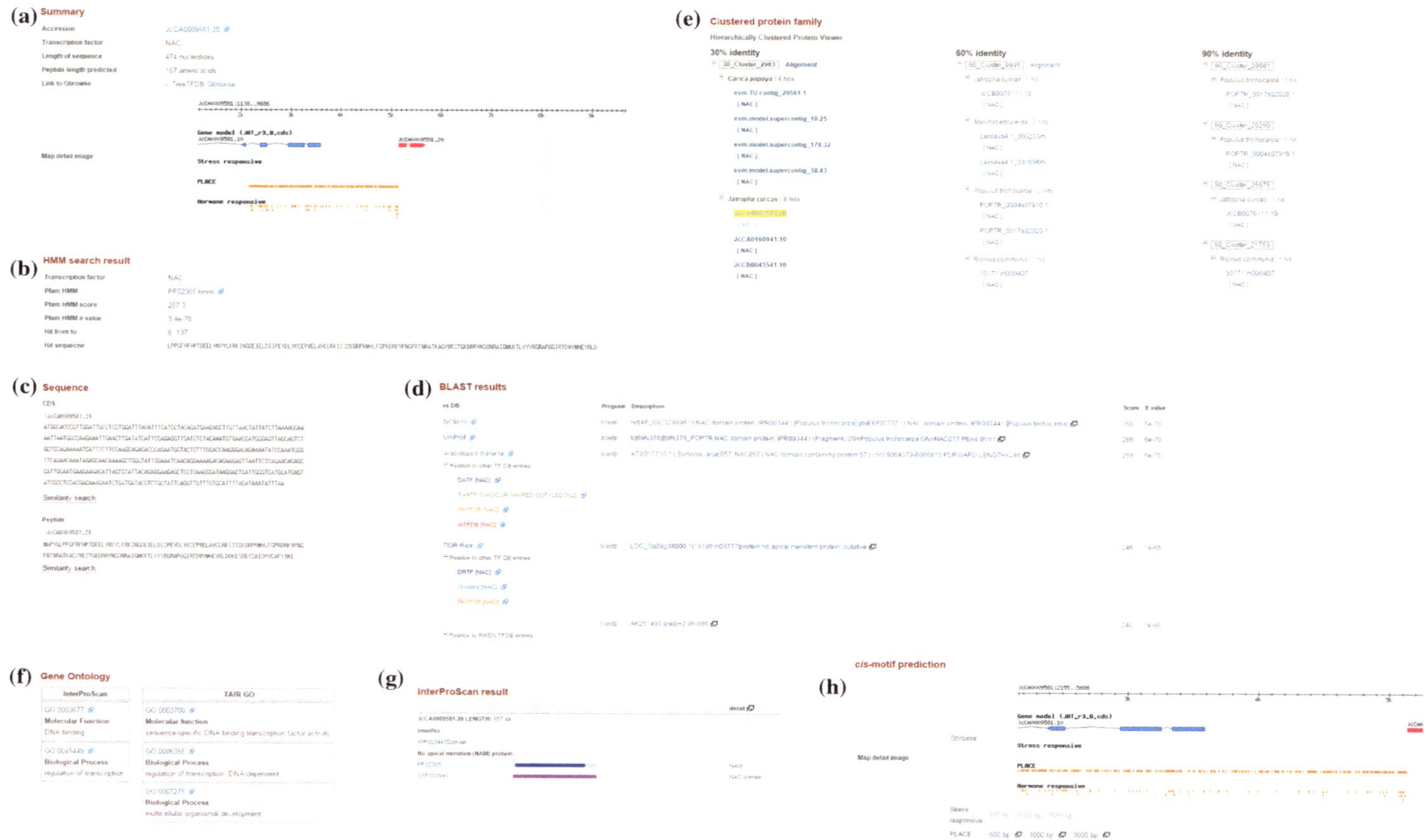

Fig. 3.3 Web-based user interface in a detailed annotation of an example of putative transcription factor (TF) encoding gene in TreeTFDB. **a** Summary provides basic information on each of the gene models annotated with gene structure with embedded image of the genome browser. **b** Hidden Markov model search result of the corresponding DNA-binding domains for the TF. **c** Sequences of the cDNA and protein. **d** Similarity search results for each of the entries against NCBI nr, UniProt, and gene models of *Arabidopsis* and other species including tree species, with detailed search results and hyperlinks to the original data. **e** Resultant hierarchical clustering of homologous TFs can be browsed with multiple alignment of each cluster. **f** Gene Ontology terms assigned to each of the entries based on InterProScan and sequence similarity search against the annotated genes of *Arabidopsis* of TAIR10. **g** Domain structure predicted by InterProScan. **h** Result of the *cis*-motif sequence pattern search of promoter regions for each gene with genomic gene structure

available information on gene functions not only from jatropha but also from model organisms is required to accelerate the understanding of gene functions. Therefore, adaptable structure of housing datasets and user-friendly interfaces are important functionalities that integrate broad range of spectrums generated from various analytical platforms. With the useful functionalities, our TreeTFDBs represent heuristic information in studies on TFs in jatropha.

References

Abdulla R, Chan ES, Ravindra P (2011) Biodiesel production from *Jatropha curcas*: a critical review. Crit Rev Biotechnol 31:53–64

Aya K, Hobo T, Sato-Izawa K, Ueguchi-Tanaka M, Kitano H, Matsuoka M (2014) A novel AP2-type transcription factor, SMALL ORGAN SIZE1, controls organ size downstream of an auxin signaling pathway. Plant Cell Physiol 55:897–912

Doebley JF, Gaut BS, Smith BD (2006) The molecular genetics of crop domestication. Cell 127:1309–1321

Donlin MJ (2009) Using the Generic Genome Browser (GBrowse). Curr Protoc Bioinf Chap 9, Unit 9 9

Eksi R, Li HD, Menon R, Wen Y, Omenn GS, Kretzler M, Guan Y (2013) Systematically differentiating functions for alternatively spliced isoforms through integrating RNA-seq data. PLoS Comput Biol 9:e1003314

Feller A, Machemer K, Braun EL, Grotewold E (2011) Evolutionary and comparative analysis of MYB and bHLH plant transcription factors. Plant J 66:94–116

Figuaroa P, Browse J (2012) The Arabidopsis JAZ2 promoter contains a G-Box and thymidine-rich module that are necessary and sufficient for jasmonate-dependent activation by MYC transcription factors and repression by JAZ proteins. Plant Cell Physiol 53:330–343

Finn RD, Bateman A, Clements J, Coggill P, Eberhardt RY, Eddy SR, Heger A, Hetherington K, Holm L, Mistry J, Sonnhammer EL, Tate J, Punta M (2014) Pfam: the protein families database. Nucleic Acids Res 42:222–230

Fojtova M, Van Houdt H, Depicker A, Kovrik A (2003) Epigenetic switch from posttranscriptional to transcriptional silencing is correlated with promoter hypermethylation. Plant Physiol 133:1240–1250

Gao G, Zhong Y, Guo A, Zhu Q, Tang W, Zheng W, Gu X, Wei L, Luo J (2006) DRTF: a database of rice transcription factors. Bioinformatics 22:1286–1287

Guo A, He K, Liu D, Bai S, Gu X, Wei L, Luo J (2005) DATF: a database of Arabidopsis transcription factors. Bioinformatics 21:2568–2569

Ha CV, Le DT, Nishiyama R, Watanabe Y, Suleiman S, Tran UT, Mochida K, Dong NV, Yamaguchi-Shinozaki K, Shinozaki K, Tran LS (2013) The auxin response factor transcription factor family in soybean: genome-wide identification and expression analyses during development and water stress. DNA Res 20:511–524

Hemmes H, Henriques R, Jang IC, Kim S, Chua NH (2012) Circadian clock regulates dynamic chromatin modifications associated with Arabidopsis CCA1/LHY and TOC1 transcriptional rhythms. Plant Cell Physiol 53:2016–2029

Higo K, Ugawa Y, Iwamoto M, Korenaga T (1999) Plant cis-acting regulatory DNA elements (PLACE) database: 1999. Nucleic Acids Res 27:297–300

Hirai MY, Sugiyama K, Sawada Y, Tohge T, Obayashi T, Suzuki A, Araki R, Sakurai N, Suzuki H, Aoki K, Goda H, Nishizawa OI, Shibata D, Saito K (2007) Omics-based identification of Arabidopsis Myb transcription factors regulating aliphatic glucosinolate biosynthesis. Proc Natl Acad Sci USA 104:6478–6483

Hirayama T, Shinozaki K (2010) Research on plant abiotic stress responses in the post-genome era: past, present and future. Plant J 61:1041–1052

Iida K, Seki M, Sakurai T, Satou M, Akiyama K, Toyoda T, Konagaya A, Shinozaki K (2005) RARTF: database and tools for complete sets of Arabidopsis transcription factors. DNA Res 12:247–256

Ji Y, Xiao J, Shen Y, Ma D, Li Z, Pu G, Li X, Huang L, Lui B, Ye H, Wang H (2014) Cloning and characterization of AabHLH1, a bHLH transcription factor that positively regulates artemisinin biosynthesis in *Artemisia annua*. Plant Cell Physiol 55:1592–1604

Jin J, Zhang H, Kong L, Gao G, Luo J (2014) PlantTFDB 3.0: a portal for the functional and evolutionary study of plant transcription factors. Nucleic Acids Res 42:1182–1187

Kasuga M, Liu Q, Miura S, Yamaguchi-Shinozaki K, Shinozaki K (1999) Improving plant drought, salt, and freezing tolerance by gene transfer of a single stress-inducible transcription factor. Nat Biotechnol 17:287–291

Kasuga M, Miura S, Shinozaki K, Yamaguchi-Shinozaki K (2004) A combination of the Arabidopsis DREB1A gene and stress-inducible rd29A promoter improved drought- and low-temperature stress tolerance in tobacco by gene transfer. Plant Cell Physiol 45:346–350

Kaufmann K, Muino JM, Osteras M, Farinelli L, Krajewski P, Angenent GC (2010a) Chromatin immunoprecipitation (ChIP) of plant transcription factors followed by sequencing (ChIP-SEQ) or hybridization to whole genome arrays (ChIP-CHIP). Nat Protoc 5:457–472

Kaufmann K, Pajoro A, Angenent GC (2010b) Regulation of transcription in plants: mechanisms controlling developmental switches. Nat Rev Genet 11:830–842

Kim JS, Mizoi J, Yoshida T, Fujita Y, Nakajima J, Ohori T, Todaka D, Nakashima K, Hirayama T, Shinozaki K, Yamaguchi-Shinozaki K (2011) An ABRE promoter sequence is involved in osmotic stress-responsive expression of the DREB2A Gene, which encodes a transcription factor regulating drought-inducible genes in Arabidopsis. Plant Cell Physiol 52:2136–2146

Le DT, Nishiyama R, Watanabe Y, Mochida K, Yamaguchi-Shinozaki K, Shinozaki K, Tran LS (2011) Genome-wide survey and expression analysis of the plant-specific NAC transcription factor family in soybean during development and dehydration stress. DNA Res 18:263–276

Lee SB, Kim J, Suh MC (2014) Overexpression of MYB94 transcription factor causes activation of Arabidopsis cuticular wax biosynthesis. Plant Cell Physiol 56:48–60

Li WZ, Godzik A (2006) Cd-hit: a fast program for clustering and comparing large sets of protein or nucleotide sequences. Bioinformatics 22:1658–1659

Li Y, Sawada Y, Hirai A, Sato M, Kuwahara A, Yan X, Hirai MY (2013) Novel insights into the function of Arabidopsis R2R3-MYB transcription factors regulating aliphatic glucosinolate biosynthesis. Plant Cell Physiol 54:1335–1344

Ma S, Bohnert HJ (2007) Integration of *Arabidopsis thaliana* stress-related transcript profiles, promoter structures, and cell-specific expression. Genome Biol 8:R49

Mangelsen E, Kilian J, Berendzen KW, Kolukisaoglu UH, Harter K, Jansson C, Wanke D (2008) Phylogenetic and comparative gene expression analysis of barley (*Hordeum vulgare*) WRKY transcription factor family reveals putatively retained functions between monocots and dicots. BMC Genom 9:194

Mitsuda N, Ohme-Takagi M (2009) Functional analysis of transcription factors in Arabidopsis. Plant Cell Physiol 50:1232–1248

Mochida K, Shinozaki K (2010) Genomics and bioinformatics resources for crop improvement. Plant Cell Physiol 51:497–523

Mochida K, Shinozaki K (2011) Advances in omics and bioinformatics tools for systems analyses of plant functions. Plant Cell Physiol 52:2017–2038

Mochida K, Shinozaki K (2013) Unlocking Triticeae genomics to sustainably feed the future. Plant Cell Physiol 54:1931–1950

Mochida K, Yoshida T, Sakurai T, Ogihara Y, Shinozaki K (2009a) TriFLDB: a database of clustered full-length coding sequences from triticeae with applications to comparative grass genomics. Plant Physiol 150:1135–1146

Mochida K, Yoshida T, Sakurai T, Yamaguchi-Shinozaki K, Shinozaki K, Tran LS (2009b) In silico analysis of transcription factor repertoire and prediction of stress responsive transcription factors in soybean. DNA Res 16:353–369

Mochida K, Yoshida T, Sakurai T, Yamaguchi-Shinozaki K, Shinozaki K, Tran LS (2010) LegumeTFDB: an integrative database of *Glycine max*, *Lotus japonicus* and *Medicago truncatula* transcription factors. Bioinformatics 26:290–291

Mochida K, Yoshida T, Sakurai T, Yamaguchi-Shinozaki K, Shinozaki K, Tran LS (2011) In silico analysis of transcription factor repertoires and prediction of stress-responsive transcription factors from six major gramineae plants. DNA Res 18:321–332

Mochida K, Uehara-Yamaguchi Y, Takahashi F, Yoshida T, Sakurai T, Shinozaki K (2013a) Large-scale collection and analysis of full-length cDNAs from *Brachypodium distachyon* and integration with pooideae sequence resources. PLoS One 8:e75265

Mochida K, Yoshida T, Sakurai T, Yamaguchi-Shinozaki K, Shinozaki K, Tran LS (2013b) TreeTFDB: an integrative database of the transcription factors from six economically important tree crops for functional predictions and comparative and functional genomics. DNA Res 20:151–162

Mulder NJ, Apweiler R (2008) The InterPro database and tools for protein domain analysis. Curr Protoc Bioinf Chap 2, Unit 2 7

Nakashima K, Tran LS, van Nguyen D, Fujita M, Maruyama K, Todaka D, Ito Y, Hayashi N, Shinozaki K, Yamaguchi-Shinozaki K (2007) Functional analysis of a NAC-type transcription factor OsNAC6 involved in abiotic and biotic stress-responsive gene expression in rice. Plant J 51:617–630

O'Maoileidigh DS, Graciet E, Wellmer F (2014) Gene networks controlling *Arabidopsis thaliana* flower development. New Phytol 201:16–30

Ohashi-Ito K, Matsukawa M, Fukuda H (2013) An atypical bHLH transcription factor regulates early xylem development downstream of auxin. Plant Cell Physiol 54:398–405

Perez-Rodriguez P, Riano-Pachon DM, Correa LG, Rensing SA, Kersten B, Mueller-Roeber B (2010) PlnTFDB: updated content and new features of the plant transcription factor database. Nucleic Acids Res 38:D822–D827

Riechmann JL, Heard J, Martin G, Reuber L, Jiang C, Keddie J, Adam L, Pineda O, Ratcliffe OJ, Samaha RR, Creelman R, Pilgrim M, Broun P, Zhang JZ, Ghandehari D, Sherman BK, Yu G (2000) Arabidopsis transcription factors: genome-wide comparative analysis among eukaryotes. Science 290:2105–2110

Sakuma S, Salomon B, Komatsuda T (2011) The domestication syndrome genes responsible for the major changes in plant form in the Triticeae crops. Plant Cell Physiol 52:738–749

Sato S, Hirakawa H, Isobe S, Fukai E, Watanabe A, Kato M, Kawashima K, Minami C, Muraki A, Nakazaki N, Takahashi C, Nakayama S, Kishida Y, Kohara M, Yamada M, Tsuruoka H, Sasamoto S, Tabata S, Aizu T, Toyoda A, Shin-I T, Minakuchi Y, Kohara Y, Fujiyama A, Tsuchimoto S, Kajiyama S, Makigano E, Ohmido N, Shibagaki N, Cartagena JA, Wada N, Kohinata T, Atefeh A, Yuasa S, Matsunaga S, Fukui K (2011) Sequence analysis of the genome of an oil-bearing tree, *Jatropha curcas* L. DNA Res 18:65–76

Stevens JD, Roalson EH, Skinner MK (2008) Phylogenetic and expression analysis of the basic helix-loop-helix transcription factor gene family: genomic approach to cellular differentiation. Differentiation 76:1006–1022

Thao NP, Tran LS (2012) Potentials toward genetic engineering of drought-tolerant soybean. Crit Rev Biotechnol 32:349–362

Tran LS, Mochida K (2010) Identification and prediction of abiotic stress responsive transcription factors involved in abiotic stress signaling in soybean. Plant Signal Behav 5:255–257

Tran LS, Nakashima K, Sakuma Y, Osakabe Y, Qin F, Simpson SD, Maruyama K, Fujita Y, Shinozaki K, Yamaguchi-Shinozaki K (2007a) Co-expression of the stress-inducible zinc finger homeodomain ZFHD1 and NAC transcription factors enhances expression of the ERD1 gene in Arabidopsis. Plant J 49:46–63

Tran LS, Nakashima K, Shinozaki K, Yamaguchi-Shinozaki K (2007b) Plant gene networks in osmotic stress response: from genes to regulatory networks. Methods Enzymol 428:109–128

Tran LSP, Nakashima K, Shinozaki K, Yamaguchi-Shinozaki K (2007c) Plant gene networks in osmotic stress response: from genes to regulatory networks. Osmosensing Osmosignaling 428:109–128

Tran LS, Nishiyama R, Yamaguchi-Shinozaki K, Shinozaki K (2010) Potential utilization of NAC transcription factors to enhance abiotic stress tolerance in plants by biotechnological approach. GM Crops 1:32–39

Tsuwamoto R, Harada T (2010) Identification of a cis-regulatory element that acts in companion cell-specific expression of AtMT2B promoter through the use of brassica vasculature and gene-gun-mediated transient assay. Plant Cell Physiol 51:80–90

Umezawa T, Urano K, Shinozaki K (2007) Molecular mechanisms of drought tolerance and signal transduction in plants. Tanpakushitsu Kakusan Koso 52:550–556

Vandepoele K, Quimbaya M, Casneuf T, de Veylder L, van de Peer Y (2009) Unraveling transcriptional control in Arabidopsis using cis-regulatory elements and coexpression networks. Plant Physiol 150:535–546

Wang Z, Libault M, Joshi T, Valliyodan B, Nguyen HT, Xu D, Stacey G, Cheng J (2010) SoyDB: a knowledge database of soybean transcription factors. BMC Plant Biol 10:14

Yamaguchi-Shinozaki K, Shinozaki K (2005) Organization of cis-acting regulatory elements in osmotic- and cold-stress-responsive promoters. Trends Plant Sci 10:88–94

Yamaguchi-Shinozaki K, Shinozaki K (2006) Transcriptional regulatory networks in cellular responses and tolerance to dehydration and cold stresses. Annu Rev Plant Biol 57:781–803

Yamasaki K, Kigawa T, Seki M, Shinozaki K, Yokoyama S (2013) DNA-binding domains of plant-specific transcription factors: structure, function, and evolution. Trends Plant Sci 18:267–276

Yan H, Jia H, Chen X, Hao L, An H, Guo X (2014) The cotton WRKY transcription factor GhWRKY17 functions in drought and salt stress in transgenic *Nicotiana benthamiana* through ABA signaling and the modulation of reactive oxygen species production. Plant Cell Physiol 55:2060–2076

Yilmaz A, Nishiyama MY Jr, Fuentes BG, Souza GM, Janies D, Gray J, Grotewold E (2009) GRASSIUS: a platform for comparative regulatory genomics across the grasses. Plant Physiol 149:171–180

Yilmaz A, Mejia-Guerra MK, Kurz K, Liang X, Welch L, Grotewold E (2011) AGRIS: the Arabidopsis gene regulatory information server, an update. Nucleic Acids Res 39:D1118–D1122

Zhang H, Jin J, Tang L, Zhao Y, Gu X, Gao G, Luo J (2011) PlantTFDB 2.0: update and improvement of the comprehensive plant transcription factor database. Nucleic Acids Res 39:1114–1117

Zhang L, Liu G, Zhao G, Xia C, Jia J, Liu X, Kong X (2014) Characterization of a wheat R2R3-MYB transcription factor gene, TaMYB19, involved in enhanced abiotic stresses in Arabidopsis. Plant Cell Physiol 55:1802–1812

Zhu JY, Sun Y, Wang ZY (2012) Genome-wide identification of transcription factor-binding sites in plants using chromatin immunoprecipitation followed by microarray (ChIP-chip) or sequencing (ChIP-seq). Methods Mol Biol 876:173–188

Zhu M, Chen G, Zhou S, Tu Y, Wang Y, Dong T, Hu Z (2014) A new tomato NAC (NAM/ATAF1/2/CUC2) transcription factor, SlNAC4, functions as a positive regulator of fruit ripening and carotenoid accumulation. Plant Cell Physiol 55:119–135

Molecular Markers in Jatropha: Current Status and Future Possibilities

4

Atefeh Alipour, Suguru Tsuchimoto and Kiichi Fukui

Abstract

Jatropha is a non-edible plant that has recently become favored worldwide for its potential as a feasible feedstock for the production of second-generation biofuels and new pharmaceutical compounds. However, the full capacity of this crop has not been realized, particularly owing to the lack of improved varieties with desirable traits such as flowering, oil content, non-toxicity, and stress tolerance. Considerable attempts in recent years have been made on exploitation of DNA polymorphism and development of molecular markers to be used for its genetic improvement. Recent progress in population genetics and comparative marker analysis of jatropha have led to the development of specific and more efficient DNA markers that may provide avenues for improving the productivity and quality of this crop. Development of the molecular markers (RFLPs, RTNs, AFLPs, and SSRs), which are linked to the genes controlling productive traits, is not only a key factor in selecting parental combinations and seedlings with desirable traits, but they are also important in detecting negative traits, such as poor flowering. This review aims to provide an overview of the pace of the progress and evolution of molecular marker technologies in jatropha. Moreover, the potential

A. Alipour (✉) · K. Fukui
Department of Biotechnology, Graduate School of Engineering, Osaka University, Suita, Osaka 565-0871, Japan
e-mail: atefeh.alipour@gmail.com

A. Alipour
Pharmaceutical Design and Bioinformatics Group, Drug Design and Bioinformatics Unit, Department of Medical Biotechnology, Pasteur Institute of Iran, No. 69 Pasteur Avenue, 13164 Tehran, Iran

S. Tsuchimoto
Plant Bioengineering for Bioenergy Laboratory, Graduate School of Engineering, Osaka University, Suita, Osaka 565-0871, Japan

S. Tsuchimoto (ed.), *The Jatropha Genome*, Compendium of Plant Genomes,
DOI 10.1007/978-3-319-49653-5_4

application of the variety of molecular marker systems to genetically enhance jatropha productivity and how to select proper molecular markers in breeding programs are discussed.

4.1 Introduction

First-generation biofuels that are derived from food crops, such as corn, sugarcane, and sugar beet place an enormous strain on world food prices and could precipitate a water crisis as well as displace food production (King et al. 2009). Therefore, vigorous research initiatives are currently aiming to find an affordable way to develop second-generation biofuels from non-edible feedstock as an alternative energy resource. Jatropha (*Jatropha curcas* L.) is a multipurpose oilseed crop that has recently attracted worldwide attention to address some of the aforementioned problems as well as contribute to energy security and a reduction of greenhouse gas emissions (Kumar and Sharma 2008; Achten et al. 2013). Moreover, jatropha oil and its by-products have pharmaceutical potential, such as immunomodulatory, antiproliferative, antifungal, and antiviral activities (Devappa et al. 2010).

Jatropha is a diploid plant, belongs to the Euphorbiaceae family, and is native to Mexico and Central America, although it is now widely propagated in tropical and subtropical areas of Asia, Africa, and Latin America (Tanya et al. 2010). High seed oil content (40–60% oil by dry weight), its physiochemical properties, the fatty acid composition of the oil, easy cultivation, and vigorous pest and drought resistance give this plant the highest potential as a raw material for biodiesel production among all oil-bearing tree species (Niu et al. 2012; Makkar et al. 1997). Apart from this potential, jatropha has multiple utilities, such as medicine, cosmetics, and animal feed (Fairless 2007). Because jatropha plantations do not need complex cultivation technologies or high capital investment, it is expected that this crop will substitute for other biofuel feedstock, especially in developing countries (Pandey et al. 2012).

However, jatropha is still a wild plant and does not have commercial reality yet, mainly owing to the lack of high-yielding varieties with robustness in agronomic and productive traits, such as high seed yield, oil content, and resistance to a variety of stresses (Kaushik et al. 2007; Montes et al. 2014). Current limited knowledge about jatropha's genetic structure is the main barrier to improving crop productivity and its commercialization (Na-ek et al. 2011). Earlier research indicated that jatropha populations are suffering from a distinctly narrow genetic base, whereas productivity of individual plants has shown high variation on various plantations (Basha et al. 2009; Ambrosi et al. 2010; Vischi et al. 2013). This observation is plausible due to continuous phenotype-based selection from local germplasms and, in all probability, cultivation from a restricted number of cuttings. The situation looks set to become rapidly worse if marker-assisted selection (MAS) is not improved or the reproductive strategies are not revised (Yi et al. 2010). In this context, it is vital to initially assess the genetic variation within and among various jatropha populations and finally determine correlations among the genetic structure, environmental factors, yield parameters, and biochemical traits in this energy crop.

Although a few studies attempt to elucidate genetic analysis for other jatropha species, owing to the incidence of natural hybridization among various species of the *Jatropha* genus, the taxonomy and genetic structure of this genus are not fully understood (Prabakaran and Sujatha 1999; Ram et al. 2008; Sudheer Pamidimarri et al. 2009b; Phumichai et al. 2011). Identification and exploitation of robust molecular markers for comprehensive genetic diversity analysis, followed by identification of distinct and divergent accessions, could accelerate breeding strategies in new crops such as jatropha (Yue et al. 2014).

This review is intended to provide an overview of the current stage of molecular marker developments and application of molecular techniques to the breeding of the energy crop jatropha. Further, each technological method is discussed together with its advantages and disadvantages in the evaluation of genetic structure of jatropha populations. In addition, we discuss further opportunities to meet the challenges of the genetic study of jatropha and how to enhance practical breeding technologies to improve yield.

4.2 Overview of Molecular Marker in Jatropha

The application of marker technologies on plants has provided significant impact on developing our genetic knowledge of them as well as characterizations of genetic variations that affect their phenotypic performance in germplasm collections (both in situ and ex situ) and breeding management (Miflin 2000). In contrast to conventional breeding approaches, most of which are time-consuming and subject to environmental conditions, progress in molecular marker systems and novel technologies to identify markers linked to eco-agronomic traits have enabled us to improve crops with higher efficiency and simplicity in a shorter time (Alipour et al. 2013). Furthermore, the markers are effective when they are used as appropriate indicators of diversity (Mittal et al. 2010). Hence, research groups are trying to identify more robust markers that can easily perform to their full potential to make the germplasm collections less expensive to store and analyze, and more useful to plant improvement.

Jatropha is a plant that is well adapted to the dry tropics, is often grown in barren regions, and can be grown under harsh conditions (Ikbal et al. 2010). However, the germplasm of jatropha is not characterized or documented well. Although it is assumed that the clonally propagated jatropha plants are heterozygous, there is an extreme narrowing of diversity among or within jatropha populations, suggesting the presence of a genetic bottleneck, particularly outside its place of origin in Mesoamerica (Popluechai et al. 2012; Sudheer Pamidimarri and Reddy 2014).

In fact, there is still no well-documented global genetic analysis of jatropha that elucidates its systematics, evolution, and geographical distribution that leads to breeding strategies for economic traits among or within accessions. One of the motivations for the application of DNA markers to jatropha is the toxicity of its seeds; the apparent lack of genetic diversity within the Asian and African accessions has resulted in higher toxicity (Vischi et al. 2013).

Various types of DNA markers have been introduced in the plant kingdom with an even greater number of variants and increasingly complex acronyms to describe them (Poczai et al. 2013). In addition to the common features of molecular makers for all plant species, there is a need to consider advantages and disadvantages of each DNA marker carefully prior to applying them to jatropha.

4.3 Random Amplified Polymorphic DNA (RAPD) Markers

The random amplified polymorphic DNA (RAPD) marker system is one of the simplest and cheapest methods and has been widely used in studies assessing genetic diversity in the plant kingdom (Ikbal et al. 2010). RAPD uses a single short primer hybridized with a low annealing temperature that allows amplification of multiple loci dispersed throughout the genome and provide a rapid technique for nucleotide sequence polymorphism. RAPD markers can suffer from poor reproducibility mainly owing to low annealing temperatures (Ram et al. 2008). Because of the ease of use, being dominant, and the ability to detect multiple loci distributed throughout the whole genome, RAPD markers are prevalently used for genetic variation studies (Kumar et al. 2009).

Reddy et al. (2007) has reported relatively low genetic variation (14–16%) using RAPD markers among 23 out of 300 selected jatropha accessions collected from all over India. Examination of 26 accessions from the same country by means of 26 RAPD primers resulted in 1859 polymorphic bands (30.92 vs. 69.08% monomorphic band) with sizes ranging from 300 to 2500 bp (Kumar et al. 2009). A high Jaccard's coefficient of similarity indicated a narrow genetic variability among these Indian populations. In accordance with this finding, Jubera et al. (2009) proved a low level of genetic variation among a set of seven Indian genotypes based on five RAPD markers. Gupta et al. (2008), Ranade et al. (2008), and Khurana-Kaul et al. (2012) also reported about the lack of genetic variation and high similarity among 13, 42, and 29 Indian jatropha accessions by 20, 7, and 47 RAPD primers, respectively. A low level of genetic divergence among 24 clones of the Chinese jatropha accessions has been affirmed with fingerprint profiles of five RAPD markers (Chen et al. 2011). Rosado et al. (2010) reported limited variation both among and within populations of 192 accessions collected across Brazil. Only 23 out of the 381 replicated RAPD markers demonstrated a polymorphic pattern, and they were not able to cluster accessions according to their geographic origins. Applying 10 RAPD markers on 17 accessions collected from Thailand, two from India, and one from Nigeria revealed narrow genetic variability in the studied populations (Popluechai et al. 2009).

In contrast, moderate-to-high genetic variability has also been detected among Indian jatropha accessions by means of 10 selected RAPD markers (Gopale and Zunjarrao 2013). Assessment of genetic diversity among 40 accessions from various parts of India by means of 44 RAPD primers revealed 93.9% polymorphism with similarity indices ranging from 0.44 to 0.92, indicating a moderate-to-high genetic variability in the Indian germplasm (Ikbal et al. 2010). A 75% polymorphism rate has also been documented for RAPD markers with higher intrapopulation genetic variances than interpopulation variability among 40 Indian jatropha accessions (Subramanyam et al. 2010). These observations partly conform to Basha and Sujatha et al. (2007), who have reported a modest level of genetic variation, with 42.0% polymorphism using 400 RAPD primers in evaluating 42 accessions representing different agro-climatic areas of India. Similarly, Rafii et al. (2012) reported a great degree of genetic diversity among 48 Malaysian accessions revealed by eight RAPD markers.

RAPD analysis has also been carried out using 26 primers to assess the genetic diversity, as well as intra- and inter-specific relationships among five *J. curcas* and seven other *Jatropha* species (*J. ramanadensis*, *J. gossypiifolia*, *J. podagrica*, *J. tanjorensis*, *J. villosa*, *J. glandulifera*, and *J. integerrima*) in India (Ram et al. 2008). Eighteen of the examined primers were able to amplify DNA fragments reproducibly, to detect a high level of genetic variation (80.2% polymorphism) among the genotypes, and to group them based on their genetic distinctness. Popluechai et al. (2009) detected moderate similarity between *J. podagrica* and *J. curcas*, which was consistent with the report by Ram et al. (2008). Sudheer Pamidimarri et al. (2009b) also have used RAPD markers to evaluate genetic variability among the aforementioned species of the *Jatropha* genus and reported a value of 68.48% polymorphism, whereas 97.74% of loci were polymorphic. In agreement with the amplified fragment length polymorphism (AFLP) data for the same set of species, they have found that *J. curcas* is highly related to *J. integerrima*.

Comparatively equal genetic variation (53 vs 47%) was inferred among and within populations for eight Kenyan *J. curcas* accessions according to the analysis of 10 RAPD markers (Machua et al. 2011). These primers have detected significantly higher levels of mean genetic diversity and polymorphisms in Kenyan accessions than those reported in other countries, although some Kenyan populations showed low variation in previous reports (Basha and Sujatha 2007; Santos et al. 2010). Moreover, molecular polymorphism has been estimated at 61.8% with 100 RAPD primers with 7.62 polymorphic bands per primer among 72 accessions from 13 countries, indicating more genetic diversity among accessions from diverse regions (Basha et al. 2009). In spite of moderate genetic variation, very low phenotypic differences were found among these populations, except in flowering date and leaf size, unlike the Mexican accessions that showed distinct levels of biochemical variation.

RAPD markers have also been utilized to discriminate toxic Indian and non-toxic Mexican *J. curcas* accessions and evaluated genetic similarities between these populations (Sujatha et al. 2005). Sudheer Pamidimarri et al. (2009c) have found 84.91% genetic similarity among the toxic and non-toxic genotypes and reported the specific RAPD markers for both germplasms. In the following study performed by the same research group on Asian, African, Central American, and European jatropha accessions, 52 out of 180 RAPD markers had more than six score-able bands, of which 39 primers resulted in a total of 66 polymorphic markers either specific to toxic or non-toxic jatropha (Sudheer Pamidimarri and Reddy 2014). Their results suggested a probable dispersal route for *J. curcas* migration from its center of origin, Mexico, to Cape Verde, to Spain or Portugal prior to its spread to other countries. Moreover, they have generated a specific sequence characterized amplified region (SCAR) marker from a polymorphic RAPD marker, which is capable of differentiating the non-toxic varieties from the conventional toxic accessions (Mastan et al. 2012). This SCAR marker reportedly provides more information for comparative genetic mapping than dominant RAPD or amplified fragment length polymorphisms (AFLP) markers, mainly owing to the locus-specific amplification, and is less sensitive to reaction conditions such as competition between primer binding sites. The reproducible amplification of defined genomic regions allows comparative mapping and synteny studies between related species and varieties. Murty et al. (2013) reported higher polymorphism (96.67%) of RAPD markers compared to inter-simple sequence repeat (ISSR) markers and direct amplification of minisatellite DNA (DAMD) markers in evaluating 15 accessions of *J. curcas* and four related species.

The RAPD fingerprinting technique has been further utilized to estimate genetic variability within mutants of *J. curcas* to help breeders distinguish the plants showing better performance in morphological characters and agro-economical traits. Assessment of genetic variability of gamma-irradiated mutants, through analysis of 23 RAPD primers, revealed 55.16% polymorphism per primer, with the level of genetic variation ranging from 0.324 to 0.397 among the mutants studied (Dhakshanamoorthy et al. 2011). Interestingly, they reported that the mutants showing differences at the DNA level revealed by RAPD markers were distinct in morphological traits too. In agreement with previous studies, Zainudin et al. (2014) reported that three random operon primers were able to figure out a considerable degree of genetic variation among 18 mutants resulting from colchicine treatment and a wild type. Thus, the RAPD markers could be considered as a simple and affordable method for detection of DNA polymorphism induced by mutation in addition to their use in genetic mapping, as well as taxonomic and population studies.

Applying 21 highly polymorphic RAPD primers to investigate possible genomic changes and somaclonal variation in micropropagated plantlets obtained by axillary shoot bud proliferation found 2.25% polymorphism in a second subculture (culture cycle), but researcher were not able to detect polymorphisms in the eighth and 16th generations (Sharma et al. 2011). Higher variation of polymorphism in RAPD as compared to AFLP in the same samples is probably due to the less effective multiplex ratio in RAPD markers. Thus, RAPD analysis would serve for testing the genetic stability and identification of possible cross-contaminations.

4.4 Amplified Fragment Length Polymorphism (AFLP) Markers

Amplified fragment length polymorphisms (AFLP) marker system is widely employed for genotype characterization, genome mapping, and marker-assisted breeding programs. AFLP, with its high multiplex ratio, relies on the PCR amplification of a subset of small restriction fragments. Polymorphisms are detected as the presence or absence of an amplified restriction fragment and are, therefore, dominant. The considerable drawbacks for the AFLP technique are its complexity and long-time operation with multiple steps comprising DNA digestion, ligation, and amplification (Poczai et al. 2013).

AFLP markers have been successfully employed to estimate genetic variations among seven species of the *Jatropha* genus (*J. curcas, J. glandulifera, J. gossypifolia, J. integerrima, J. multifida, J. podagrica*, and *J. tanjorensis*) and to establish the phylogenetic relationships among them (Sudheer Pamidimarri et al. 2009b). The data obtained in this study was inconsistent with prevailing idea about *J. tanjorensis* as a natural interspecific hybrid between *J. curcas* and *J. gossypifolia* (Prabakaran and Sujatha 1999). It was also partly in conflict with the recent findings based on RAPD, ISSR, and DAMD markers demonstrating *J. gossypiifolia* as one of the ancestors of *J. tanjorensis* (Gautam Murty et al. 2013), but confirmed that *J. curcas* is not a parent of *J. tanjorensis*. On the other hand, AFLP analysis of 114 accessions from 15 populations of four *Jatropha* species in Costa Rica revealed that *J. curcas* is more related to *J. stevensii* than others, whereas having the highest distance with *J. gossypifolia* (Avendano et al. 2015). Thirty-two AFLP markers and two combinatorial tubulin-based polymorphism (cTBP) analyses in 38 *J. curcas* accessions from 13 countries on three continents, in addition to six other *Jatropha* species, indicated limited genetic variation in the *J. curcas* populations (Popluechai et al. 2009). Appreciable genetic diversity among six *Jatropha* species demonstrated that interspecific breeding could be considered for improving *J. curcas* productivity.

AFLP-based molecular characterization of 56 Chinese and two Malaysian *J. curcas* accessions revealed a high genetic similarity (0.96) among these populations (Sun et al. 2008). Unlike SSR markers that could only separate the Chinese and Malaysian varieties, AFLP markers were able to distinguish all accessions based on their geographical origin in spite of the low genetic variation. In contrast, seven effective AFLP marker combinations have revealed a high level of polymorphism (88%) among 48 accessions collected from six provinces (states) of India, of which 8.7 and 15.9% of fragments were accession-specific and rare, respectively (Tatikonda et al. 2009). The AFLP-based clustering was also in accordance with the geographical locations. Santos et al. (2010) declared high genetic variation among 12 Brazilian jatropha accessions through analysis of 164 loci generated from 17 AFLP primer combinations.

Pecina-Quintero et al. (2011) has found great genetic diversity among and within populations of 18 Mexican *J. curcas* genotypes by means of six AFLP primer combinations. They have reported some rare fragments in 10% of accessions, which earlier exhibited genetic diversity in agronomic traits. These rare alleles would be used for developing a marker system linked to agronomic traits such as the rate of pistillated flowers and seed yield. Similarly, screening five populations of *J. curcas* in the state of Chiapas, Mexico, using AFLP markers affirmed higher genetic variation than in the rest of the world suggesting Mesoamerica as its center of origin and diversity (Ovando-Medina et al. 2011), even though lower diversity values than those obtained by SSR markers previously (Ambrosi et al. 2010). It is consistent with the data obtained by Sanou et al. (2015) in populations from Chiapas with the expected heterozygosity between 0.34 and 0.54 in contrast to low genetic diversity in African, Asian, and even other Mexican populations from the regions of Veracruz, Puebla, and Morelos. AFLP fingerprinting of 18 combinations of AFLP selective primers has resulted 667 polymorphic markers, which indicated 0.67 to 1.00 genetic similarities with a low percentage of polymorphism (31.46) among 42 accessions from Cape Verde, India, Madagascar, Mexico, and Spain (Sudheer Pamidimarri and Reddy 2014). Although some intermixing was observed within the Indian accessions, the clustering pattern of the germplasms based on AFLP data was correlated with the RAPD analysis and consistent with their geographical distribution. This AFLP data confirms the earlier findings on the highest genetic variation in Mexican populations based on analysis with RAPD, SSR, retrotransposons, morphological, and other AFLP data (Basha et al. 2009; Sudheer Pamidimarri et al. 2009b; Wen et al. 2010; Ambrosi et al. 2010; He et al. 2011; Ovando-Medina et al. 2011; Kanchanaketu et al. 2012; Alipour et al. 2013). Moreover, AFLP marker analysis demonstrated that Indian accessions were genetically very close to the Cape Verde samples followed by the Mexican varieties.

Florescence AFLP (fAFLP) was not able to detect more than 0.1 polymorphism within and among five African, Asian, and Mesoamerican populations of *J. curcas* that had been collected from China, Indonesia, Suriname, Tanzania, and India whereas methylation-sensitive florescence AFLP (MfAFLP) analysis showed more than 25% polymorphic bands with variations in the CCGG methylation pattern among and within the aforementioned populations (Yi et al. 2010). Moreover, a parallel plantation study affirmed that various collections had some climate- and practice-independent differences in agronomic traits, such as seed yield, oil content, branching pattern, and flowering time. According to the significant epigenetic diversity within and among jatropha accessions collected from three continents, Popluechai et al. (2009) suggested that some stable epigenetic events may have occurred during jatropha development. These epigenetic markers are almost inheritable, whereas epialleles follow Mendelian segregation pattern (Yi et al. 2010).

The methylation-sensitive amplification polymorphism (MSAP) fingerprinting system, considered as a kind of AFLP, has also detected very high genetic similarity (0.95 to 1.0) among 56 toxic and non-toxic accessions of *J. curcas* with high-yield ability collected from India, Indonesia, Mexico, Myanmar, Cambodia, Peru, Sri lanka, Thailand, and USA comprising some hybrids and mutant varieties (Kanchanaketu et al. 2012). Combining AFLP and MSAP data could distinguish all non-toxic, mutant, and hybrid jatropha from other accessions (Mastan et al. 2014). Similar results of the capability of DNA markers to recognize non-toxic from toxic jatropha accessions were reported by Sudheer Pamidimarri et al. (2009c) using seven accessions. Altogether, they indicated that the MSAP technique is a suitable method for evaluating nucleotide polymorphisms as well as the differences in the levels of DNA methylation in the populations suffering low genetic diversity. These results are consistent with Kanchanaketu et al. (2012) who found different levels of DNA methylation in the *PAO* gene located in the matured chloroplasts among various jatropha accessions. AFLP markers have also been exploited for the assessment of genetic stability of micropropagated jatropha clones from the same mother plant in three generations but demonstrated less ability to detect polymorphisms in the second generation as compared to the RAPD markers (Sharma et al. 2011).

4.5 Simple Sequence Repeats (SSR) Markers

Simple sequence repeats (SSRs) or microsatellite markers are short pieces of DNA that contain tandem repeats of di-, tri-, or tetra-nucleotide units (Cai et al. 2010). Potential of the SSRs as codominant PCR-based markers with uniform genome coverage and a high level of polymorphism makes microsatellites well poised for mapping, linkage analysis, and tracing inheritance patterns in plant breeding programs (Poczai et al. 2013). In spite of the rather high expense for the isolation of these markers and the time-consuming process, this technique is routinely used in major crop plants as a powerful tool to identify a large number of alleles at a specific locus even among closely related accessions (Vischi et al. 2013). Although SSR markers detect lower genetic diversity than RAPD in the same set of jatropha accessions, SSR data is positively correlated with the data obtained by RAPD as verified by the Mantel test (Diniz-Filho et al. 2013).

Most microsatellite markers that have been developed in jatropha were found to be monomorphic or having a small number of alleles in the populations other than Mesoamerica. In a study carried out by Sun et al. (2008), only one out of 17 genomic SSR markers that have been developed by using the fast isolation by AFLP of sequences containing repeats

(FIASCR) method was able to produce a polymorphic pattern among the 58 Chinese and Malaysian jatropha accessions with only two alleles. The low level of genetic diversity among them has also been verified by AFLP markers.

A screening by 17 SSR markers in 72 accessions from 13 countries has revealed admirable genetic variability for Mexican genotypes and high genetic similarity among Asian accessions (Basha et al. 2009). Analyzing with 12 SSR markers, developed using the FIASCR method, has led to a very low estimation of genetic diversity among 44 diverse genotypes sampled from distinct geographical locations in India, although high allele polymorphism was observed (Sudheer Pamidimarri et al. 2010). Further analysis demonstrated that five of the isolated microsatellite loci had significant deviation from Hardy–Weinberg equilibrium, whereas no significant linkage disequilibrium was found among any of the loci. In addition, the clustering pattern of the Indian germplasms based on the studied SSR loci was not consistent with the geographical area of collection.

Na-ek et al. (2011) demonstrated low levels of average genetic diversity within and among populations of 32 accessions collected from Cambodia, China, India, Laos, Myanmar, Mexico (non-toxic), and Thailand (both wild type and gamma-irradiated seeds). The average number of alleles per SSR locus (1.4) in this study was the least compared to other reports on SSR analysis (Basha et al. 2009; Sudheer Pamidimarri et al. 2010; Tanya et al. 2010; Wen et al. 2010). Genetic evaluation of 26 Mexican, three Chinese, three Thai, and four Vietnamese jatropha accessions using 17 out of 25 microsatellite markers exhibited a monomorphic pattern, whereas eight markers were polymorphic (Tanya et al. 2011). The polymorphic set of SSR markers was capable of distinguishing non-toxic Mexican accessions from 10 toxic Asian genotypes. In genetic evaluation of 103 Senegal accessions using 33 microsatellite markers, only one out of 33 SSR markers showed a polymorphic pattern (Oauttara et al. 2014). Genotyping of 48 progenies derived from selfed seeds of an Indian plant using 31 SSR markers showed low level of heterozygosity (Maurya and Yadav 2016).

Bressan et al. (2012) have reported nine microsatellite markers showing a polymorphic allelic pattern (two to eight alleles per locus), which are suited for analyzing genetic diversity and population dynamics, mating systems, and gene flow studies among 41 Brazilian jatropha accessions, while 31 primer sets demonstrated a monomorphic pattern. This allele frequency for SSR markers was significantly higher than those reported in the prevailing literatures (Basha et al. 2009; Sudheer Pamidimarri et al. 2010). However, this finding was inconsistent with a previous survey on 192 Brazilian accessions, which exhibited a homozygous pattern at all except one of six examined SSR markers (Rosado et al. 2010).

Assessment of the 64 germplasms, collected from seven geographical locations in Brazil, Cape Verde, Cuba, Mozambique, and Senegal, with 32 SSR markers and two gene-specific primers (ISPJ-1 and *Curcin-P2* gene promoter) has led to the identification of distinct genotypes and population-specific molecular markers (Ricci et al. 2012). Moreover, accessions collected from the islands of Cuba and Cape Verde were very different from genotypes of the mainland in Brazil, Mozambique, and Senegal at several loci. In agreement with the studies conducted in India, China, and Brazil, all these populations displayed the least amount of within-population variation (Basha et al. 2009; Wen et al. 2010; Bressan et al. 2012). No significant polymorphism has been detected for the promoter region of the *Curcin* gene and *ISPJ1*, which are supposed to be associated with the toxicity.

No genetic diversity was detected by all the 29 microsatellites loci, which had been applied to 276 accessions of jatropha collected from nine locations in five countries in South America, Asia, and Africa (Yue et al. 2014). They found homozygous patterns at all studied SSR loci in spite of high phenotypic variation among these accessions. Vischi et al. (2013) also could not find any polymorphism for a large subset of 10 accessions collected from Africa and South America using 40 SSR markers, although a high level of genetic diversity was found both among and within 19 wild Mexican populations. This is in agreement with a previous report from Ambrosi et al. (2010) indicating high degrees of genetic variation in Mexican varieties but low genetic diversity among varieties from other regions of the globe.

The monomorphic pattern of 41 SSR markers in populations from eight different countries (Cape Verde, India, Indonesia, Malaysia, Philippines, South Africa, Thailand, and Vietnam) also confirmed earlier reports on the existence of a narrow genetic base and restriction for genetic improvement and breeding (Siju et al. 2015). Osorio et al. (2014) found that Mexican and Central American jatropha accessions have higher genetic variation than populations from Africa, Asia, and South America.

A study of a global set of jatropha consisting of 70 accessions from 14 countries, including Mexico, reported that 38 of 54 assessed SSR markers exhibit a rather high level of polymorphism and are particularly suited to classifying germplasm (Montes et al. 2014). Further, they concluded that the most common mating system in jatropha was possibly self-fertilization with low outcrossing rates. Their approach led to the identification of four SSR markers that flank responsible loci for phorbol ester biosynthesis. SSR markers are considered as powerful DNA markers for identification and classification of the non-toxic varieties from the conventional toxic jatropha according to the level of phorbol ester.

Sudheer Pamidimarri et al. (2009a) have studied cross-species amplification of the 12 microsatellites that were isolated from *J. curcas,* in the other six species of the genus *Jatropha*; *J. glandulifera, J. gossypifolia, J. integerrima, J. multifida, J. podagrica*, and *J. tanjorensis*. Moreover, 31 out of 49 microsatellite markers isolated from *J. curcas* were found to be able to infer the genetic relationship of this crop with the six sister taxa (Sudheer Pamidimarri et al. 2011). The highest and least genetic similarities have been observed between *J. integerrima*/*J. tanjorensis* and *J. curcas*/*J. multifidi*, with the rate of 0.96 and 0.76, respectively, whereas the highest percentage of polymorphism (24.00) was detected between *J. curcas*/*J. podagrica*. These findings are more consist with results on RAPD and AFLP markers than those of nrDNA in their earlier report. Phumichai et al. (2011) have reported 11 out of 55 isolated SSR markers to be able to detect polymorphism among 26 *J. curcas* accessions collected from different provinces in Thailand, whereas 34 markers were found to be useful for investigating related species including *J. integerrima*, *J. gossypifolia*, and *J. podagrica*. Six markers identified from *J. curcas* by Bressan et al. (2012) showed good transferability rates and were able to detect polymorphism in several other *Jatropha* species, such as *J. podagrica*, *J. pohliana*, and *J. gossypifolia*, but not in other more distant members of the Euphorbiaceae family, such as *Hevea brasiliensis*, *Manihot esculenta*, and *Ricinus communis*.

Assessment of the genetic relationships among 10 different Indian accessions of *J. curcas* using a novel set of chloroplast microsatellite markers demonstrated high to moderate amounts of genetic diversity among them (Mittal et al. 2010). This observation is in agreement with previous report that pointed out the higher ability of the chloroplast microsatellite markers to detect population genetic diversities because they are haploid and good indicators of historical bottlenecks (Asif et al. 2010). In fact, ISSR markers have detected lower genetic divergence among the studied populations.

Expressed sequence tags (EST)-derived SSR (EST-SSR) markers have been reported to be more useful than other genomic markers as they are highly conserved, represent the transcriptome, and directly link to the functional genes (Jain et al. 2014). Moreover, the transferability of the SSRs developed from the ESTs across species has accelerated comparative studies among closely related species. EST-SSRs or microsatellites from comparative genomics (G-SSRs) have also been exploited widely for studying genetic relationships and phylogenetic construction in *J. curcas*. Thirty-six EST-SSR and 20 G-SSR transferable markers from cassava, which also belongs to the Euphorbiaceae family, could successfully be used for evaluating genetic variations and clustering 45 accessions collected from America, Indonesia, Grenada, and two regions of China into the six distinct groups

(Wen et al. 2010). They concluded that domestication of *J. curcas* may have occurred partly in its origin, South America, and partly in China, and that higher inter-group gene diversity than the intra-group diversity would be evidence for partial domestication of this crop in America and in Hainan, China.

Transferability of EST-SSRs was higher than G-SSRs that reflects lower mutation frequency of EST sequences than genomic sequences. Wang et al. (2011) constructed a first-generation linkage map comprising 216 SSRs and generated a comparative map between *J. curcas* and *A. thaliana* containing 192 marker loci derived from ESTs. They have found 18 quantitative trait loci (QTLs) underlying oil traits using a map panel containing two backcross populations between *J. curcas* and *J. integerrima*. Twenty-one SSRs have also been identified by searching 13,201 ESTs of jatropha databases and were applied for the assessment of genetic diversity among 25 accessions collected throughout India (Yadav et al. 2011). Their dendrogram analysis depicted different clustering patterns of *J. curcas*, independent of their geographical distribution. Further investigation on mapping the same jatropha backcross populations of Wang et al. (2011) led to identify 28 novel QTL for 11 growth and seed traits (Sun et al. 2012). Moreover, in silico mining of the microsatellites using *J. curcas* ESTs from various tissues reported in NCBI public domain led to the identification of EST-SSRs including JES35, a novel informative marker for genetic improvement in *J. curcas*, which has generated from EST of a fatty acid biosynthesis pathway gene (Jain et al. 2014).

4.6 Inter-simple Sequence Repeat (ISSR) Markers

Inter-simple sequence repeats (ISSRs) are DNA segments lying between two identical microsatellite repeats oriented in opposite directions using primers designed from microsatellite core regions (Gupta et al. 2008). Although ISSR markers are dominant, they are more stable and reproducible and have a higher capacity to detect polymorphisms than RAPD, particularly in closely related species (Noor Camellia et al. 2012). Due to these properties, ISSR markers have recently been found to be used extensively for finger printing, phylogenetic studies, and population structure analysis in *Jatropha* (Basha et al. 2009; Vijayanand et al. 2009; Murty et al. 2013).

A study to elucidate genetic background of 72 accessions encompassing 13 countries showed a polymorphic pattern with 35.5% of 586 bands from 48 ISSR primers (Basha et al. 2009). This report also confirmed a narrow genetic base for Asian populations compared to the high genetic variation for Mexican germplasms. A combination of 19 morphological traits and 21 ISSR primers to evaluate the genetic relationships of five *J. curcas* accessions and seven other *Jatropha* species native to India deciphered maximum diversity between *J. villosa* and *J. integerrima* and the low variation among *J. curcas* accessions (Vijayanand et al. 2009). This report has pointed out a higher capability of ISSR markers in clustering jatropha accessions compared to the morphological markers. Fingerprint profiles of 12 ISSR markers for 24 clones of *J. curcas* collected from China (Chen et al. 2011) were consistent with low genetic diversity among Asian populations examined by another set of ISSR markers in *J. curcas* accessions collected from China, Indonesia, and Thailand (Duan and Gue 2010). Noor Camelia et al. (2012) have reported a narrow genetic base for 16 accessions of *J. curcas* from Malaysia, Brunei, India, and Thailand based on analysis of eight ISSR primers. A molecular study of 48 accessions of *J. curcas* by means of 10 ISSR markers demonstrated high genetic similarities among populations collected from three states of Malaysia (Arolu et al. 2012). Applying 12 polymorphic ISSR primers have revealed lower genetic variation among 10 different accessions sampled across India compared to cp SSR markers (Mittal and Dubay 2010).

These results are opposed to a high degree of genetic variation at the species level detected by ISSR markers in eight *J. curcas* accessions collected from Yunnan province, China (Xiang et al.

2007). In the same way, Ou et al. (2009) have reported high polymorphism among 120 accessions revealed by ISSR markers. Cai et al. (2010) also reported that 15 ISSR primers detected a high level of genetic variability in 225 Chinese and five Myanmarian accessions. Khurana-Kaul et al. (2012) have observed more than 50% polymorphism for 25 ISSR markers when applied to 29 *J. curcas* accessions collected from across India. This observation is different from previous observations for Indian populations (Basha and Sujatha 2007). Similarly, 10 ISSR markers have revealed 31.23% variability among 78 *J. curcas* accessions originating from eight countries, cultivated in two locations in Taiwan, whereas 68.77% variation occurs within these populations (Mavuso et al. 2016).

Studying 332 accessions of *J. curcas* from eight states in Brazil using seven ISSR primers proved the high efficiency and reliability of these markers and showed a rather high level of genetic diversity among these accessions (Grativol et al. 2011). The authors concluded that despite finding close intra- and inter-relations between these populations, Brazilian germplasms, predominantly the accessions from the Natal area, harbored a higher level of genetic diversity compared to earlier reports on accessions from other countries. ISSR fingerprinting of 144 accessions from six *J. curcas* populations in Malaysia, Indonesia, the Philippines, and India with 10 primer combinations has revealed wide intrapopulation variation (87%), whereas only a 13% variation corresponded to the interpopulation genetic dissimilarity (Biabani et al. 2013). This finding was opposed to the phenotypic variation data reported earlier by the same research group, based on the morphological traits in the studied populations of *J. curcas* (Biabani et al. 2012).

Genetic evaluation of 15 Indian *J. curcas* accessions and four other species of *Jatropha* genus (*J. gossypiifolia, J. tanjorensis, J. podagrica*, and *J. integerrima*) using 11 ISSR markers revealed high genetic variability in populations from Ranchi and Assam states but low diversity in those from Gujarat (Murty et al. 2013). ISSR markers were shown to be capable of separating the wild species from *J. curcas* accessions and interspecific hybrids, although they exhibited very low correlation with RAPD and DAMD markers for the same set of samples.

4.7 Single Nucleotide Polymorphism (SNP) Markers

Single nucleotide polymorphism (SNP) is a single base change in a DNA sequence that can be effectively used for germplasm fingerprinting, genetic linkage analysis, population studies, and studying marker-trait associations mainly owing to their high conservation throughout evolution and within a population. SNP analysis has been recognized as a powerful marker system that is a more efficient, faster, and relatively inexpensive method compared to other DNA marker detection approaches.

The first linkage map including SNP markers of jatropha was generated by mapping 506 markers, including 290 SNPs and 216 microsatellites from ESTs (Wang et al. 2011). This study clarified that 79.2, 86.5, and 91.0% of jatropha ESTs were homologous to counterpoints in Arabidopsis, poplar, and castor bean, respectively. Popluechai et al. (2012) have found alleles of *JcOle3* (the Oleosin gene) with SNPs in its intron in jatropha accessions, related species and hybrids, which have high potential to be used as markers in phylogenetic studies or breeding programs.

Ricci et al. (2012) have identified a population-specific SNP marker through sequencing of highly conserved monomorphic fragments obtained by the SSR markers that have been used for studying the genetic structure of 64 jatropha populations from seven locations on two continents. The single base substitution at the *Jcps9* locus efficiently distinguished genotypes of the islands of the Pacific Ocean from their mainland counterparts. This marker has deciphered a low level of genetic variation within the studied populations.

Genetic analysis of a worldwide set of jatropha germplasms including 16 accessions from

Asia, Africa, and South and Central America, using 120 SNP markers, affirmed that 14 SNP markers were heterozygous and have a good potential to cluster the accessions into distinct groups (Monte et al. 2014). In accordance with the SSR data for the same set of germplasms from all continents, most of the SNP markers were not able to detect multiple alleles per locus, indicating a rather high level of homozygosis in jatropha accessions worldwide. However, SNP markers were capable of apparently discriminating toxic and non-toxic varieties and could group them based on the presence or absence of phorbol esters.

4.8 Ribosomal DNA Markers

DNA sequence of the internal transcribed spacer (ITS) region of the 18S–5.8S–26S nuclear ribosomal cistron consists of three components—the ITS1, ITS2 spacers, and the highly conserved 5.8S nrDNA exon—and is most widely used as a noncoding DNA marker system in the plant kingdom, suited for studying phylogenetic inference at the specific, generic, or even family levels (Mastan et al. 2012). Nuclear ribosomal DNA internal transcribed spacer (nrDNA-ITS) markers have been reported as a reproducible, cheap, and rapid assay for deciphering phylogenetic reconstruction, patterns of reticulate evolution, hybrid formation, and parentage verification (Calonje et al. 2009). These characteristics are particularly due to their biparental inheritance, technical simplicity, and inter-genomic variability in addition to having rather high copy numbers.

The nrDNA-ITS sequence analysis among 42 jatropha accessions from Asia, Africa, and Europe, as well as its center of origin, Mexico, has revealed a very low genetic distance (0.03), far less than the mean genetic diversity estimated with RAPD and AFLP markers for the same set of germplasms. The maximum diversity was found for ITS1 and the least for 5.8S nrDNA (Sudheer Pamidimarri and Reddy 2014). Further, in contrast to the AFLP data, the nrDNA-ITS sequence phylogram showed a close relation between the African and Madagascar varieties.

4.9 Sequence-related Amplified Polymorphism (SRAP) Markers

Sequence-related amplified polymorphism (SRAP) is a simple marker technique designed for the amplification of open reading frames (ORFs). The SRAP system has been developed for many crop plants such as potatos, rice, lettuce, rapeseed, garlic, apples, citrus, and celery (Li and Quiros 2001). Shen et al. (2010) have used 45 SRAP primer pairs for assessing genetic diversity and studying genetic relationships of eight Asian jatropha populations from six provinces of China and one from Indonesia. Although 28% of the loci were polymorphic, SRAP analysis revealed low genetic diversity in the Asian accessions of this crop. Identified markers were able to separate the Indonesian germplasm from the Chinese counterparts.

4.10 Directed Amplified Minisatellite DNA (DAMD) Markers

Directed amplified minisatellite DNA (DAMD) is a cheap, PCR-based method to amplify minisatellites, tandemly repeated DNA segments of eukaryotic genomes that have high allelic length variability owing to the differences in number of repeated units (Karca and Ince 2008). It is highly reproducible and has been introduced in a variety of crops, such as wheat and bermudagrass, in mapping and genetic diversity studies (Karca and Ince 2008). Ranade et al. (2008) have been first to employ four DAMD primers to jatropha for assessment of genetic relations among 18 Indian accessions. The DAMD markers detected an adequate genetic diversity among the accessions and were able to discriminate them according to the place of origin.

Characterization of 15 *J. curcas* accessions from different geographical areas of India in addition to four other species of *Jatropha* genus (*J. gossypiifolia* L., *J. podagrica* H., *J. integerrima* J., and *J. tanjorensis*) using four DAMD primers resulted in specific groupings in accordance with the location of origin (Murty et al.

2013). These primers generated, in total, 36 species-specific DAMD markers, out of which 10 were specific to *J. curcas* accessions, whereas 26 were specific to related species and hybrids. Although these markers demonstrated a high level of polymorphism, genetic diversity values were disclosed to be significantly lower for DAMDs as compared to the ISSR markers in a comparative study. Altogether, it has been suggested that DAMD markers exhibit a high potential for exploitation as species-specific diagnostic markers.

4.11 Retrotransposon-based Markers

Retrotransposons constitute a class of mobile genetic elements. They frequently copy and paste themselves via RNA intermediates and create additional new copies in the host genome. Retrotransposons are the largest components of the eukaryotic genome and are supposed to contribute to genome evolution by causing a wide range of mutations, genome reorganization, as well as increasing the physical size of the genome (Kalendar et al. 2010). Due to the high occurrence and irreversibility of retrotransposition events and dispersion of these repeat elements throughout the plant genome, verifying the presence or absence of retrotransposons has been introduced as a powerful tool for investigating biodiversity and phylogenetic relationships among and within species, particularly for tracing the evolution (Alipour et al. 2013). Retrotransposons have also been exploited widely to decipher evolutionary histories, as well as to perform gene tagging, mutagenesis, and marker-assisted selection (Alipour et al. 2014).

The availability of the jatropha genome sequence also facilitated the development of retrotransposon markers through data mining and comparative analysis in this crop (Sato et al. 2011). Based on the sequence of the reverse transcriptase (RT) gene, *Ty-copia* retrotransposon families were identified and investigated to elucidate the structure, phylogenetic diversity, and chromosomal distribution in this biodiesel crop (Alipour et al. 2013). Comparative analysis of the copy number and presence of the genes in the flanking sequences of them were performed. Presence and absence of their members in 12 jatropha accessions from Asia, Africa, and the center of origin, Mexico, were examined, and a specific recently amplified retrotransposon-based insertional polymorphism (RBIP) marker, *JC7-1*, which is able to distinguish certain Mexican and Guatemalan accessions from other populations, was identified. This report has suggested that the retrotransposition event of the identified markers at their loci took place in the center of origin prior to migration to other continents. Moreover, the data affirmed Asian and African populations have very close genetic relationship, which shows that most of them share the same origin in Mesoamerica.

A comparative phylogenetic analysis of *Ty3-gypsy* retrotransposons in jatropha with retrotransposon families from other plants has led to identification of two jatropha-specific families (Alipour et al. 2014). Although *Ty3-gypsy* markers have detected a high genetic similarity between Philippine and Chinese jatropha accessions, a high degree of diversity among *gypsy*-type retrotransposons in the jatropha genome has indicated their potential for genetic variation and evolutionary studies. To identify a greater number of recently amplified *copia*- or *gypsy*-type retrotransposons to generate RBIP markers, as well as to conduct a comprehensive genetic study on various genotypes worldwide will lead to elucidate the history of jatropha, especially how it was transmitted from its Mesoamerican origin to Africa and Asia.

4.12 QTL Identification and Mapping in Jatropha

Various desirable traits in jatropha have been reported to be under polygenic control, for example, phorbol ester content, plant growth, and oil yield (King et al. 2013). Considering the low genetic variability of jatropha populations, the ability to evaluate complex phenotypes at the seedling stage would significantly accelerate breeding of new jatropha cultivars. Thus, identification of the quantitative trait loci (QTLs) will serve as an effective starting point in the establishment of new varieties of jatropha with improved performance for seed and oil productivity. Genetic mapping of QTL involves characterizing and assessing the degree of association between dominant or recessive traits and a set of genetic markers (King et al. 2015). For accurate QTL identification, it is necessary to construct a saturated genetic map covering the entire genome. Codominant markers such as single nucleotide polymorphisms (SNP) and simple sequence repeats (SSR) are the most useful markers for the construction of linkage maps as they supply required information on both alleles present in a diploid species such as jatropha. In spite of genetic analyses among jatropha populations, there are still not enough number of codominant markers for jatropha, and they are in the form of SSR in most reports (Basha et al. 2009; Yadav et al. 2011; Phumichai et al. 2011; Montes et al. 2014). Liu et al. (2011) initially have identified 18 QTLs underlying the oil traits and three eQTLs of the oleosin genes. King et al. (2013) have discovered some additional SNP markers from the jatropha genome for the construction of a linkage map. These comparative mapping approaches have led to the identification of the locus responsible for phorbol ester biosynthesis and paved the way to breed new non-toxic varieties in a shorter time. Further, recent comprehensive positioning of a large number of genes involved in the biosynthesis of lipid storage onto the genetic map of jatropha has been carried out using a candidate gene approach and integrating physical mapping data from new release of the jatropha genome (Hirakawa et al. 2012; King et al. 2015). The QTL identification for traits associated with seed oil yield in three mapping populations of jatropha in conjunction with phenotypic selection can be used as a reliable system to create genetically stable, high-yield jatropha cultivars.

4.13 Discussion and Future Prospects

Jatropha is an earth-friendly non-food bioenergy crop that can grow in a wide range of climates with lower requirements for water, fertilizer, and herbicides compared to other oleaginous crops (Kumar et al. 2008). However, several agro-technological contexts about its productivity still remain to be clarified prior to the commercial utilization as a feasible biofuel crop (Pandey et al. 2012). Most jatropha plantation studies launched in Brazil, China, India, Mexico, and the Philippines have shown lower productivity than commercial needs, indicating the necessity for urgent genetic improvement (Ovando-Medina et al. 2011). In this context, the genetic potential of jatropha genotypes worldwide with regard to different biochemical traits in different environments is important for screening and assessing affordable crop production. Genetic improvement for increased yield and stress resistance needs comprehensive genetic analysis of the jatropha germplasm using robust markers that could be a good basis for research for marker-assisted selection, as well as breeding (Sudheer Pamidimarri and Reddy 2014). The genetic improvement of crops suffering from a low genetic base requires individuals with distinct differences in their inheritable traits. The development and use of molecular markers for the detection and exploitation of DNA polymorphism are the most important in the field of molecular genetics required for this purpose.

Introduction of high-throughput DNA marker systems would be advantageous in identifying new jatropha germplasms carrying desired traits. They could be used not only for genetic diversity assessment but also to expand existing linkage maps, as well as to map agronomic traits, all of which are required for crop development (Yue et al. 2014). Moreover, availability of the jatropha genome database could accelerate to develop further high-throughput marker platforms such as SNP arrays (Hirakawa et al. 2012).

Higher levels of genetic diversity, as well as better agronomic traits, have been reported within Mesoamerican jatropha accessions compared to those collected from rest of the world (He et al. 2011; Kanchanaketu et al. 2012). This suggests that Mexico and Central America may be a provenance of this energy crop (Basha et al. 2009; Sudheer Pamidimarri et al. 2009b; Wen et al. 2010; Ambrosi et al. 2010; Ovando-Medina et al. 2011; Alipour et al. 2013; Vischi et al. 2013). These findings are stimulating further efforts to avail required polymorphisms in Mexican populations for marker-based breeding. Hence, using new accessions from Mesoamerica, which exhibit robustness in agronomic traits such as high oil content, non-toxicity, and 100% pistillate flowers, could be applied as resources for breeding programs and genetic improvement of jatropha (Pecina-Quintero et al. 2011).

Reports on genetic variation among accessions of jatropha out of the center of origin are sometimes controversial. The partial inconsistencies among the results of diversity analysis obtained for common regions in Asia, Africa, and South America seem to depend on the type of markers used and eco-geographic cultivation differences.

In spite of performing enormous studies to evaluate the intra- and inter-specific relationships in *J. curcas* and among different *Jatropha* species, further investigation on applying molecular markers for gene tagging, mapping, inter-specific hybridization, marker-assisted selection, or map-based cloning of genes coding for desirable traits is necessary. A combination of the data obtained from molecular genetic technologies and the integration of several omics resources can drive and sustain successful breeding strategies in jatropha, thus finding of new and diverse accessions with high levels of both phenotypic and genetic diversity is desirable to improve the germplasm collection and finally to enhance the productivity of this economic crop. The presence of *Ty-copia* retrotransposons in the gene-rich regions of the jatropha genome has been shown by data mining and fluorescence in situ hybridization (FISH) assay (Alipour et al. 2013). Thus, further investigation of these retrotransposons in further jatropha accessions, particularly from the center of origin, Mesoamerica, would result in new useful DNA markers.

Obvious phenotypic variation in productivity and agronomic traits such as seed oil content, toxicity, and stress resistance in the absence of polymorphism at the DNA level could indicate possible effects of epigenetic variations and changes in DNA methylation, which has recently been pointed out as a key factor involved in heterosis or hybrid vigor (Yi et al. 2010; Shen et al. 2012). Therefore, applying whole-genome bisulfite sequencing of multiple individuals to assess epigenetic patterns could be considered to address the current challenges on the effects of eco-geographic factors on the diversity of jatropha (Lou et al. 2014). Considering the effect of biotic and abiotic stresses on gene expression through epigenetic-mediated mechanisms (Popluechai et al. 2012), comparative expression analysis of the genes essential for agronomic and productive traits in different agro-climates should be targeted as molecular screens as well.

Acknowledgments Plant Bioengineering for Bioenergy Laboratory was supported by Corporate Social Responsibility (CSR) Foundation of the Sumitomo Electric Industries (SEI), Ltd. This work was conducted and funded in part under the Joint Research Program of the Arid Land Research Center, Tottori University, Japan.

References

Achten WMJ, Trabucco A, Maes WH, Verchot LV, Aerts R, Mathijs E, Vantomme P, Singh VP, Muy B (2013) Global greenhouse gas implications of land conversion to biofuel crop cultivation in arid and semi-arid lands-Lessons learned from Jatropha. Arid Environ 98:135–145

Alipour A, Tsuchimoto S, Sakai H, Ohmido N, Fukui K (2013) Structural characterization of copia-type retrotransposons leads to insights into the marker development in a biofuel crop, *Jatropha curcas* L. Biotechnol Biofuel 6:129

Alipour A, Cartagena JA, Tsuchimoto S, Sakai H, Ohmido N, Fukui K (2014) Identification and characterization of novel gypsy-type retrotransposons in a biodiesel crop, *Jatropha curcas* L. Plant Mol Biol Rep 10:11105–013

Ambrosi DG, Galla G, Purelli M, Barbi T, Fabbri A, Lucretti S, Sharbel TF, Barcaccia G (2010) DNA markers and FCSS analyses shed light on the genetic diversity and reproductive strategy of *Jatropha curcas* L. Diversity 2:810–836

Arolu IW, Rafii MY, Hanafi MM, Mahmud TMM, Latif MA (2012) Molecular characterization of *Jatropha curcas* germplasm using inter simple sequence repeat (ISSR) markers in Peninsular Malaysia. Aust J Crop Sci 6(12):1666–1673

Asif M, Mantri S, Sharma A, Srivastava A, Trivedi I, Gupta P, Mohanty C, Sawant S, Tuli R (2010) Complete sequence and organisation of the *Jatropha curcas* (Euphorbiaceae) chloroplast genome. Tree Genet Genome 6:941–952

Avendaño R, García Díaz A, Valdez-Melara M, Chaves Solano N, Mora Villalobos A, Aguilar Cascante F, Williamson Benavides B, Solís-Ramos L (2015) Genetic diversity analysis of jatropha species from Costa Rica using AFLP markers. Am J Plant Sci 6:2426–2438

Basha SD, Sujatha M (2007) Inter and intra-population variability of *Jatropha curcas* (L.) characterized by RAPD and ISSR markers and development of population-specific SCAR markers. Euphytica 156:375–386

Basha SD, Francis G, Becker K, Makkar HPS, Sujatha M (2009) A comparative study of biochemical traits and molecular markers for assessment of relationships between *Jatropha curcas* L. germplasm from different countries. Plant Sci 176:812–823

Biabani A, Rafii MY, Saleh GB, Shabanimofrad M, Latif MA (2012) Phenotypic and genetic variation of *Jatropha curcas* L. populations from different countries. Maydica 57:164–174

Biabani A, Rafii MY, Saleh GB, Latif M (2013) Inter- and intra-population genetic variations in *Jatropha curcas* populations revealed by inter-simple sequence repeat molecular markers. Maydica 58:111–118

Bressan AE, Scotton DC, Ferreira RR, Jorge EC, Sebbenn AM, Gerald LTS, Figueira A (2012) Development of microsatellite primers for *Jatropha curcas* (Euphorbiaceae) and transferability to congeners. Am J Bot 99(6):237–239

Cai Y, Sun D, Wu G, Peng J (2010) ISSR based genetic diversity of *Jatropha curcas* germplasm in China. Biomass Bioenergy 34:1739–1750

Calonje M, Martin-Bravo S, Dobes C, Gong W, Jordon-Thaden I, Kiefer C, Kiefer M, Paule J, Schmickl R, Koch M (2009) Non-coding nuclear DNA markers in phylogenetic reconstruction. Plant Syst Evol 282(3–4):257–280

Chen K, Ren P, Ying C, Jiang Q, Jia X (2011) Genetic relationships among *Jatropha curcas* L. clones from Panzhihua, China as revealed by RAPD and ISSR. Afr J Biotechnol 6(11):2582–2585

Devappa RK, Makkar HP, Becker K (2010) Nutritional, biochemical, and pharmaceutical potential of proteins and peptides from jatropha: review. J Agric Food Chem 58(11):6543–6555

Dhakshanamoorthy D, Selvaraj R, Chidambaram ALA (2011) Induced mutagenesis in *Jatropha curcas* L. using gamma rays and detection of DNA polymorphism through RAPD marker. Crit Rev Biol 334:24–30

Diniz-Filho JAF, Soares TN, Lima JS, Dobrovolski R, Landeiro VL, de Campos Telles MP, Rangel TF, Bini LM (2013) Mantel test in population genetics. Genet Mol Biol 36(4):475–485

Duan YB, Guo HC (2010) Seed quality and ISSR analysis of *Jatropha curcas* L. from different origins. J Yunnan Agric Univ 25:458–465

Fairless D (2007) Biofuel: the little shrub that could: maybe. Nature 499:652–655

Gautam Murty S, Patel F, Punwar BS, Patel M, Singh AS, Fougat RS (2013) Comparison of RAPD, ISSR, and DAMD markers for genetic diversity assessment between accessions of *Jatropha curcas* L. and its related species. J Agric Sci Technol 15:1007–1022

Gopale KD, Zunjarro RS (2013) Evaluation of genetic diversity of *Jatropha curcas* L. using RAPD marker in Maharashtra. Int J Pure Appl Sci Technol 14(2):12–24

Grativol C, da Fonseca Lira-Medeiros C, Hemerly AS, Ferreira PCG (2011) High efficiency and reliability of inter-simple sequence repeats (ISSR) markers for evaluation of genetic diversity in Brazilian cultivated *Jatropha curcas* L accessions. Mol Biol Rep 38:4245–4256

Gupta S, Srivastava M, Mishra GM, Naik PK, Chauhan RS, Tiwari SK, Kumar M, Singh R (2008) Analogy of ISSR and RAPD markers for comparative analysis of genetic diversity among different *Jatropha curcas* genotypes. Afr J Biotechnol 7 (23):4230–4243

He W, King AJ, Awais Khan M, Cuevas JA, Ramiaramanana D, Graham IA (2011) Analysis of seed phorbol-ester and curcin content together with genetic diversity in multiple provenances of *Jatropha curcas* L. from Madagascar and Mexico. Plant Physiol Biochem 49:1183–1190

Hirakawa H, Tsuchimoto S, Sakai H, Nakayama S, Fujishiro T, Kishida Y, Kohara M, Watanabe A, Yamada M, Aizu T, Toyoda A, Fujiyama A, Tabata S, Fukui K, Sato T (2012) Upgraded genomic information of *Jatropha curcas* L. Plant Biotechnol 29:123–130

Ikbal Boora KS, Dhillon RS (2010) Evaluation of genetic diversity in *Jatropha curcas* L. using RAPD markers. Indian J Biotechnol 9:50–57

Jain N, Patil GB, Bhargava P, Nadgauda RS (2014) In silico mining of EST-SSRs in *Jatropha curcas* L. towards assessing genetic polymorphism and marker development for selection of high oil yielding clones. Am J Plant Sci 5(11):1521–1541

Jubera MA, Janagoudar BS, Biradar DP, Ravikumar RL, Koti RV, Patil SJ (2009) Genetic diversity analysis of elite *Jatropha curcas* (L.) genotypes using randomly amplified polymorphic DNA markers. Karnataka J Agric Sci 22:293–295

Kalendar R, Flavell AJ, Ellis TH, Sjakste T, Moisy C, Schulman AH (2010) Analysis of plant diversity with retrotransposon-based molecular markers. Heredity 106:520–530

Kanchanaketu T, Sangduen N, Toojinda T, Hongtrakul V (2012) Genetic diversity analysis of *Jatropha curcas*

L. (*Euphorbiaceae*) based on methylation-sensitive amplification polymorphism. Genet Mol Res 11 (2):944–955

Karca M, Ince AJ (2008) Minisatellites as DNA markers to classify bermudagrasses (Cynodon spp.): confirmation of minisatellite in amplified products. J Genet 87 (1):83–86

Kaushik N, Kumar K, Kumar S, Kaushik N, Roy S (2007) Genetic variability and divergence studies in seed traits and oil content of Jatropha (*J. curcas* L.) accessions. Biomass Bioenergy 31:497–502

Khurana-Kaul V, Kachhwaha S, Kothari SL (2012) Characterization of genetic diversity in *Jatropha curcas* L. germplasm using RAPD and ISSR markers. Indian J Biotechnol 11:54–61

King K, He W, Cuevas JA, Freudenberger M, Ramiaramanana DL, Graham IA (2009) Potential of *Jatropha curcas* as a source of renewable oil and animal feed. J Exp Bot 60:2897–2905

King AJ, Montes LR, Clarke JG, Affleck J, Li Y, Witsenboer H, van der Vossen E, van der Linde P, Tripathi Y, Tavares E, Shukla P, Rajasekaran T, van Loo EN, Graham IM (2013) Linkage mapping in the oilseed crop *Jatropha curcas* L. reveals a locus controlling the biosynthesis of phorbol esters which cause seed toxicity. Plant Biotechnol J 11(8):986–996

King AJ, Montes LR, Clarke JG, Itzep J, Perez CAA, Jongschaap REE, Visser RGF, van Loo EN, Graham IA (2015) Identification of QTL markers contributing to plant growth, oil yield and fatty acid composition in the oilseed crop *Jatropha curcas* L. Biotechnol Biofuels 8:160

Kumar A, Sharma S (2008) An evaluation of multipurpose oil seed crop for industrial uses (*Jatropha curcas* L.): a review. Ind Crop Prod 28:1–10

Kumar RV, Tripathi YK, Shukla P, Ahlawat SP, Gupta VK (2009) Genetic diversity and relationships among germplasm of *Jatropha curcas* L. revealed by RAPDs. Trees 23:1075–1079

Li G, Quiros CF (2001) Sequence-related amplified polymorphism (SRAP), a new marker system based on a simple PCR reaction: its application to mapping and gene tagging in Brassica. Theor Appl Genet 103:455–461

Liu P, Wang CM, Li L, Sun F, Liu P, Yue GH (2011) Mapping QTLs for oil traits and QTLs for oleosin genes in jatropha. BMC Plant Biol 11:132

Lou S, Lee H, Qin H, Li J, Gao Z, Liu X, Chan LL, Lam VKL, So W, Lok S, Wang J, Ma RC, Tsui SK, Chan JCN, Chan T, Yip KY (2014) Whole-genome bisulfite sequencing of multiple individuals reveals complementary roles of promoter and gene body methylation in transcriptional regulation. Genome Biol 15:408

Machua J, Muturi G, Omondi SF, Gicheru J (2011) Genetic diversity of *Jatropha curcas* L. populations in Kenya using RAPD molecular markers: Implication to plantation establishment. Afr J Biotechnol 10 (16):3062–3069

Makkar HPS, Becker K, Sporer F, Wink M (1997) Studies on nutritive potential and toxic constituents of different provenances of *Jatropha curcas*. J Agric Food Chem 45:3152–3157

Mastan S, Sudheer P, Rahman H, Reddy M, Chikara J (2012) Development of SCAR marker specific to non-toxic *Jatropha curcas* L. and designing a novel multiplexing PCR along with nrDNA ITS primers to circumvent the false negative detection. Mol Biotechnol 50(1):57–61

Mastan S, Rathore MS, Bhat VD, Chikara J, Ghosh A (2014) DNA methylation and methylation polymorphism in ecotypes of *Jatropha curcas* L. using methylation-sensitive AFLP markers. Mol Biol Rep 41(12):8261–8271

Maurya R, Yadav HK (2016) Microsatellite markers based heterozygosity assessment in *Jatropha curcas* L.: a potential bioenergy crop. Trop Plant Res 3 (1):191–198

Mavuso C, Wu Y, Chen F, Huang B, Lin S (2016) Genetic diversity analysis of *Jatropha curcas* L. accessions cultivated in Taiwan using inter simple sequence repeats (ISSR) markers. Agrofor Syst 90 (3):417–431

Miflin B (2000) Crop improvement in the 21st century. J Exp Bot 51(342):1–8

Mittal N, Dubey AK (2010) A novel set of highly polymorphic chloroplast microsatellite and ISSR markers for the biofuel crop *Jatropha curcas*. EurAsia J BioSci 4:119–131

Montes J, Technow F, Martin M, Becker K (2014) Genetic diversity in *Jatropha curcas* L. assessed with SSR and SNP markers. Diversity 6:551–566

Murty SG, Patel F, Punwar BS, Patel M, Singh AS, Fougat RS (2013) Comparison of RAPD, ISSR, and DAMD markers for genetic diversity assessment between accessions of *Jatropha curcas* L. and its related species. J Agric Sci Technol 15:1007–1022

Na-ek Y, Wongkaew A, Phumichai T, Kongsiri N, Kaveeta R, Reewongchai T, Phumichai C (2011) Genetic diversity of physic nut (*Jatropha curcas* L.) revealed by SSR markers. J Crop Sci Biotechnol 14 (2):105–110

Niu G, Rodriguez D, Mendoza M, Jifon J, Ganjegunte G (2012) Responses of *Jatropha curcas* to salt and drought stresses. Int J Agron 2012:6320–6326

Noor Camellia NA, Thohirah Lee A, Abdullah NAP (2012) Genetic relationships and diversity of *Jatropha curcas* accessions in Malaysia. Afr J Biotechnol 11:3048–3054

Oauttara B, Ndir KN, Gueye MC, Diedhiou I, Bernaud A, Fonceka D, Cisse N, Akpo EL, Diouf D (2014) Genetic diversity of *Jatropha curcas* L. in Senegal compared with exotic accessions based on microsatellite markers. Genet Resour Crop Evol 61(6):1039–1045

Osorio LRM, Salvador AFT, Jongschaap REE, Perez CAA, Sandoval JEB, Trindade LM, Visser RGF, van Loo EN (2014) High level of molecular and phenotypic biodiversity in *Jatropha curcas* from Central America compared to Africa, Asia and South America. BMC Plant Biol 14:77

Ou WJ, Wang WQ, Li KM (2009) Molecular genetic diversity analysis of 120 accessions of *Jatropha curcas* L. germplasm. Chin J Trop Crop 30:284–292

Ovando-Medina I, Sánchez-Gutiérrez A, Adriano-Anaya L, Espinosa-García F, Núñez-Farfán J, Salvador-Figueroa M (2011) Genetic diversity in *Jatropha curcas* populations in the state of Chiapas, Mexico. Diversity 3:641–659

Pandey VC, Singh K, Singh JS, Kumar A, Singh B, Singha RP (2012) *Jatropha curcas*: a potential biofuel plant for sustainable environmental development. Renew Sustain Energy Rev 16:2870–2883

Pecina-Quintero V, Anaya-Lopez JL, Colmenero AZ, Garcia NM, Colin CAN, Solis Bonilla JL, Aguilar-Rangela MR, Langarica HRG, Bustamante DJM (2011) Molecular characterization of *Jatropha curcas* L. genetic resources from Chiapas, Mexico through AFLP markers. Biomass Bioenergy 35:1897–1905

Phumichai C, Phumichai T, Kongsiri N, Wongkaew A, Sripichit P, Kaveeta R (2011) Isolation of 55 microsatellite markers for *Jatropha curcas* and its closely related species. Biol Plant 55(2):387–390

Poczai P, Varga I, Laos M, Cseh A, Bell N, Valkonen JPT, Hyvönen J (2013) Advances in plant gene-targeted and functional markers: a review. Plant Methods 9:6

Popluechai S, Breviario D, Mulpuri S, Makkar HPS, Reddy MRAR, Palchetti E, Gatehouse AMR, Syers JK, O'Donnell AG, Kohli A (2009) Narrow genetic and apparent phenetic diversity in *Jatropha curcas*: initial success with generating low phorbol ester interspecific hybrids. Nat Preced 3:1–44

Popluechai S, Froissard M, Jolivet P, Breviario D, Gatehouse AMR, O'Donnell AG, Chardot L, Kohli T (2012) *Jatropha curcas* oil body proteome and oleosins: L-form JcOle3 as a potential phylogenetic marker. Plant Physiol Biochem 49:352–356

Prabakaran AJ, Sujatha M (1999) *Jatropha tanjorensis* Ellis & Saroja, a natural interspecific hybrid occurring in Tamil Nadu, India. Genet Resour Crop Evol 46:213–218

Rafii MY, Shabanimofrad M, Edaroyati MWP, Latif MA (2012) Analysis of the genetic diversity of physic nut, *Jatropha curcas* L. accessions using RAPD markers. Mol Biol Rep 39:6505–6511

Ram SG, Parthiban KT, Kumar RK, Thiruvengadam M, Paramathma M (2008) Genetic diversity among Jatropha species as revealed by RAPD markers. Genet Resour Crop Evol 55:803–809

Ranade AS, Srivastava AP, Rana TS, Srivastava J, Tuli R (2008) Easy assessment of diversity in *Jatropha curcas* L. plants using two single-primer amplification reaction (SPAR) methods. Biomass Bioenergy 32:533–540

Reddy MP, Chikara J, Patolia JS, Ghosh A (2007) Genetic improvement of *J. curcas* adaptability and oil yield. In: Fact seminar on *J. curcas* L., agronomy and genetics, Wageningen

Ricci A, Chekhovskiy K, Azhaguvel P, Albertini E, Falcinelli M, Saha M (2012) Molecular characterization of *Jatropha curcas* resources and identification of population- specific markers. Bioenerg Res 5:215–224

Rosado TB, Laviola BG, Faria DA, Pappas MR, Bhering LL, Quirino B, Grattapaglia D (2010) Molecular marker reveal limited genetic diversity in large germplasm collection of the biofuel crop *Jatropha curcas* L. in Brazil. Crop Sci 50:2372–2382

Sanou H, Angulo-Escalanteb MA, Martínez-Herrerac J, Konéa S, Nikiemad A, Kalinganire A, Hansen JK, Kjær ED, Graudal L, Nielsen LR (2015) Loss of genetic diversity of *Jatropha curcas* L. through domestication: implications for its genetic improvement. Crop Sci 55(2):749–759

Santos CAF, Drumond MA, Rodrigues MA, Evangelista MRV (2010) Genetic similarity of *Jatropha curcas* accessions based on AFLP markers. Crop Breed Appl Biotechnol 10:364–369

Sato S, Hirakawa H, Isobe S, Fukai E, Watanabe A, Kato M, Kawashima K, Minami C, Muraki A, Nakazaki N, Takahashi C, Nakayama S, Kishida Y, Kohara M, Yamada M, Tsuruoka H, Sasamoto S, Tabata S, Aizu T, Toyoda A, Shin-I T, Minakuchi Y, Kohara Y, Fujiyama A, Tsuchimoto S, Kajiyama S, Makigano E, Ohmido N, Shibagaki N, Cartagena JA, Wada N, Kohinata T, Alipour A, Yuasa S, Matsunaga S, Fukui K (2011) Sequence analysis of the genome of an oil-bearing tree, *Jatropha curcas* L. DNA Res 18:65–76

Sharma S, Sudheer DVNP, Anand KGV, Reddy MP (2011) Assessment of genetic stability in micropropagules of *Jatropha curcas* genotypes by RAPD and AFLP analysis. Indust Crop Prod 34:1003–1009

Shen JL, Ni HQ, Chen XY, Huang SW (2010) Genetic diversity of *Jatropha curcas* with SRAP molecular markers. J Zhejiang Forest Coll 27:347–353

Shen H, He H, Li J, Chen W, Wang X, Guo L, Peng Z, He G, Zhong S, Qi Y, Terzaghi W, Deng XW (2012) Genome-wide analysis of DNA methylation and gene expression changes in two *Arabidopsis* ecotypes and their reciprocal hybrids. Plant Cell 24:875–892

Siju S, Ismanizan I, Wickneswari R (2015) Genetic homogeneity in *Jatropha curcas* L individuals as revealed by microsatellite markers: implication to breeding strategies. Br J Bot. doi:10.1007/s40415-014-0117-7

Subramanyam K, Muralidhararao D, Devanna N (2010) Genetic diversity assessment of wild and cultivated varieties of *Jatropha curcas* L in India by RAPD analysis. Afr J Biotechnol 8:1900–1910

Sudheer Pamidimarri DVN, Sinha R, Kothari P, Reddy MP (2009a) Isolation of novel microsatellites from *Jatropha curcas* L. and their cross species amplification. Mol Ecol Resour 9:431–433

Sudheer Pamidimarri DVN, Singh S, Mastan SG, Patel J, Reddy MP (2009b) Molecular characterization and identification of markers for toxic and non-toxic varieties of *Jatropha curcas* L. using RAPD, AFLP and SSR markers. Mol Biol Rep 36(6):1357–1364

Sudheer Pamidimarri DVN, Pandya N, Reddy MP, Radhakrishnan T (2009c) Comparative study of interspecific genetic divergence and phylogenic analysis of genus jatropha by RAPD and AFLP. Mol Biol Rep 36:901–907

Sudheer Pamidimarri DVN, Rahman H, Mastan SG, Reddy MP (2010) Isolation of novel microsatellites using FIASCO by dual probe enrichment from *Jatropha curcas* L. and study on genetic equilibrium and diversity of Indian population revealed by isolated microsatellites. Mol Biol Rep 37:3785–3793

Sudheer Pamidimarri DVN, Mastan SG, Rahman H, Prakash CR, Singh S, Reddy MP (2011) Cross species amplification ability of novel microsatellites isolated from *Jatropha curcas* and genetic relationship with sister taxa. Cross species amplification and genetic relationship of *Jatropha* using novel microsatellitess. Mol Biol Rep 38:1383–1388

Sudheer Pamidimarri DVN, Reddy MP (2014) Phylogeography and molecular diversity analysis of *Jatropha curcas* L. and the dispersal route revealed by RAPD, AFLP and nrDNA-ITS analysis. Mol Biol Rep 41(5):3225–3234

Sujatha M, Makker HPS, Becker K (2005) Shoot bud proliferation from axillary nodes and leaf sections of non-toxic *Jatropha curcas* L. Plant Growth Regul 47:49–83

Sun Q, Li L, Li Y, Wu G, Ge X (2008) SSR and AFLP markers reveal low genetic diversity in the biofuel plant *Jatropha curcas* in China. Crop Sci 48(5):1865–1871. doi:10.2135/cropsci2008.02.0074

Sun F, Liu P, Ye J, Chuan Lo L, Cao S, Li L, Hua Yue G, Ming Wang C (2012) An approach for jatropha improvement using pleiotropic QTLs regulating plant growth and seed yield. Biotechnol Biofuels 5:42

Tanya P, Taeprayoon P, Hadkam Y, Srinives P (2010) Genetic diversity among Jatropha and Jatropha - related species based on ISSR markers. Plant Mol Biol Rep 29:252–264

Tanya P, Dachapak S, Tar MM, Srinives P (2011) New microsatellite markers classifying nontoxic and toxic *Jatropha curcas*. J Genet 90:76–78

Tatikonda L, Wani SP, Kannan S, Beerelli N, Sreedevi TK, Hoisington DA, Devi P, Varshney RK (2009) AFLP-based molecular characterization of an elite germplasm collection of *Jatropha curcas* L., a biofuel plant. Plant Sci 176:505–513

Vijayanand V, Senthil N, Vellaikumar S, Paramathma M (2009) Genetic diversity of Indian Jatropha species as revealed by morphological and ISSR markers. J Crop Sci Biotechnol 12:115–120

Vischi M, Raranciuc S, Baldini M (2013) Evaluation of genetic diversity between toxic and non-toxic *Jatropha curcas* L. accessions using a set of simple sequence repeat (SSR) markers. Afr J Biotechnol 12:265–274

Wang CM, Liu P, Yi C, Gu K, Sun F, Li L, Lo LC, Liu X, Feng F, Lin G, Cao S, Hong Y, Yin Z, Yue GH (2011) A first generation microsatellite-and SNP-based linkage map of Jatropha. PLoS One 6:e23632

Wen M, Wang H, Xia Z, Zou M, Lu C, Wang W (2010) Development of EST-SSR and genomic-SSR markers to assess genetic diversity in *Jatropha curcas* L. BMC Res Note 3:42

Xiang ZY, Song SQ, Wang GJ, Chen MS, Yang CY, Long CL (2007) Genetic diversity of *Jatropha curcas* (Euphorbiaceae) collected from Southern Yunnan, detected by inter-simple sequence repeat (ISSR). Acta Bot Yunnanica 29:619–624

Yadav HK, Ranjan A, Asif MH, Mantri S, Sawant SV, Tuli R (2011) EST-derived SSR markers in *Jatropha curcas* L.: development, characterization, polymorphism, and transferability across the species/genera. Tree Genet Genomes 7:207–219

Yi C, Zhang S, Liu X, Bui HT, Hong Y (2010) Does epigenetic polymorphism contribute to phenotypic variances in *Jatropha curcas* L.? BMC Plant Biol 10:259

Yue GH, Lo LC, Sun F, Cao SY, Yi CX, Hong Y, Sun WB (2014) No variation at 29 microsatellites in the genome of *Jatropha curcas*. J Genom 2:59–63

Zainudin A, Maftuchah Fitriani H (2014) Analysis of genetic diversity on mutants *Jatropha curcas* using RAPD. Energy Procedia 47:1–6

Part II

Metabolomics and Physiology

Jatropha Metabolomics

5

Daisuke Shibata, Ryosuke Sano and Takeshi Ara

Abstract

Jatropha, *Jatropha curcas* L., has been used for materials in traditional medicine, and recently its oil in the seed kernel has received much attention as a renewable energy source, biodiesel. Various biological activities such as antimicrobial, insecticidal, and anti-inflammatory have been found in all parts of the plant. Therefore, search for compounds with bioactivities in this plant and also its related species has been conducted for pharmaceutical and agricultural applications. The property of the seed oil has been well characterized. The oil quality meets the standards of the USA and European Union as diesel fuel. A major advantage of the oil production over other oil crops is that the jatropha plant cultivation does not compete with food production as the plant grows well in wastelands. Given the jatropha genome sequences have been determined, a systematic research in conjunction with the analysis of the whole metabolites (metabolome) could be a key for pharmaceutical or agricultural application. This chapter summarizes metabolomics approaches for jatropha, including integration of information of transcriptome and metabolome on jatropha metabolic pathway maps.

5.1 Introduction

In traditional medicine, all parts of Jatropha, *Jatropha curcas* L., (hereafter referred as Jatropha, unless otherwise mentioned), have been used to cure various ailments in Africa, Asia, and Latin America, and therefore with the intention of finding novel useful chemicals, much attention has been paid to analyze the jatropha extracts to search compounds with cytotoxicity toward mammalian cell lines and anti-inflammatory

D. Shibata (✉)
Department of Research & Development, Kazusa DNA Research Institute, 2-6-7 Kazusa-Kamatari, Kisarazu, Chiba 292-0818, Japan
e-mail: shibata@kazusa.or.jp

R. Sano
Nara Institute of Science and Technology, 8916-5 Takayama, Ikoma, Nara 630-0192, Japan
e-mail: r-sano@bs.naist.jp

T. Ara
Kyoto University, Gokasho, Uji, Kyoto 611-0011, Japan
e-mail: ara@kais.kyoto-u.ac.jp

S. Tsuchimoto (ed.), *The Jatropha Genome*, Compendium of Plant Genomes,
DOI 10.1007/978-3-319-49653-5_5

activity for pharmaceutical purpose and insecticidal property and antimicrobial activity for agricultural application. The plant has also been attractive as a new feedstock for biofuel because the seeds contain oils that can be converted easily to biodiesel, and the plant is well adapted to wastelands that are not suitable for production of vegetables and agronomic crops, so that the energy source may not compete with food production (Achten et al. 2007; Johnson et al. 2011; Maghuly and Laimer 2013). The recent concern for carbon emissions to atmosphere has accelerated the cultivation of jatropha worldwide so that a huge amount of by-products (stems, kernel husk, and so on) after extraction of oil would be a good source for preparation of bioactive compounds for pharmaceutical and agricultural applications (Kumar and Sharma 2008).

Metabolomics is an area of genomics in which the metabolism of an organism is understood in terms of genomic information and the whole metabolites (called "metabolome") of an organism (Saito and Matsuda 2010). These biological research areas are called "omics". Since the completion of genome sequences of *Arabidopsis thaliana*, the most widely used model plant, with high accuracy in 2000 (The Arabidopsis Genome Initiative 2000), highly accurate sequences of rice (International Rice Genome Sequencing Project 2005) and tomato (Tomato Genome Consortium 2012) have been determined as additional plant reference genomic sequences, following genome sequencing of various plant species with moderate accuracy, including *Jatropha curcas* (Hirakawa et al. 2012) (see the Chap. 1 in this book). The availability of the genomic sequences has triggered analyses of the whole transcripts (called "transcriptome") and proteins (called "proteome"). Jatropha transcriptome (Jiang et al. 2012; Pan et al. 2014; Zhang et al. 2014) and proteome (Liu et al. 2013) have been investigated. Metabolome analysis has still been under technical development, mainly because of difficulty of dealing with the structural complexity of various types of metabolites, especially secondary metabolites and of a broad dynamic range of concentrations from a few molecules to millions of molecules per cell (Misra et al. 2014), which requires distinct types of state-of-the-art instruments such as gas chromatography (GC) coupled mass spectrometer (MS), liquid chromatography (LC) coupled MS, and nuclear magnetic resonance (NMR). Bioinformatics is indispensable to omics research due to the need of huge data processing. Here, we summarize previous researches on jatropha concerning biological activities and metabolites and recent progresses in metabolomics approaches. This chapter includes metabolites of other species of the genus *Jatropha* because metabolome analysis of *J. curcas* may provide useful information on the structural information of metabolites of other closely related species, which is one of the advantages of metabolomics approaches.

5.2 Metabolism in Jatropha

5.2.1 Biological Activities in Jatropha

Information of biological activities in an organism is start-up for identifying active compounds for pharmaceutical or agricultural application. Here, we summarize literal information of various biological activities of jatropha and related species, which includes uncertain description as folk stories.

5.2.1.1 Antimicrobial Activity

Extracts prepared from dried fruit parts of *J. curcas* were shown to exhibit activity against pathogenic microorganisms *Pseudomonas aeruginosa, Staphylococcus aureus*, and *Escherichia coli* when examined using agar-well diffusion method, in which the methanolic extracts showed the maximum inhibition zone (Rachana et al. 2012). Antimicrobial and antifungal potential of the crude methyl acetate or methanol extracts from four *Jatropha* species, *J. curcas, J. integerrima, J. podagrica, and J. multifida* as well as *Ricinus communis* (castor bean) toward six gram-positive and six gram-negative strains and one fungus (*Candida albicans*) had been examined in comparison with the activity of tetracycline as positive control by measuring the diameter of inhibition zones (Rampadarath et al. 2014). The results showed that most of the crude extracts of the plants had significant antibacterial and antifungal potential in varying degrees. They also showed that the crude extracts had varied effects on the two Diptera larvae, *Bactrocera zonata* and *B. cucurbitae*, which cause economic losses to local fruits in the island of Mauritius. Antibacterial activity of leaf extracts of *J. gossypifolia* toward *E. coli* and *Bacillus subtilis*, which have been used traditionally as a therapeutic bathing agent effective for wounds, sores, sprains, rash, and so on, and known to possess pesticidal property, was shown (Seth and Sarin 2010).

5.2.1.2 Molluscicidal Potential

Schistosomiasis is a chronic disease of great importance for public health in developing countries and transmitted through the host snails *Biomphalaria glabrata, B. tenagophila,* or *B. straminea.* The extract of leaves from *J. gossypiifolia* was shown to be a strong molluscicidal agent, causing 100% mortality of *B. glabrata*, even in 25 ppm, the lowest concentration tested (Pereira Filho et al. 2014).

5.2.1.3 Anti-Inflammatory Activity

The root powder in paste form of *J. curcas* exhibited anti-inflammatory activity in TPA (12-*O*-tetradecanoylphorbol-13-acetate)-induced ear inflammation in albino mice, and the methanol extract of the roots exhibited systemic and significant anti-inflammatory activities in acute carrageenan-induced Wistar albino rat paw edema (Mujumdar and Misar 2004). The alcoholic extract of root, stem, and leaf of *J. curcas* also exhibited systemic anti-inflammatory activity in acute carrageenan-induced rat paw edema (Nayak and Patel 2010). The methanolic extract of the leaves of *J. curcas* also exhibited the anti-inflammatory activity in egg albumin-induced paw edema of the Wistar albino rat (Uche and Aprioku 2008). The methanolic extracts of the leaves of *J. curcas* also exhibited anti-inflammatory activity in formaldehyde-induced arthritic albino mice paw edema (Ogunnaike et al. 2013). The ethanolic extract of the underground parts of *J. isabellei* and its alkaloid-rich fraction presents antinociceptive and anti-inflammatory effects in a rat gout model, similar to that observed after treatment with colchicine (Silva et al. 2013). These results support the traditional use and future applications of this plant for the inflammatory disorders.

5.2.1.4 Tumor Promoter

The seed oil of *J. curcas* was shown to contain skin tumor promoters in a two-stage mouse carcinogenesis experiment. The irritant fraction was partially purified from the methanol extract of the seed oil, and this fraction induced tumors in the skin of the mice ear tested in 30 weeks after initiation with 7,12-dimethylbenz [alpha] anthracene (DMBA) (Horiuchi et al. 1987). A new type of phorbol ester, 13,16-diester of 12-deoxy-16-hydroxyphorbol (DHPB), was isolated from the seed oil of *J. curcas* and showed a weak tumor-promoting activity in a two-stage carcinogenesis experiment on mouse skin. Along with the comparison of biological and biochemical activities between DHPB and 12-O-tetradecanoylphorbol-13-acetate (TPA), the authors indicated DHPB is a tumor promoter with weaker activity than TPA (Hirota et al. 1988).

5.2.2 Metabolites Found in Jatropha

5.2.2.1 Jatropha Seed Oil

Oil is the major concern as a renewable fuel feedstock of the jatropha seeds. Oil contents in the seed (kernel) vary among jatropha accessions and are affected by cultivation conditions. The contents are reported as from 44 to 62% in kernel (see review; King et al. 2009). Fatty acid composition of Jatropha oil affects largely the quality of biodiesel and fatty acid methyl esters (FAMES) that are produced by transesterification of the vegetable oil with methanol. The quality is evaluated with cetane number, cold-flow and cloud point properties, and kinetic viscosity and oxidative stability, in which the cetane number is the most important factor for biodiesel, and the actual cetane values of Jatropha biodiesel have so far been within the range of 50–57, which meets the Standard Specification for Biodiesel Fuel

Blend Stock (B100) for Middle Distillate Fuels ATSM D6751 for the USA and the stricter European Union requirement (EN14214) (see King et al. 2009). A comparison of Jatropha FAMES and the European standards was shown in the review article of Makkar and Becker (2009).

5.2.2.2 General Secondary Metabolites

Secondary metabolites of *J. curcas*, alkaloids, tannins, saponins, flavonoids, phenolics, HCN, and phytate in roots, bark, leaves, and seeds were quantified (Harry-Asobara et al. 2014). Comparison of secondary metabolites, alkaloids, tannins, flavonoids, saponins, and phenols was made among *J. curcas, J. gossypifolia, J. multifida*, and *J. podagrica* (Nwokocha et al. 2011). In these researches, metabolites were measured as compound groups, but not as individual chemicals. Six metabolites, fraxidin, fraxetin, scoparone, 3-acetylaleuritolic acid, beta-sitosterol, and sitosterone of *J. podagrica* were found by comparing their spectral data with previously reported values (Rumzhum et al. 2012).

5.2.2.3 Terpenoids

Terpenes are the largest group of phytochemicals, many of them exhibit diverse functions in mediating antagonistic and beneficial interactions in, and among, organisms. Each *Jatropha* species is a rich source of terpenoid compounds (Devappa et al. 2011). The diterpenes isolated from *Jatropha* species contain the skeletal structures of rhamnofolane, daphnane, lathyrane, tigliane, dinorditerpene, deoxy preussomerin, and pimarane. The diterpene compounds from *Jatropha* species exhibited diverse bioactivities such as cytotoxicity, antitumor activity, or antimicrobial activity. Jatrophone, jatrophatrione, spruceanol, cleistanthol, curcusones (A and B), and japodagrol possess in vitro antitumor activities. Many diterpenes (jatrophalactam, faveline derivatives, multifolone, curcusone, jatrophone derivatives, etc.) are cytotoxic when examined in vitro, while japodagrin, jatrogrossidione, and jatropholone derivatives exhibited antimicrobial activities. Jatropha diterpenes having a wide spectrum of bioactivity could form lead compounds for the synthesis of new compounds with better biological activity for utilization in the pharmaceutical industries. The other terpenoids (sesquiterpenoids and triterpenoids) are also isolated from *Jatropha* species, and two triterpenoids showed the cytotoxic activity (García and Delgado 2006a, b; Sutthivaiyakit et al. 2009; Yang et al. 2013). The derivatives of one of the isolated triterpenoid (calenduladiol) exhibited the antiretroviral activity to HIV-1 infection in permissive cells (Barroso-González et al. 2009).

The phorbol-type diterpenes (called phorbol esters, such as Jatropha factor C1–C6 and jatropherol) isolated have a wide range of toxicity (rodenticidal, piscicidal, molluscicidal, and insecticidal activities) and tumor-promoting activity (Hirota et al. 1988; Goel et al. 2007; Gaudani et al. 2009; Devappa et al. 2011). This toxicity limits the use of the plants and agricultural by-products containing phorbol esters to be used as animal feed. Therefore, various chemical and physical treatments have been evaluated to extract or inactivate phorbol esters so that protein-rich seed meals could be used as feed resources. More details of phorbol esters in *Jatropha* species are described in the Chap. 6 in this issue.

5.2.2.4 Cyclic Peptides

Cyclic peptides have been isolated from latex, seeds, and roots of *Jatropha* species and shown to possess biological activities such as cytotoxic activity, immunosuppressive activity, antimalarial activity, vasorelaxant activity, and inhibitory activities to some enzymes (see the review of Devappa et al. 2010). These authors reviewed the properties of 19 cyclic peptides (8 heptapeptides: cyclogossine A, integerrimides (A and B), mahafacyclin (A and B), podacyclin B, and pohlianins (A and B); 6 octapeptides: chevalierins (A and B), curcacycline A, cycloglossine B, jatrophidin, and pohlianins C; 4 nonapeptides: biobollein, chevalierins C, curcacycline B, and podacyclin A; 1 decapeptide: labaditin) isolated from Jatropha species including *J. curcas, J. mahafalensis, J. multifida, J. chevalieri, J. gossypifolia, J. podagrica, J. pohliana*, and *J. integerrima*. As biological activity, chevalierin A, curcacycline B, mahafacyclin (A and B), pohlianins (A to C) exhibited antimalarial activity against *Plasmodium falciparum*. Jatrophidin had weak antifungal effect against the strains of *Candida albicans, C. krusei, C. parapsilosis*, and *Cryptococcus neoformans*. Integerrimides (A and B) significantly inhibited neurite outgrowth of E7 chicken spinal cord neurons. Labaditin and curcacycline A inhibited the classical pathway of human complement activation in vitro. Molecular targets of these compounds are also reported. Jatrophidin had moderate activity as an acetylcholinesterase inhibitor, and curcacycline B enhances peptidyl prolyl *cis-trans* isomerase activity. Labaditin and biobollein bind to aggregated and antigen-bound IgG.

5.2.2.5 Novel Metabolites

Preussomerins and deoxypreussomerins, which are known to possess a wide range of biological properties including antibacterial, antifungal, herbicidal, antibiotic, and antitumor activities, were found in the *Jatropha* stems (Ravindranath et al. 2004). These chemicals are generally reported as fungal metabolites. The authors argued against the possibility that their occurrence in an endophytic fungus present in the plant body, since appreciable amount as 172 mg of preussomerins was purified from 3 kg of jatropha stems. These compounds were shown to exhibit significant antibacterial activity against the bacteria *Staphylococcus aureus*. Their activity was comparable to that of the standard antibiotics, penicillin-G. Staubmann and colleagues (1999) isolated pyrimidinedione and 5-hydroxy-2-pyrrolidinone from the leaves of *J. curcas*, although the latter compound was previously isolated from the leaves of *Hyptis verticillata* in the Lamiaceae family.

5.3 Metabolomics Approaches in Jatropha

5.3.1 Metabolome Analysis

5.3.1.1 Techniques of Metabolome Analysis

The total number of metabolites in Jatropha has not yet been estimated. It has been generally believed that the number of metabolites in a plant species is more than 5,000 (Trethewey 2004), although no definitive evidence for the number has been provided. The sum of all metabolites of all plant species on the earth is estimated to be one million by statistical analysis using a metabolite-plant species database, KNApSAcK (Afendi et al. 2012). Due to the huge diversity of the metabolites and the large dynamic range of the amounts of metabolites in an organ, neither a single detection condition nor a single detection machine is available for comprehensive metabolite analysis. The most prevalent analytical methods for metabolome comprise of a combination of some mass spectrometers (MS). Gas chromatography coupled MS (GC–MS) and liquid chromatography coupled MS (LC–MS) are commonly used for metabolome analysis. Capillary electrophoresis MS (CE–MS) is used for detection and quantification of ionic compounds (Soga 2007). Liquid chromatography coupled nuclear magnetic resonance (LC–NMR) is also used for metabolites that exist in a large amount, as the limitation of the detection sensitivity (Timmers and Urban 2011). Liquid chromatography Fourier transform ion cyclotron resonance mass spectrometry (LC–FTICR–MS) is a powerful tool to analyze metabolites. LC–FTICR–MS enables acquisition of high-resolution mass spectra with high accuracy. The chemical composition of each compound separated by LC can be determined from the exact mass value (*m/z*). The information of chemical composition facilitates to annotate metabolites. LC–FTICR–MS also gives information of MS/MS fragmentation patterns of some but not all compounds, which is crucial to identify the compound.

5.3.1.2 Bioinformatics for Metabolome Analysis

It is essential to analyze a huge data set produced from a MS machine with informatics techniques for extracting meaningful metabolome information. Several informatics tools are available to process MS data sets such as MetAlign (Tikunov et al. 2005), XCMS (Smith et al. 2006), and MZmine (Katajamaa et al. 2006). For metabolome annotation from exact mass values obtained by LC–FTICR–MS, Sakurai and colleagues (2014) developed a bioinformatics tool PowerGet, which consists of two programs, PowerFT and PowerMatch. The PowerFT module processes all metabolite peaks detected with high-accurate mass values, by which ^{13}C-isotope peaks are assigned into corresponding ^{12}C-isotope peaks, ion adducts assigned to the $^{13}C/^{12}C$ ion groups, and elemental compositions of each ion group are calculated. The PowerMatch module aligns metabolite peaks assigned by the PowerFT module of multiple samples with high accuracy. The graphical user interface (GUI) of the module integrates the information of metabolite features (retention time, *m/z* value, MS/MS fragmentation pattern, PDA spectra, molecular formula, metabolite database search result, metabolite annotation, etc.) for the metabolomics analysis. It is noteworthy that the calculation speed of chemical composition of thousand compounds obtained from the MS is the fastest among other calculation tools, by which all calculations can be done within a few minutes while others need a few days for that many compounds (Sakurai et al. 2013). The MatchedIonsFinder tool is for realignment of the alignment produced by the PowerMatch tool to correct a mis-alignment (Yamamoto et al. 2012). The tool ShiftedIonsFinder is a search program to find peaks having specified mass differences from such as stable isotopes and/or chemical modifications (hydration, glycosylation) (Kera et al. 2014). LC–Orbitrap MS also gives exact mass values as comparable as LC–FTICR–MS. The PowerGet tool is applicable to the data sets obtained from LC–Orbitrap MS.

5.3.1.3 Mass Spectrometry-Based Analysis of Jatropha

The changes of metabolites during jatropha fruit maturation were analyzed with LC–Orbitrap MS (Sano et al. 2012). Metabolites of maturing fruits from stage 1 to stage 4 were extracted with methanol and separated with reverse-phased C18 column for exact mass measurement (Fig. 5.1a). In total, 6,778 metabolite peaks were detected for the four stages. The changes in each metabolite peak were analyzed by batch-learning self-organization mapping (BL-SOM), in which the patterns of quantitative changes of metabolites were grouped and assigned in a 30×20 matrix with their similarity (Fig. 5.1b). A drastic change of SOM pattern was observed between stages 2 and 3, suggesting occurrence of some physiological events conferring such metabolic changes. Interestingly, total metabolite peak numbers decreased from stages 1–3, and slightly increased at stage 4 (Fig. 5.2). Flavonoid oligomers were annotated in all stages, and the annotation result suggested that the peak numbers of oligomers (dimer, trimer, and tetramer) decreased from stages 1–3, and slightly increased at stage 4 (Fig. 5.3). The authors argued that polymerization or incorporation of low molecular metabolites into insoluble materials, which were not detectable by mass-spectrometry under the analytical conditions applied in the study, could occur during fruit maturation. Further investigations are required for a general perspective of the metabolite changing during fruit maturation.

The metabolome of the seed kernels from phorbol ester (PE)-containing and PE-free accessions of *J. curcas* is compared by using LC–Orbitrap MS (Ohtani et al. 2012). Among more than 12,000 metabolites detected, only 18 ions were specific to the PE-containing accession, and four ions were specific to the PE-free accession. These results indicate that PE-containing and PE-free Jatropha are broadly similar in their metabolism. The chemical composition of the callus of *J. curcas* under salicyclic acid elicitation was studied by GC–MS (Mahalakshmi et al. 2013). In total, 51 compounds were detected and the major constituents are 1-docosene, 1-octadecene, 1-hexadecene (E)-3-eicosene, (E)-5-eicosene, 1-hexacene, nonahexacontaoic acid, and methyl ester-3-oxocyclohexane carboxylic acid. Salicylic acid treatment resulted in the production of higher percentage of alkanes and fatty acid than those in control plant.

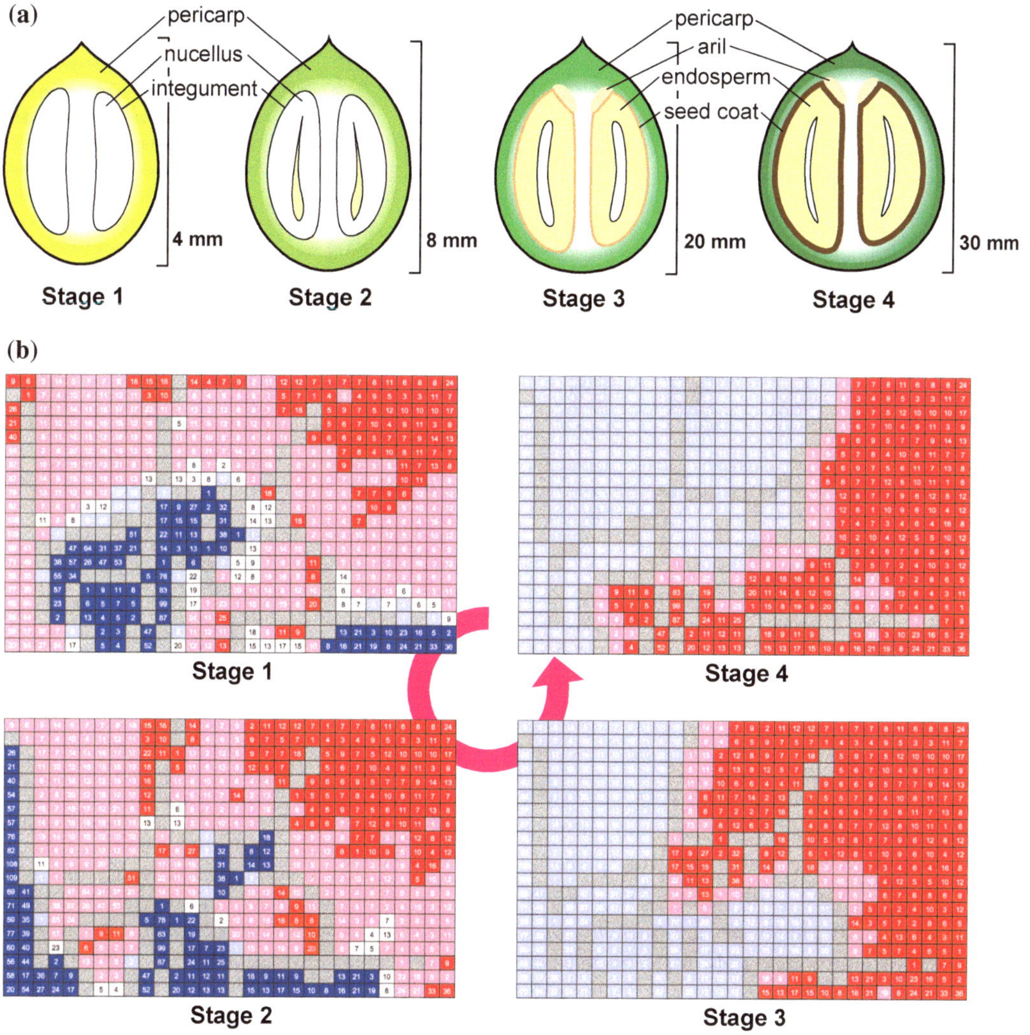

Fig. 5.1 Metabolic changes in fruit metabolites during maturation. **a** Schematic drawings of fruit development (longitudinal section). *Scale bar* for each stage of fruit indicates an approximate length. Note that the seed coat of stage 4 has already hardened (and deposited pigments) to a considerable extent. **b** BL-SOM representation using metabolome data sets. *Colors* of lattices represent the intensity of ion peaks: *red* (highest), *pink* (high), *light blue* (low), and *blue* (lowest). *Number* in each lattice denotes the number of objects (metabolites) included, and *hatched lattices* represent blank ones that include no objects

5.3.1.4 NMR-Based Analysis of Jatropha

The changes in metabolic profiles in *J. curcas*, infected with *Jatropha* mosaic virus (JMV), have been investigated using magnetic resonance imaging (MRI) and proton high-resolution magic angle spinning (HR-MAS) NMR spectroscopy (Sidhu et al. 2010). The metabolic analysis indicated that viral infection significantly affected the plant metabolism such as TCA cycle, amino acid, and sugar metabolism. The qualitative analysis of the seeds stored at different conditions and the metabolism of *J. curcas* during its initial growth stage are investigated using ^{13}C-glucose

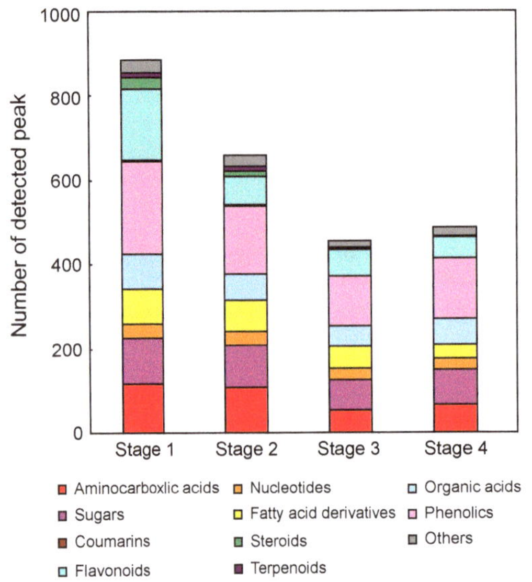

Fig. 5.2 Number of annotated metabolites in each development stage. Annotated metabolites are classified in 11 groups (aminocarboxylic acids, sugars, nucleotides, fatty acid derivaives, organic acids, phenolics, coumarins, flavonoids, steroids, terpenoids, and others)

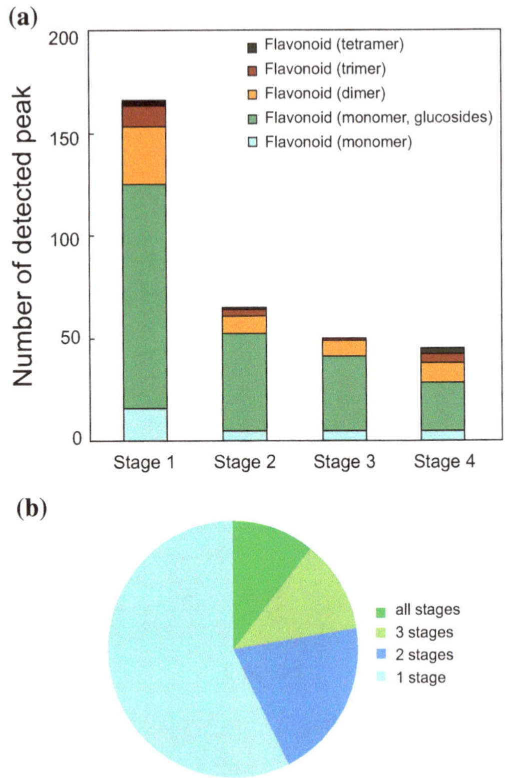

Fig. 5.3 **a** Number of flavonoid chemicals. The number of annotated flavonoid chemicals as monomer (monomer as *blue*, monomer with glucosides as *green*) and oligomers (dimer as *orange*, trimer as *brown*, and tetramer as *black*) in each development stage are represented. **b** Distribution of flavonoid chemicals. The ratio of the number of flavonoids detected in all stages (*green*), three stages (*light green*), two stages (*blue*), and one stage (*light blue*) is represented. Around 10% of annotated flavonoids are detected in all stages, whereas 57% of them are only detected in one of the stages

and ^{15}N-nitrate (Komatsu et al. 2014). The NMR analysis identified positive (sucrose and raffinose family oligosaccharides) and negative (gluconic acid) markers of seed germination. The analysis also indicated that glutamine and arginine were the major organic compounds for nitrogen and carbon transfer from roots to leaves. To analyze the effects of different conditions of torrefaction on *Jatropha* biomass, six different types of Jatropha tissues (seed coat, kernel, stem, xylem, bark, and leaf) were torrefied at four different temperature conditions, and changes in the metabolite composition of the torrefied products were determined by Fourier transform–infrared spectroscopy (FT–IR) and NMR analyses (Watanabe et al. 2014). The study showed that the degradation patterns of cellulose, hemicellulose, glucuronoxylan, starch, and lignin depending on the tissue and temperature conditions. These results suggest that torrefaction is a feasible treatment for further processing of residual biomass to biorefinery stock or fertilizer.

5.3.1.5 Database Deposition of Jatropha Metabolome Data

The raw data sets of maturing Jatropha fruit obtained by LC–Orbitrap MS were deposited in the database MassBase (http://webs2.kazusa.or.jp/massbase/), which provides thousands of raw mass chromatograms of GC–MS, LC–MS, and CE–MS from various biological samples of more than 100 plant species. All data sets in the MassBase are freely available from the site.

The analytical information (metadata) of Jatropha obtained from the LC–Orbitrap MS analysis is available at the metabolome metadata database Metabolonote as the accession number SE15 (http://metabolonote.kazusa.or.jp/SE15:/).

5.3.2 Integrated Approaches of Metabolomics Information of Jatropha

5.3.2.1 Omics Data sets of Jatropha

A wealth of information for genome, transcriptome, proteome, and metabolome of Jatropha would be a key to utilize the plant fully for pharmaceutical, agricultural, and environmental purposes. Given that the genome sequences of *Jatropha curcas* have been determined (Sato et al. 2011; Hirakawa et al. 2012), the whole gene expression (transcriptome) data of seedlings exposed to salt stress (Zhang et al. 2014), meristems treated with cytokinin (Pan et al. 2014), and developing seeds (Jiang et al. 2012) obtained by using high-performance sequencing technologies have became available. Proteomic analysis of the seed development in Jatropha was also reported to reveal carbon flux along with the lipid accumulation (Liu et al. 2013). As mentioned in the previous section, metabolome analyses of the plant are undergoing. Although gene expression and metabolite accumulation may not be directly correlated, integration of information of omics data sets could lead to a new way for applied research (Hirai et al. 2004; Tohge et al. 2005; Saito et al. 2008; Ritter et al. 2014). As genetic transformation protocols for *Jatropha curcas* have been established (Misra and Misra 2010; Tsuchimoto et al. 2012; Kajikawa et al. 2012), finding key genes for genetic control of metabolisms has been expected in integrated approaches of omics data sets.

5.3.2.2 Integration of Transcriptome and Metabolome Data of Jatropha

The KaPPA-View4 system (http://kpv.kazusa.or.jp) is provided to integrate quantitative transcriptome and metabolome data to overlay gene-to-gene and/or metabolite-to-metabolite relationships as curves on a metabolic pathway map, or on a combination of up to four maps (Sakurai et al. 2011). The system includes KaPPA-View4 Classic, KaPPA-View4 KEGG, and KaPPA-View4 Jatropha (Sakurai et al. 2012). The Classic was created originally for integration of the large data sets of Arabidopsis transcriptome and metabolome on 110 metabolic pathways (Tokimatsu et al. 2005), and then the maps have been added for major plant species, soybean (*Glycine max*), barley (*Hordeum vulgare*), *Lotus japonicus*, rice (*Oryza sativa*), poplar (*Populus trichocarpa*), tomato (*Solanum lycopersicum*), wheat (*Triticum aestivum*), grape (*Vitis vinifera*), and maize (*Zea mays*). To extend the tool function of the Classic, the KEGG version has been incremented, in which the wealth of maps of the KEGG (Kyoto Encyclopedia of Genes and Genomes) (http://www.genome.jp/kegg/) for human (*Homo sapiens*), rat (*Rattus norvegicus*), mouse (*Mus musculus*), *Drosophila melanogaster, Caenorhabditis elegans, Escherichia coli* K, *Saccharomyces cerevisiae, Arabidopsis thaliana, Physcomitrella patens* subsp. *patens*, rice, maize, *Sorghum bicolor*, poplar, grape, and castor bean (*Ricinus communis*) are available. Here we introduce the KaPPA-View4 Jatropha briefly.

The KaPPA-View4 Jatropha (http://kpv2.kazusa.or.jp/kpv4-jat/) has some of KEGG maps. Out of 40,929 predicted genes, 8,085 genes were mapped on the KEGG Metabolism maps, other KEGG maps, or gene category maps that were generated from gene classification data of KEGG BRITE. Two transcriptome data sets (RNA-Seq data from hermaphrodite flowers and early stage of developing fruits of *J. curcas* accession IBP-1 grown in an experimental field), four metabolome data sets (nontargeted metabolome data by LC–Orbitrap MS from four stages of developing fruits of *J. curcas*), and one gene co-expression data set were registered in the viewer as sample data set. The gene co-expression data set was calculated by 6 transcriptome data sets for developing flowers and fruits (Akashi et al. unpublished data) using the cosine correlation coefficients (0.99) between the 6,898 genes assigned on the Metabolism pathway maps of

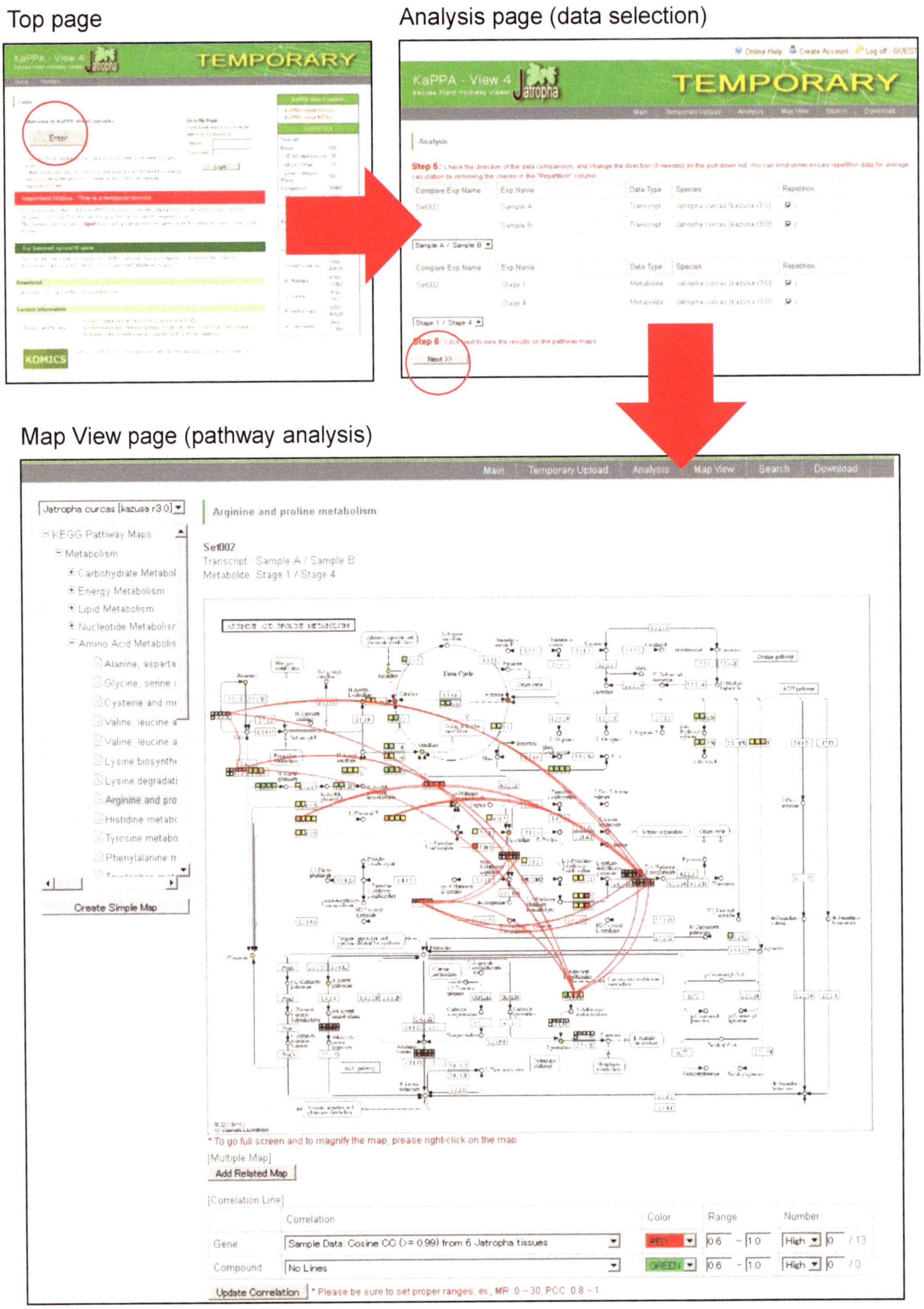
Top page
TEMPORARY
Enter
KOMICS
Analysis page (data selection)
KaPPA - View 4 Jatropha
TEMPORARY
Analysis
Next >>
Map View page (pathway analysis)
Main
Temporary Upload
Analysis
Map View
Search
Download
Jatropha curcas [kazusa r3.0]
KEGG Pathway Maps
Metabolism
Carbohydrate Metabol
Energy Metabolism
Lipid Metabolism
Nucleotide Metabolisr
Amino Acid Metabolis
Alanine, asparta
Glycine, serine
Cysteine and me
Valine, leucine a
Valine, leucine a
Lysine biosynthe
Lysine degradati
Arginine and pro
Histidine metabc
Tyrosine metabo
Phenylalanine m
Create Simple Map
Arginine and proline metabolism
Set002
Transcript : Sample A / Sample B
Metabolite : Stage 1 / Stage 4
* To go full screen and to magnify the map, please right-click on the map
[Multiple Map]
Add Related Map
[Correlation Line]
Correlation
Color
Range
Number
Gene
Sample Data: Cosine CC (>= 0.99) from 6 Jatropha tissues
RED
0.6
1.0
High
0
/ 13
Compound
No Lines
GREEN
0.6
1.0
High
0
/ 0
Update Correlation
* Please be sure to set proper ranges. ex., MR: 0 ~ 30, PCC: 0.8 ~ 1

◀ **Fig. 5.4** KaPPA-View Jatropha integrates transcriptome and metabolome data sets for the prediction of gene functions in functional genomics studies. A workflow is represented by the "top page", "analysis page", and "map view page". Transcriptome and metabolome data sets selected in the analysis page are overlaid on the pathway maps in the map view page. A *colored circle* represents a detected metabolite, and a *colored square* represents a detected transcript. A *red arc* represents the correlation of a pair of transcripts. The name, identifier, and values of metabolites and transcripts are displayed by mouseover action of the *colored circles* and *squares* in each pathway map. The detailed manual can be found in the Online Help page

KEGG. From the metabolome analysis data, 71 compounds that were annotated and attributed to the corresponding KEGG IDs were selected and registered in the system. Figure 5.4 shows an overview of the KaPPA-View4 Jatropha. The instructions are as follows: The Main page for the analysis is displayed after clicking Enter button in the top page. To select transcriptome and metabolome data sets for the analysis, click "Analysis" in menu bar. Otherwise you can click "Temporary Upload" in menu bar to upload your own data sets. (i) After the selection of both transcriptome and metabolome data sets, click Next button in the Analysis page to go to the map view page. (ii) In the map view page, the pathway of your interest can be selected from left side menu. The detected metabolites and transcripts in each metabolic pathway are indicated as colored circles and squares, respectively. The correlation of gene expression between a pair of transcripts is represented as a red arc. The value of a ratio of each metabolite accumulation and transcript expression is displayed by mouseover action of a colored circle and square, respectively. The detailed manual can be found in the Online Help page (http://kpv2.kazusa.or.jp/kpv4-jat/help/en/index.htm).

5.4 Conclusions

The Jatropha plant is a useful resource for the production of biodiesel from seed oil as renewable energy and of bioactive chemicals for pharmaceutical and agricultural applications. The seed oil meets the standards of the USA and the European Union as diesel fuel. All parts of the plant have been used in traditional medicine, and various bioactivities such as antimicrobial, insecticidal, and anti-inflammatory are known. Several uncommon metabolites are found in the plant. Metabolomics approaches, in which the whole metabolites (metabolome) are investigated in conjunction with the whole transcripts (transcriptome), may open an opportunity of developing new technology for further utilization of the plant. Given that the genome sequences of Jatropha have been determined, metabolomics approaches on Jatropha, in which the metabolism of the plant is investigated with the information of metabolome and transcriptome, will accelerate the applied research of oil and biological active compounds. In this context, integration of information obtained from transcriptome and metabolome analyses could be a key to understand the metabolism of useful metabolites such as oil and bioactive compounds. It is worth noticing that the expansion of Jatropha plantation in tropical regions, where the renewable oil production is expected from the plant grown in wastelands that are not suitable for vegetables or crops, is preferable to produce useful chemicals for pharmaceutical and agricultural purposes as a stable source for supplying the materials.

References

Achten WMJ, Mathijs E, Verchot L, Singh VP, Aerts R, Muys B (2007) *Jatropha* biodiesel fueling sustainability? Biofuels Bioprod Biorefin 1:283–291

Afendi FM, Okada T, Yamazaki M, Hirai-Morita A, Nakamura Y, Nakamura K, Ikeda S, Takahashi H, Altaf-Ul-Amin M, Darusman LK, Saito K, Kanaya S (2012) KNApSAcK family databases: integrated metabolite–plant species databases for multifaceted plant research. Plant Cell Physiol 53:e1

Barroso-González J, El Jaber-Vazdekis N, García-Expósito L, Machado J-D, Zárate R, Ravelo ÁG, Estévez-Braun A, Valenzuela-Fernández A (2009) The lupane-type triterpene 30-oxo-calenduladiol is a CCR5 antagonist with anti-HIV-1 and anti-chemotactic activities. J Biol Chem 284:16609–16620

Devappa RK, Makkar HPS, Becker K (2010) Nutritional, biochemical, and pharmaceutical potential of proteins and peptides from Jatropha: review. J Agric Food Chem 58:6543–6555

Devappa RK, Makkar HPS, Becker K (2011) *Jatropha* diterpenes: a review. J Am Oil Chem Soc 88:301–322

García A, Delgado G (2006a) Uncommon sesquiterpenoids and new triterpenoids from *Jatropha neopauciflora* (Euphorbiaceae). Helv Chim Acta 89:16–29

García A, Delgado G (2006b) Cytotoxic *cis*-fused bicyclic sesquiterpenoids from *Jatropha neopauciflora*. J Nat Prod 69:1618–1621

Gaudani H, Gupta M, Gupta N, Trivedi S, Patil P, Gupta G, Krishna KV, Reddy MP, Sethiya BD, Rathod MR (2009) Isolation and characterization of Phorbol esters (toxin) from the *Jatropha curcas* L. Int J Microbiol Res 1:1–7

Goel G, Makkar HPS, Francis G, Becker K (2007) Phorbol esters: structure, biological activity, and toxicity in animals. Int J Toxicol 26:279–288

Harry-Asobara JL, Eno-Obong SO (2014) Comparative study of the phytochemical properties of *Jatropha curcas* and *Azadirachta indica* plant extracts. J Poison Med Plants Res 2:020–024

Hirai MY, Yano M, Goodenowe DB, Kanaya S, Kimura T, Awazuhara M, Arita M, Fujiwara T, Saito K (2004) Integration of transcriptomics and metabolomics for understanding of global responses to nutritional stresses in *Arabidopsis thaliana*. Proc Natl Acad Sci USA 101:10205–10210

Hirakawa H, Tsuchimoto S, Sakai H, Nakayama S, Fujishiro T, Kishida Y, Kohara M, Watanabe A, Yamada M, Aizu T, Toyoda A, Fujiyama A, Tabata S, Fukui K, Sato S (2012) Upgraded genomic information of *Jatropha curcas* L. Plant Biotechnol 29:123–130

Hirota M, Suttajit M, Suguri H, Endo Y, Shudo K, Wongchai V, Hecker E, Fujiki H (1988) A new tumor promoter from the seed oil of *Jatropha curcas* L., an intramolecular diester of 12-deoxy-16-hydroxyphorbol. Cancer Res 48:5800–5804

Horiuchi T, Fujiki H, Hirota M, Suttajit M, Suganuma M, Yoshioka A, Wongchai V, Hecker E, Sugimura T (1987) Presence of tumor promoters in the seed oil of *Jatropha curcas* L. from Thailand. Jpn J Cancer Res 78:223–226

International Rice Genome Sequencing Project (2005) The map-based sequence of the rice genome. Nature 436:793–800

Jiang H, Wu P, Zhang S, Song C, Chen Y, Li M, Jia Y, Fang X, Chen F, Wu G (2012) Global analysis of gene expression profiles in developing physic nut (*Jatropha curcas* L.) seeds. PLoS One 7:e36522

Johnson TS, Eswaran N, Sujatha M (2011) Molecular approaches to improvement of *Jatropha curcas* Linn. as a sustainable energy crop. Plant Cell Rep 30:1573–1591

Kajikawa M, Morikawa K, Inoue M, Widyastuti U, Suharsono S, Yokota A, Akashi K (2012) Establishment of bispyribac selection protocols for *Agrobacterium tumefaciens*- and *Agrobacterium rhizogenes*-mediated transformation of the oil seed plant *Jatropha curcas* L. Plant Biotechnol 29:145–153

Katajamaa M, Miettinen J, Orešič M (2006) MZmine: toolbox for processing and visualization of mass spectrometry based molecular profile data. Bioinformatics 22:634–636

Kera K, Ogata Y, Ara T, Nagashima Y, Shimada N, Sakurai N, Shibata D, Suzuki H (2014) ShiftedIonsFinder: A standalone Java tool for finding peaks with specified mass differences by comparing mass spectra of isotope-labeled and unlabeled data sets. Plant Biotechnol 31:269–274

King AJ, He W, Cuevas JA, Freudenberger M, Ramiaramanana D, Graham IA (2009) Potential of *Jatropha curcas* as a source of renewable oil and animal feed. J Exp Bot 60:2897–2905

Komatsu T, Ohishi R, Shino A, Akashi K, Kikuchi J (2014) Multi-spectroscopic analysis of seed quality and ^{13}C-stable-iotopologue monitoring in initial growth metabolism of *Jatropha curcas* L. Metabolites 4:1018–1033

Kumar A, Sharma S (2008) An evaluation of multipurpose oil seed crop for industrial uses (*Jatropha curcas* L.): a review. Ind Crops Prod 28:1–10

Liu H, Wang C, Komatsu S, He M, Liu G, Shen S (2013) Proteomic analysis of the seed development in *Jatropha curcas*: from carbon flux to the lipid accumulation. J Proteomics 91:23–40

Maghuly F, Laimer M (2013) *Jatropha curcas*, a biofuel crop: Functional genomics for understanding metabolic pathways and genetic improvement. Biotechnol J 8:1172–1182

Mahalakshmi R, Eganathan P, Parida AK (2013) Salicylic acid elicitation on production of secondary metabolite by cell cultures of *Jatropha curcas* L. Int J Pharm Pharm Sci 5:655–659

Makkar HPS, Becker K (2009) *Jatropha curcas*, a promising crop for the generation of biodiesel and value-added coproducts. Eur J Lipid Sci Technol 111:773–787

Misra BB, Assmann SM, Chen S (2014) Plant single-cell and single-cell-type metabolomics. Trends Plant Sci 19:637–646

Misra M, Misra AN (2010) Jatropha: the biodiesel plant biology, tissue culture and genetic transformation—a review. Int J Pure Appl Sci Technol 1:11–24

Mujumdar AM, Misar AV (2004) Anti-inflammatory activity of *Jatropha curcas* roots in mice and rats. J Ethnopharmacol 90:11–15

Nayak BS, Patel KN (2010) Anti-inflammatory screening of *Jatropha curcas* root, stem and leaf in albino rats. Rom J Biol Plant Biol 55:9–13

Nwokocha Blessing A, Agbagwa IO, Okoli BE (2011) Comparative phytochemical screening of *Jatropha* L. species in the Niger Delta. Res J Phytochem 5:107–114

Ogunnaike BF, Okutachi IR, Anucha ES, Gbodi OO, Shokunbi OS, Onajobi FD (2013) Comparative anti-inflammatory activities of *Jatropha curcas*, *Ocimum gratissimum* and *Solanum scabrum* leaves. J Nat Prod Plant Resour 3:59–66

Ohtani M, Nakano Y, Usami T, Demura T (2012) Comparative metabolome analysis of seed kernels in phorbol ester-containing and phorbol ester-free accessions of *Jatropha curcas* L. Plant Biotechnol 29:171–174

Pan B-Z, Chen MS, Ni J, Xu Z-F (2014) Transcriptome of the inflorescence meristems of the biofuel plant *Jatropha curcas* treated with cytokinin. BMC Genom 15:974

Pereira Filho AA, França CRC, Oliveira DDS, Mendes RJDA, Gonçalves JDRS, Rosa IG (2014) Evaluation of the molluscicidal potential of hydroalcoholic extracts of *Jatropha gossypiifolia* Linnaeus, 1753 ON *Biomphalaria glabrata* (Say, 1818). Rev Inst Med Trop Sao Paulo 56:505–510

Rachana S, Tarun A, Rinki R, Neha A, Meghna R (2012) Comparative analysis of antibacterial activity of *Jatropha curcas* fruit parts. J Pharm Biomed Sci 15:1–4

Rampadarath S, Puchooa D, Ranghoo-Sanmukhiya M (2014) Antimicrobial, phytochemical and insecticidal properties of *Jatropha* species and wild *Ricinus communis* L. found in Mauritius. Int J Pharm Phytochem Res 6:831–840

Ravindranath N, Reddy MR, Mahender G, Ramu R, Kumar KR, Das B (2004) Deoxypreussomerins from *Jatropha curcas*: are they also plant metabolites? Phytochemistry 65:2387–2390

Ritter A, Dittami S, Goulitquer S, Correa J, Boyen C, Potin P, Tonon T (2014) Transcriptomic and metabolomic analysis of copper stress acclimation in *Ectocarpus siliculosus* highlights signaling and tolerance mechanisms in brown algae. BMC Plant Biol 14:116

Rumzhum NN, Sohrab MH, Al-Mansur MA, Rahman MS, Hasan CM, Rashid MA (2012) Secondary metabolites from *Jatropha podagrica* Hook. J Phys Sci 23:29–37

Saito K, Matsuda F (2010) Metabolomics for functional genomics, systems biology, and biotechnology. Annu Rev Plant Biol 61:463–489

Saito K, Hirai MY, Yonekura-Sakakibara K (2008) Decoding genes with coexpression networks and metabolomics—'majority report by precogs'. Trends Plant Sci 13:36–43

Sakurai N, Ara T, Ogata Y, Sano R, Ohno T, Sugiyama K, Hiruta A, Yamazaki K, Yano K, Aoki K, Aharoni A, Hamada K, Yokoyama K, Kawamura S, Otsuka H, Tokimatsu T, Kanehisa M, Suzuki H, Saito K, Shibata D (2011) KaPPA-View4: a metabolic pathway database for representation and analysis of correlation networks of gene co-expression and metabolite co-accumulation and omics data. Nucleic Acids Res 39:D677–D684

Sakurai N, Ogata Y, Ara T, Sano R, Akimoto N, Hiruta A, Suzuki H, Kajikawa M, Widyastuti U, Suharsono S, Yokota A, Akashi K, Kikuchi J, Shibata D (2012) Development of KaPPA-View4 for omics studies on Jatropha and a database system KaPPA-Loader for construction of local omics databases. Plant Biotechnol 29:131–135

Sakurai N, Ara T, Kanaya S, Nakamura Y, Iijima Y, Enomoto M, Motegi T, Aoki K, Suzuki H, Shibata D (2013) An application of a relational database system for high-throughput prediction of elemental compositions from accurate mass values. Bioinformatics 29:290–291

Sakurai N, Ara T, Enomoto M, Motegi T, Morishita Y, Kurabayashi A, Iijima Y, Ogata Y, Nakajima D, Suzuki H, Shibata D (2014) Tools and databases of the KOMICS web portal for preprocessing, mining, and dissemination of metabolomics data. Biomed Res Int 2014:11

Sano R, Ara T, Akimoto N, Sakurai N, Suzuki H, Fukuzawa Y, Kawamitsu Y, Ueno M, Shibata D (2012) Dynamic metabolic changes during fruit maturation in *Jatropha curcas* L. Plant Biotechnol 29:175–178

Sato S, Hirakawa H, Isobe S, Fukai E, Watanabe A, Kato M, Kawashima K, Minami C, Muraki A, Nakazaki N, Takahashi C, Nakayama S, Kishida Y, Kohara M, Yamada M, Tsuruoka H, Sasamoto S, Tabata S, Aizu T, Toyoda A, Shin-i T, Minakuchi Y, Kohara Y, Fujiyama A, Tsuchimoto S, Kajiyama S, Makigano E, Ohmido N, Shibagaki N, Cartagena JA, Wada N, Kohinata T, Atefeh A, Yuasa S, Matsunaga S, Fukui K (2011) Sequence analysis of the genome of an oil-bearing tree, *Jatropha curcas* L. DNA Res 18:65–76

Seth R, Sarin R (2010) Analysis of the phytochemical content and anti-microbial activity of *Jatropha gossypifolia* L. Arch Appl Sci Res 2:285–291

Sidhu OP, Annarao S, Pathre U, Snehi SK, Raj SK, Roy R, Tuli R, Khetrapal CL (2010) Metabolic and histopathological alterations of *Jatropha mosaic begomovirus*-infected *Jatropha curcas* L. by HR-MAS NMR spectroscopy and magnetic resonance imaging. Planta 232:85–93

Silva CR, Frohlich JK, Oliveira SM, Cabreira TN, Rossato MF, Trevisan G, Froeder AL, Bochi GV, Moresco RN, Athayde ML, Ferreira J (2013) The antinociceptive and anti-inflammatory effects of the crude extract of *Jatropha isabellei* in a rat gout model. J Ethnopharmacol 145:205–213

Smith CA, Want EJ, O'Maille G, Abagyan R, Siuzdak G (2006) XCMS: processing mass spectrometry data for metabolite profiling using nonlinear peak alignment, matching, and identification. Anal Chem 78:779–787

Soga T (2007) Capillary electrophoresis–mass spectrometry for metabolomics. In: Weckwerth W (ed) Metabolomics: methods and protocols. Humana Press, New York, pp 129–137

Staubmann R, Manfred S-Z, Hiermann A, Kartnig T (1999) A complex of 5-hydroxypyrrolidin-2-one and pyrimidine-2,4-dione isolated from *Jatropha curcas*. Phytochemistry 50:337–338

Sutthivaiyakit S, Mongkolvisut W, Prabpai S, Kongsaeree P (2009) Diterpenes, sesquiterpenes, and a sesquiterpene– coumarin conjugate from *Jatropha integerrima*. J Nat Prod 72:2024–2027

The Arabidopsis Genome Initiative (2000) Analysis of the genome sequence of the flowering plant *Arabidopsis thaliana*. Nature 408:796–815

The Tomato Genome Consortium (2012) The tomato genome sequence provides insights into fleshy fruit evolution. Nature 485:635–641

Tikunov Y, Lommen A, de Vos CHR, Verhoeven HA, Bino RJ, Hall RD, Bovy AG (2005) A novel approach for nontargeted data analysis for metabolomics. Large-scale profiling of tomato fruit volatiles. Plant Physiol 139:1125–1137

Timmers M, Urban S (2011) On-line (HPLC–NMR) and off-line phytochemical profiling of the Australian plant, *Lasiopetalum macrophyllum*. Nat Prod Commun 6:1605–1616

Tohge T, Nishiyama Y, Hirai MY, Yano M, Nakajima J, Awazuhara M, Inoue E, Takahashi H, Goodenowe DB, Kitayama M, Noji M, Yamazaki M, Saito K (2005) Functional genomics by integrated analysis of metabolome and transcriptome of Arabidopsis plants over-expressing an MYB transcription factor. Plant J 42:218–235

Tokimatsu T, Sakurai N, Suzuki H, Ohta H, Nishitani K, Koyama T, Umezawa T, Misawa N, Saito K, Shibata D (2005) KaPPA-View. A web-based analysis tool for integration of transcript and metabolite data on plant metabolic pathway maps. Plant Physiol 138:1289–1300

Trethewey RN (2004) Metabolite profiling as an aid to metabolic engineering in plants. Curr Opin Plant Biol 7:196–201

Tsuchimoto S, Cartagena J, Khemkladngoen N, Singkaravanit S, Kohinata T, Wada N, Sakai H, Morishita Y, Suzuki H, Shibata D, Fukui K (2012) Development of transgenic plants in jatropha with drought tolerance. Plant Biotechnol 29:137–143

Uche FI, Aprioku JS (2008) The phytochemical constituents, analgesic and anti-inflammatory effects of methanol extract of *Jatropha curcas* leaves in mice and Wister albino rats. J Appl Sci Environ Manag 12:99–102

Watanabe T, Shino A, Akashi K, Kikuchi J (2014) Chemical profiling of *Jatropha* tissues under different torrefaction conditions: application to biomass waste recovery. PLoS One 9:e106893

Yamamoto N, Suzuki T, Ara T, Sakurai N, Shinpo S, Morishita Y, Sasaki R, Tsugane T, Suzuki H, Shibata D (2012) *MatchedIonsFinder*: a software tool for revising alignment matrices of spectrograms from liquid chromatography-mass spectrometry. Plant Biotechnol 29:109–113

Yang YF, Liu JQ, Li XY, Liu EQ, Li ZR, Qiu MH (2013) New terpenoids from the roots of *Jatropha curcas*. Chin Sci Bull 58:1115–1119

Zhang L, Zhang C, Wu P, Chen Y, Li M, Jiang H, Wu G (2014) Global analysis of gene expression profiles in physic nut (*Jatropha curcas* L.) seedlings exposed to salt stress. PLoS One 9:e97878

Toxic Substances in Jatropha Seeds: Biosynthesis of the Most Problematic Compounds, Phorbol Esters

6

Misato Ohtani, Yoshimi Nakano, Ryosuke Sano, Tetsuya Kurata and Taku Demura

Abstract

Jatropha (*Jatropha curcas* L.) is considered to be an important oilseed crop. Its seed oil can be used as a biodiesel in existing diesel engines without special modifications, making this plant an attractive bioresource for the production of sustainable bioenergy. Additionally, it is possible that, after extracting the oil from Jatropha seed cakes, they can be used as animal feed because they contain high amounts of proteins. However, the toxic and antinutritional substances in the seeds and oils, such as a lectin, protease inhibitors, saponins, and phorbol esters, have generated a serious safety concern for further utilization of Jatropha. In this chapter, we review the toxic substances found in Jatropha seeds, focusing on the most problematic toxic substance, phorbol ester, one of the characteristic tetracyclic diterpenoids in Thymelaeaceae and Euphorbiaceae plants. Recent advances in genome-wide Jatropha research using toxic and non-toxic varieties have identified the genetic locus related to the phorbol ester content and the genes responsible for casbane-type diterpenoid biosynthesis. Based on these findings, Jatropha genomic research should help to achieve the practical utilization of Jatropha as a multipurpose crop.

Misato Ohtani and Yoshimi Nakano contributed equally.

M. Ohtani · Y. Nakano · R. Sano · T. Kurata · T. Demura (✉)
Graduate School of Biological Sciences, Nara Institute of Science and Technology, 8916-5 Takayama-cho, Ikoma 630-0192, Japan
e-mail: demura@bs.naist.jp

M. Ohtani · T. Demura
Biomass Engineering Research Division, RIKEN Center for Sustainable Resource Science, Yokohama, Japan

Present Address:
Y. Nakano
National Institute of Advanced Industrial Science and Technology (AIST), Tsukuba, Japan

Present Address:
T. Kurata
Graduate School of Life Sciences, Tohoku University, Sendai, Japan

S. Tsuchimoto (ed.), *The Jatropha Genome*, Compendium of Plant Genomes,
DOI 10.1007/978-3-319-49653-5_6

6.1 Introduction

Jatropha (*Jatropha curcas* L.) produces seeds rich in oil (44–62%), making this plant a favorable oilseed crop. Since the biodiesel made from Jatropha seed oil has properties similar to conventional petrodiesel, it can be used in existing diesel engines (El Diwani et al. 2009; Singh and Padhi 2009; Raja et al. 2011), and thus, Jatropha is expected to be a potential resource for the production of sustainable bioenergy. Jatropha is also traditionally used as a medicine, insecticide, and garden hedge (Thomas et al. 2008; Islam et al. 2011; Sabandar et al. 2013). Additionally, the seed kernels are known to contain a high content (22–35%) of proteins (Makkar and Becker 1997; Makkar et al. 1998). The amino acid score of proteins in Jatropha seed kernels indicates that the seeds have good balanced proteins, i.e., the level of essential amino acids tends to be higher than those of the FAO reference protein (Makkar and Becker 1997; Makkar et al. 1998). Therefore, Jatropha has the potential to be feedstock not only for biodiesel, but also for protein. However, one serious drawback to the use of the plant and its by-products is the presence of toxic components in its seeds and/or oil. The Jatropha seed contains several different toxic and/or antinutritional substances, such as a lectin, protease inhibitors, saponins, and phorbol esters (PEs) (Makkar and Becker 1997; Makkar et al. 1998; Martínez-Herrera et al. 2006). This safety concern makes the utilization of Jatropha seeds challenging, especially for the use of Jatropha seed cakes as animal feed after extracting the oil. Importantly, a non-toxic variety, i.e., an edible variety, of Jatropha is reported to grow naturally in Mexico and Central America (Makkar and Becker 1997; Makkar et al. 1998). Comparative analysis of the toxic factors present in the toxic and non-toxic varieties found that the non-toxic variety contains very low amounts of PEs, a group of characteristic tetracyclic diterpenoids found in the Thymelaeaceae and Euphorbiaceae families (Makkar and Becker 2009; Makkar et al. 2011) (Fig. 6.1), suggesting that the major contributor to the toxicity of Jatropha seeds is PEs.

In this chapter, we summarize the toxic substances in Jatropha, particularly those contained in the seeds. After reviewing the well-studied toxic substances, we will focus on the current research into the biosynthesis of PEs, the most problematic toxic compounds, and a summary of the future direction of Jatropha genome research in terms of the utilization of Jatropha as a multipurpose crop.

6.2 Toxic Compounds Found in Jatropha

Jatropha is known to be toxic when consumed by animals and humans. The toxic and antinutritional compounds contained in Jatropha are PEs, lectins, trypsin inhibitors, phytic acids, and saponins, which are summarized in this chapter.

6.2.1 Phorbol Esters (PEs)

Phorbol esters (PEs) are tetracyclic diterpenoids found in the seeds, sap, and latex of the Thymelaeaceae and Euphorbiaceae families. Six PE derivatives have been identified from Jatropha oil and named as Jatropha factors C1–C6 (Haas et al. 2002; Fig. 6.1). All these Jatropha factors contain a 12-hydroxy-16-deoxyphorbol structure (Adolf et al. 1984; Haas et al. 2002), tetracyclic diterpene (tigliane), and two fatty acids that are esterified at the C13 and C16 positions of the tigliane skeleton (Haas et al. 2002; Fig. 6.1). Hence, PEs are lipophilic and easily dissolved in oil. Jatropha seeds contain 0.03–4 mg/g kernel of PEs (Makkar and Becker 1997; Makkar et al. 1998; Martínez-Herrera et al. 2006; Ohtani et al. 2012). The non-toxic varieties contain up to 0.11 mg/g PEs in the seed kernels, and the values differ depending on the variety (Makkar and Becker 1997; Makkar et al. 1998; Martínez-Herrera et al. 2006), indicating that in some cases, low amounts of PEs can be tolerated for human consumption. The toxicity of Jatropha seeds has been studied in several animal species including goats, sheep, mice, rats, and fish

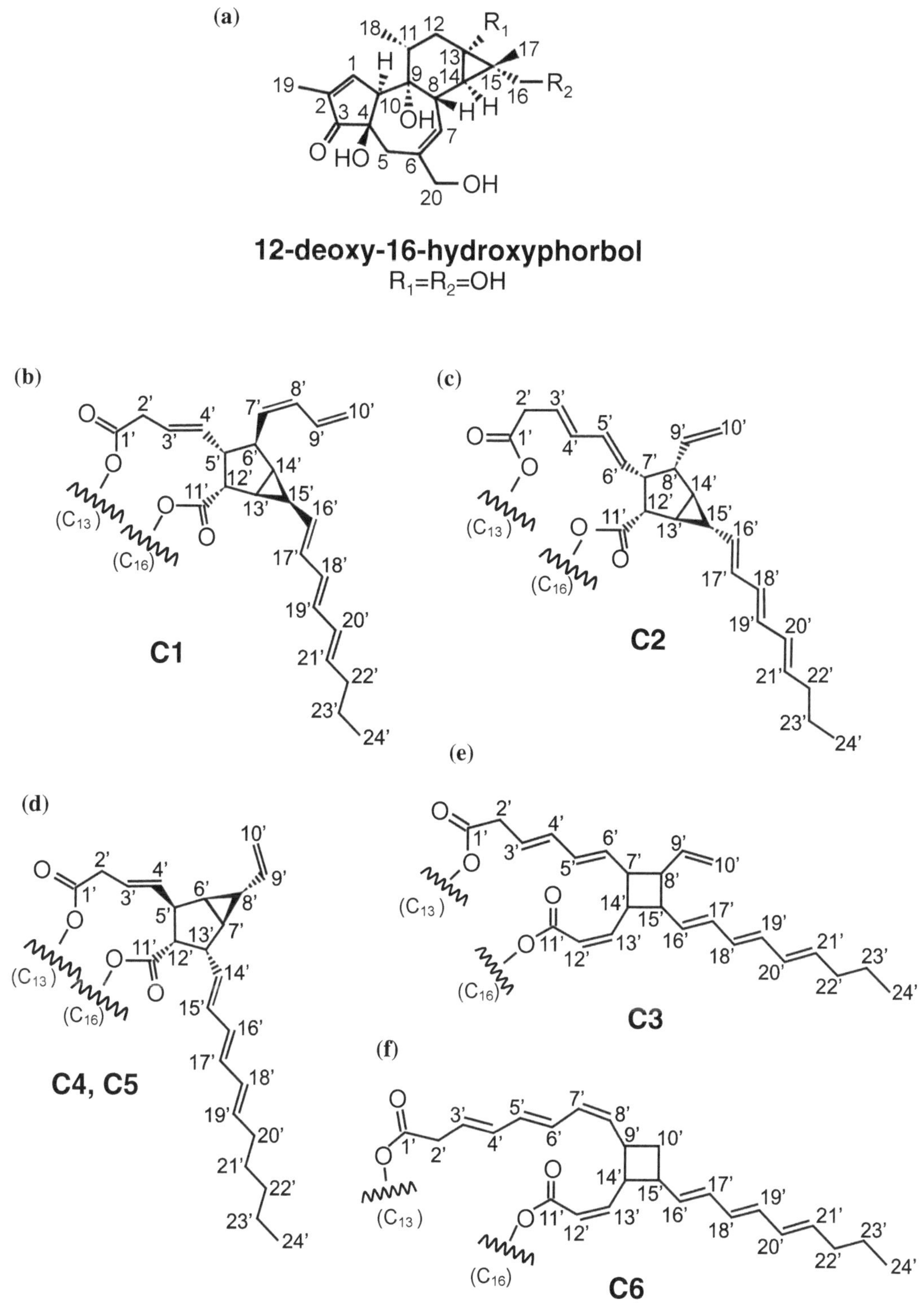

Fig. 6.1 Molecular structures of phorbol esters (PEs) found in *Jatropha curcas* L. (adopted from Haas et al. 2002). **a** The common diterpene, 12-deoxy-16-hydroxyphorbol. **b–f** The acid moiety of C1 (**b**), C2 (**c**), C3 (**e**), C4 and C5 (**d**), and C6 (**f**)

(Adam 1974; Adam and Magzoub 1975; Bourin et al. 1982; Gandhi et al. 1995; Makkar and Becker 1999). In most cases, PE-containing feeds showed negative effects on animal health and growth, despite the differences in species-specific toxicity. PE toxicity has also been observed for snails (Liu et al. 1997), silkworm (Jing et al. 2005), and several insects (Wink et al. 1997).

PEs are reported to promote tumors and inflammation as well as act as a purgative agent (Adolf et al. 1984; Hirota et al. 1988). The promotion of tumors by PE has been studied at the molecular level, indicating that PEs mimic the action of the lipid second messenger, diacylglycerol, which activates multiple signaling pathways via binding to cysteine-rich modules of signaling-related proteins called C1 domains (Goel et al. 2007). The C1 domains were originally identified in protein kinase C, but are also found in the protein kinase D family of proteins, chimaerins, RasGRPs, diacylglycerol kinases, and others (Colón-González and Kananietz 2006). All are thought to be targets for PE toxicity in animals (Goel et al. 2007).

6.2.2 Lectins

Lectins are proteins with binding activities to specific carbohydrate structures. Curcin, a toxic lectin, was isolated from Jatropha seeds by Felke (1914). Curcin functions as a type I ribosome-inactivating protein with high *N*-glycosidase activity (Barbieri et al. 1993; Lin et al. 2003). The *N*-glycosidic bond of adenine A_{4234} of eukaryotic 28S rRNA is cleaved, resulting in the disturbance of binding between the ribosome and elongation factors, consequently arresting protein synthesis (Endo and Tsurugi 1987). Since the non-toxic Jatropha varieties contain similar amounts of lectins to the toxic varieties (Makkar and Becker 1997; Aregheore et al. 1998; Makkar et al. 1998), lectins are likely not responsible for the toxicity, at least for the quick-impact toxicity. Recently, in addition to the identification of curcin in Jatropha seeds, the stress-induced lectin, curcin-L (also known as curcin 2), was identified in the leaves of Jatropha (Qin et al. 2005, 2009). The genes encoding curcin and curcin-L are specifically expressed in seeds and in stress-treated leaves, respectively, and curcin and curcin-L have different antifungal activities (Qin et al. 2010), suggesting that each lectin has different physiological functions. Ribosome-inactivating proteins have the potential to be developed as antitumor drugs that selectively target tumor cells, and thus, Jatropha curcin may have therapeutic applications (Zheng et al. 2013).

6.2.3 Trypsin Inhibitors

Several trypsin inhibitors were reported in Jatropha seeds, but there were no differences between the toxic and non-toxic varieties (Makkar and Becker 1997; Makkar et al. 1998). Additionally, the trypsin inhibitor activity is easily inactivated by moist heat treatment (Martínez-Herrera et al. 2006), indicating that trypsin inhibitors have only a minor contribution to Jatropha toxicity. Recently, the novel trypsin inhibitor JcTI-I was purified from Jatropha seed cakes (Costa et al. 2014). JcTI-I has antibacterial activity, as shown for Xb-KTI, a Kunitz trypsin inhibitor present in *Xanthosoma blandum* (Lima et al. 2011), and for fistulin, a serine protease inhibitor present in *Cassia fistula* (Arulpandi and Sangeetha 2012). These findings support the potential of plant-derived protease inhibitors as a novel class of antimicrobial agents.

6.2.4 Phytic Acids

Phytic acid, also known as inositol hexaphosphate, is a saturated cyclic acid and is the principal storage form of phosphorus in many plant tissues, such as seeds and brans (Coulibaly et al. 2011). The Jatropha seeds contain around 9% weight in dry seed materials regardless of their edibility (Makkar and Becker 1997; Makkar et al. 1998; Martínez-Herrera et al. 2006). It is known that phytic acid can chelate multivalent metal ions, especially zinc, calcium, and iron, and together with its protein residue, it is considered to be an antinutrient (Coulibaly et al. 2011). The content of phytic acid in Jatropha seeds is higher than those found in common food sources (Coulibaly et al. 2011), and the moist heating treatment does not affect the level of phytic acids in Jatropha seeds (Martínez-Herrera et al. 2006). Therefore, continuous consumption of Jatropha seeds will result in potentially problematic levels of phytic acid.

However, many beneficial effects of phytic acid have been reported. For example, phytate, the calcium salt form of phytic acid, can protect against dietary lead ions, and thus, phytate has the ability to counteract acute oral lead toxicity (Wise and Gilbert 1981). Moreover, phytic acid functions in mineral balancing in animals, such as zinc-copper metabolism, and further related to several diseases (Coulibaly et al. 2011). Phytic acid is one of a few available chelating therapies used for uranium removal (Cebrian et al. 2007), suggesting that well-controlled ingestion of phytic acid can be an effective remedy for a wide range of heavy-metal pollution. Therefore, Jatropha seed cakes are a possible source of phytate for therapeutic use.

6.2.5 Saponins

Saponins are a class of glycosides composed of a steroid or triterpenoid and a saccharide identified from various plant species. As saponins are amphipathic glycosides, they can permeabilize the cell membrane. Additionally, saponin forms complexes with cholesterols and/or proteins, and thus, saponins can greatly affect growth, feed intake, and reproduction in animals (Francis et al. 2002). Moreover, specific species of saponins have been utilized as traditional medicines (Podolak et al. 2010).

The amounts of saponins in Jatropha seeds are lower than soybean meal samples, and the amounts of saponins in the non-toxic varieties are higher than those in the toxic varieties (Makkar and Becker 1997; Makkar et al. 1998; Martínez-Herrera et al. 2006). This suggests that saponins are not significant contributors to the toxicity of Jatropha. Saponins are not degraded by ordinary cooking procedures (Reddy and Pierson 1994), but the ethanol extraction treatment of Jatropha seeds is known to reduce the saponin content (Martínez-Herrera et al. 2006).

6.3 Biosynthesis of Phorbol Esters (PEs) in Jatropha

6.3.1 Putative PE Biosynthesis Pathway

PE is composed of a tetracyclic diterpene, named tiglian, and two fatty acids, which are esterified to the C13 and C16 of the tiglian skeleton. Our knowledge of PE biosynthesis is still limited, perhaps because of their low abundance, instability, and susceptibility to oxidation, hydrolysis, and transesterification (Haas et al. 2002). The first cyclization step in the biosynthesis of tigliane has been proposed to involve the biosynthesis of casbane-type diterpenoids catalyzed by casbene synthase (CS), which converts geranylgeranyl diphosphate into casbane-type diterpenoids through a "head-to-tail" condensation (Schmidt 1987; Fig. 6.2). To date, several *CS* genes have been identified from the Euphorbiaceae family of plants, and some have been cloned (Mau and West 1994; Kirby et al. 2010; Nakano et al. 2012; Zerbe et al. 2013; King et al. 2014). Analysis of recombinant CS proteins and/or heterologous transient expression of *CS* genes demonstrated that some of the identified *CS* genes do not encode proteins with CS activity, but with neocembrene synthase activity (Huang et al. 1998; Kirby et al. 2010; King et al. 2014). Neocembrene is a nomocyclic diterpene isolated from pine and spruce resin (Shmidt et al. 1970).

Prior to finishing the genome sequencing of Jatropha in 2011 (Sato et al. 2011), we had identified the CS gene from a toxic variety collected from the Philippines based on the conserved amino acid sequences among known CS proteins. Shortly thereafter, we successfully identified the *JcCSH* gene (Nakano et al. 2012), which was derived from an identical gene locus to *JcCAS1* reported by King et al. (2014).

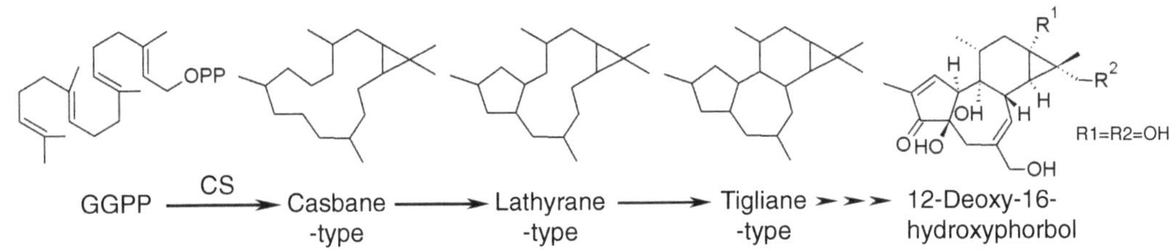

Fig. 6.2 Proposed biosynthesis pathway of the tigliane skeleton. A geranylgeranyl pyrophosphate (GGPP) is cyclized into a casbane-type diterpene by casbene synthase (CS). The phorbol structure is based on a tetracyclic diterpene known as tigliane, which is assumed to be synthesized from a casbane-type diterpene through a lathylane-type diterpene as an intermediate. Only the casbene synthesis step catalyzed by CS has been confirmed experimentally

JcCSH/*JcCAS1* encodes a protein with a chloroplast transit peptide and a DDxxD motif. Indeed, when the JcCSH/JcCAS1 protein was fused with a yellow fluorescent protein, it was found to be localized to the chloroplast, consistent with the hypothesis that CS functions in plastids where geranylgeranyl diphosphate and diterpenoids are synthesized (Chen et al. 2011). The DDxxD motif, which is conserved among known terpene cyclases, is presumed to be a catalytic domain for the cyclization reaction (Prisic et al. 2007). The expression of *JcCSH*/*JcCAS1* is detected in seedling, mature leaves, and the flesh of fruits, while no expression is found in developing seeds. Thus, the precursors of PEs or PEs themselves may be synthesized in distinct organs from seeds, and then transported and accumulated into the seeds. Alternatively, CSs other than JcCSH/JcCAS1 may be involved in PE biosynthesis in seeds because at least nine genes are predicted to be *CS* genes in the Jatropha genome (Sato et al. 2011). Further gene expression analysis of these Jatropha *CS* genes and functional analysis of their products will reveal the sites of casbene biosynthesis for PEs accumulated in Jatropha seeds.

The tigliane skeleton is thought to be biosynthesized through a series of intramolecular cyclization and oxidation steps from casbane-type diterpenoids (Schmidt 1987). These modification steps are presumed to be catalyzed by the cytochrome P450 family of enzymes, and in fact, the cytochrome P450 enzymes isolated from castor bean and *Euphorbia peplus* are known to catalyze the oxidation of casbane-type diterpenoids (King et al. 2014). After the construction of the tigliane skeleton, two fatty acids are esterified to C13 and C16 of the tigliane skeleton, a process presumably catalyzed by acyltransferases. The BAHD (BEAT, AHCT, HCBT, and DAT) acyltransferase family is thought to be involved because the BAHD family of proteins are reported to participate in the biosynthesis of a wide-range secondary metabolites, such as alkaloids, terpenoids, and phenolics in plants (D'Auria 2006), and some BAHD family proteins use fatty acids as substrates (Molina and Kosma 2015).

6.3.2 Genetic Approaches to Identify PE Biosynthesis-Related Genes

The detection of non-toxic varieties of Jatropha in Latin America supported the development of genetics approaches to identify the genes responsible for PE biosynthesis (Sujatha et al. 2005; Basha et al. 2009; Sudheer Pamidimarri et al. 2009). Because Jatropha has unisexual flowers and the non-toxicity of seeds is a stable heritable trait suggested that the seed toxicity, i.e., the PE content of the seeds, should be a maternal trait. Indeed, crossing studies indicated that the PE content of seeds is determined by the traits of the maternal plant (Sujatha et al. 2005).

To further develop molecular breeding of Jatropha, the establishment of a genetic linkage map is one of the most important strategies. Recently, a high-density genetic linkage map of *J. curcas* was reported (King et al. 2013). This linkage map was created from four F_2 mapping populations, which were generated by crossing toxic and non-toxic varieties, open- and closed-branching varieties, high- and low-oil content varieties, and high- and low-oleic acid content varieties, and contained more than 500 codominant genetic markers including simple sequence repeat (SSR) and single nucleotide polymorphism (SNP). One-third of the F_2 populations derived from the crossing between the toxic and non-toxic varieties produced non-toxic seeds, indicating that the genes responsible for seed toxicity should be located at a single genetic locus. King et al. (2013) mapped this locus to 37.6–39.9 cM on linkage group 8. However, interestingly, the *CS* gene clusters that are evolutionally conserved among the Euphorbiaceae are not located in this region; three *CS* genes (*JcCSH*/*JcCAS1*, *JcCAS2*, and *JcCAS3*) and two cytochrome P450 genes, the CYP726A genes, are in a single locus on linkage group 1 (King et al. 2014). The non-toxic variety used by King et al. (2014) did not accumulate PEs in the seeds, but can produce diterpenoids in the roots (King et al. 2014). Thus, the conserved gene cluster for diterpene biosynthesis is not directly related to determining the trait of PE content in Jatropha seeds. This is consistent with our observation that *JcCSH*/*JcCAS1* is expressed in organs other than seeds (Nakano et al. 2012). Additional information on the identified genetic locus on linkage group 8 will uncover what kinds of genes are involved in determining the trait of PE content in seeds.

6.3.3 Transcriptomic Approach to Identify PE Biosynthesis-Related Genes

In addition to the genetic approaches described above, a number of transcriptomic studies on Jatropha have been performed (Costa et al. 2010; Natarajan et al. 2010; King et al. 2011; Jiang et al. 2012; Wang et al. 2013). Combined with the genomic sequence information (Sato et al. 2011; Hirakawa et al. 2012), these studies have greatly developed our molecular biological understanding of Jatropha. However, transcriptome data for a non-toxic variety of Jatropha has not been reported.

In an attempt to identify novel molecular markers to distinguish the seed toxicity at the seedling stages, we conducted comparative transcriptome analysis using the toxic variety (NBF-1; Nakano et al. 2012) and non-toxic variety (NBF-7; Ohtani et al. 2012) of Jatropha. First, in order to establish reference contigs de novo, we generated the full-length cDNA library for sequencing by Roche 454 GL FLX+. The mRNA fractions were isolated from the mature leaves of each variety, and the mRNA samples of the toxic and non-toxic varieties were mixed equally. The mixed mRNA samples were used to construct the normalized full-length cDNA library. By normalizing the cDNA library, the prevalence of highly abundant transcripts should be decreased and the efficiency of sequencing for rare and/or specific transcripts should be increased (Natarajan et al. 2010). The cDNA library was sequenced with the Roche 454 GL FLX+ by the sequencing service at RIKEN LSA support (GeNAS, http://www.osc.riken.jp/genas/faq/index.html). Next, the RNA-sequence analysis was performed using an Illumina GAIIx platform. The total RNA samples were isolated from two individual toxic and non-toxic varieties (in total four samples) and subjected to mRNA-sequence analysis using an mRNA-seq 8-Sample Prep Kit (Illumina Inc. San Diego, CA USA). The resulting long reads obtained by 454 GL FLX+ and short reads obtained by GAIIx sequencing were assembled using Trinity (ver. Trinityrnaseq_r20131110, Grabherr et al. 2011). The assembled sequences containing a total of 32,091 unique contigs were used as reference contigs in our analysis. The longest contig was 11,065 bp, the average length of contigs was 763,064 bp, and the N50 length was 1095 bp. A total of 75.39% of constructed contigs (24,194 contigs) have high similarities with the *Arabidopsis thaliana* proteins annotated in TAIR10 (BLASTX search; *E* value < 0.00001).

For the comparative analysis between the toxic and non-toxic varieties, the sequence tags obtained by GAIIx were mapped to the reference contigs using TopHat2 (ver. 2.0.13, Kim et al. 2013), and then, the Reads Per Kilobase per Million Mapped reads (RPKM) were calculated for each contig by Cuffdiff2 (ver. 2.2.1, Trapnell et al. 2013). Significantly different expression was found for 1462 contigs between the toxic and non-toxic varieties using the TCC package on the R console (Sun et al. 2013); 599 and 863 contigs showed higher and lower expression in the toxic variety than in the non-toxic variety, respectively. These transcripts may be biomarker candidates useful to select low PE content plants at the seedling stages using the leaf samples. The gene ontology (GO) enrichment analysis was performed by BiNGO (Maere et al. 2005) using the annotation information on the *A. thaliana* homologs (ver. TAIR10) of these differentially expressed contigs. The results of GO Slim analysis showed that the terms "catalytic activity," "metabolic process," "secondary metabolic process," "cellular amino acid and derivative metabolic process," and "extracellular region" were significantly over-represented in the sets of relatively highly expressed contigs in the toxic variety (corrected p value < 0.05) (Table 6.1). In contrast, the terms "response to stress," "oxygen binding," "response to abiotic stimulus," "pollen–pistil interaction," and "death" were significantly enriched in the sets of relatively highly expressed contigs in the non-toxic variety (corrected p value < 0.05) (Table 6.1). These results suggest that metabolism regulation and stress response greatly varied between the toxic and non-toxic varieties.

Table 6.1 Gene ontology (GO) term analysis on differentially expressed contigs between the toxic and non-toxic varieties

GO term	GO term ID	Description	Corrected p-value
Relatively highly-expressed contigs in the toxic variety			
Biological process	8152	Metabolic process	3.42E-13
	19748	Secondary metabolic process	1.78E-11
	6519	Cellular amino acid and derivative metabolic process	9.51E-09
	6629	Lipid metabolic process	3.99E-08
	6950	Response to stress	1.03E-05
	5975	Carbohydrate metabolic process	1.61E-05
	9607	Response to biotic stimulus	5.43E-04
	9058	Biosynthetic process	2.74E-03
	9987	Cellular process	3.58E-03
	16265	Death	1.37E-02
	8219	Cell death	1.37E-02
	9719	Response to endogenous stimulus	2.39E-02
	9628	Response to abiotic stimulus	3.58E-02
Cellular component	5576	Extracellular region	9.51E-09
	30312	External encapsulating structure	6.69E-08
	5618	Cell wall	6.69E-08
	5623	Cell	1.61E-05
	5773	Vacuole	1.97E-04
Molecular function	3824	Catalytic activity	9.67E-21
	16740	Transferase activity	6.16E-06

(continued)

Table 6.1 (continued)

GO term	GO term ID	Description	Corrected *p*-value
	16301	Kinase activity	8.60E-05
	30246	Carbohydrate binding	1.03E-04
	19825	Oxygen binding	1.27E-04
	16787	Hydrolase activity	4.88E-04
Relatively highly-expressed contigs in the non-toxic variety			
Biological process	6950	Response to stress	5.53E-12
	9628	Response to abiotic stimulus	1.70E-06
	9875	Pollen-pistil interaction	2.79E-06
	16265	Death	6.46E-05
	8219	Cell death	6.46E-05
	9607	Response to biotic stimulus	9.84E-05
	9856	Pollination	1.08E-03
	9719	Response to endogenous stimulus	1.90E-03
	7154	Cell communication	5.72E-03
	6464	Protein modification process	7.80E-03
	19748	Secondary metabolic process	9.90E-03
	8150	Biological process	1.44E-02
	9838	Abscission	2.34E-02
	6519	Cellular amino acid and derivative metabolic process	2.76E-02
Molecular function	19825	Oxygen binding	1.48E-06
	30246	Carbohydrate binding	2.68E-03
	16301	Kinase activity	7.13E-03
	3700	Transcription factor activity	8.01E-03
	5488	Binding	9.71E-03

In the 32,091 contigs finally assembled, we identified one contig corresponding to *JcCSH/JcCAS1*, 171 contigs corresponding to cytochrome P450 genes, and 33 contigs corresponding to BAHD/BAHD-like genes, all of which are presumed to be involved in PE biosynthesis (see Sect. 3.1). A certain subset of these contigs had differences in expression between the toxic and non-toxic varieties (Table 6.2). The contig JcLf4rcHa26552 with the highest similarities to the *JcCSH/JcCAS1* gene was more highly expressed in the toxic variety than in the non-toxic variety. A total of 15 and 17 contigs related to cytochrome P450 were expressed at higher and lower levels in the toxic variety and non-toxic variety, respectively. The contig JcLf4rcHa10528, which is similar to the *CYP726A20* gene and was reported to be involved in the oxidation of diterpenes by King et al. (2014), was more highly expressed in the non-toxic variety than in the toxic variety. Interestingly, among the 33 contigs related to the BAHD/BAHD-like genes, the contig JcLf4rcHa09206 was specifically expressed in the toxic variety. The BAHD proteins are enzymes possibly involved in the esterification of two fatty acids to C13 and C16 of the tigliane skeleton. Thus, the lack of BAHD/BAHD-like gene expression could be related to the toxicity, i.e., the PE content in Jatropha. Our comparative transcriptome data was obtained from the leaves of toxic and non-toxic varieties, and thus, further

Table 6.2 Subsets of differentially expressed contigs between the toxic and non-toxic varieties

				RPKM[a]						
			Contig length	Toxic		Non-toxic		Fold Change		
Contig ID	*A. thaliana* Gene	Short description for *A. thaliana* Gene	(bp)	NBF-1_1	NBF-1_2	NBF-7_1	NBF-7_2	(log2)	*p* value	*q* value
CSH/CAS1 gene										
JcLf4rcHa26552	AT5G23960	Terpene synthase 21	386	22.54	13.35	1.24	3.53	−2.91	3.67E-04	1.16E-02
Cytochrome P450 genes										
JcLf4rcHa03679	AT2G46960	Cytochrome P450, family 709, subfamily B, polypeptide 1, CYP709B1	1661	110.30	42.87	0.55	0.99	−6.64	1.56E-04	5.88E-03
JcLf4rcHa19593	AT5G07990	Cytochrome P450 superfamily protein, CYP75B1	599	20.79	34.27	1.91	1.87	−3.86	1.39E-06	1.15E-04
JcLf4rcHa08841	AT4G15440	Hydroperoxide lyase 1, CYP74B2	1996	117.40	155.22	10.36	15.86	−3.38	1.73E-11	4.43E-09
JcLf4rcHa26677	AT4G31950	Cytochrome P450, family 82, subfamily C, polypeptide 3, CYP82C3	221	47.88	54.02	6.47	5.09	−3.14	2.74E-05	1.39E-03
JcLf4rcHa07144	AT2G30490	Cinnamate-4-hydroxylase, CYP73A5	1765	920.75	801.35	164.36	88.95	−2.77	1.33E-09	2.39E-07
JcLf4rcHa13270	AT4G00360	Cytochrome P450, family 86, subfamily A, polypeptide 2, CYP86A2	711	44.96	49.09	12.53	2.92	−2.61	2.10E-08	2.90E-06
JcLf4rcHa03014	AT2G45970	Cytochrome P450, family 86, subfamily A, polypeptide 8, CYP86A8	1213	20.95	25.42	6.91	1.15	−2.53	4.14E-08	5.47E-06
JcLf4rcHa20021	AT4G31940	Cytochrome P450, family 82, subfamily C, polypeptide 4, CYP82C4	824	8.49	8.09	1.46	1.42	−2.52	1.22E-05	7.18E-04
JcLf4rcHa23386	AT1G12740	Cytochrome P450, family 87, subfamily A, polypeptide 2, CYP87A2	477	10.96	15.26	1.52	3.12	−2.50	1.53E-04	5.79E-03
JcLf4rcHa21521	AT5G07990	Cytochrome P450 superfamily protein, CYP75B1	350	29.38	24.59	4.85	6.09	−2.30	1.27E-04	5.00E-03
JcLf4rcHa22737	AT2G23180	Cytochrome P450, family 96, subfamily A, polypeptide 1, CYP96A1	459	12.24	8.63	3.55	1.27	−2.12	3.66E-04	1.16E-02
JcLf4rcHa27201	AT4G39480	Cytochrome P450, family 96, subfamily A, polypeptide 9, CYP96A9	482	6.40	5.99	2.17	0.75	−2.09	2.17E-03	4.67E-02

(continued)

Table 6.2 (continued)

				RPKM[a]						
			Contig length	Toxic		Non-toxic		Fold Change		
Contig ID	*A. thaliana* Gene	Short description for *A. thaliana* Gene	(bp)	NBF-1_1	NBF-1_2	NBF-7_1	NBF-7_2	(log2)	*p* value	*q* value
JcLf4rcHa17100	AT1G12740	Cytochrome P450, family 87, subfamily A, polypeptide 2, CYP87A2	829	16.72	18.92	3.45	6.00	−1.92	3.66E-04	1.16E-02
JcLf4rcHa02027	AT3G26310	Cytochrome P450, family 71, subfamily B, polypeptide 35, CYP71B35	1839	72.14	52.77	18.08	20.87	−1.68	2.51E-04	8.42E-03
JcLf4rcHa06006	AT4G37340	Cytochrome P450, family 81, subfamily D, polypeptide 3, CYP81D3	1547	34.31	24.54	8.52	11.60	−1.55	1.01E-03	2.60E-02
JcLf4rcHa11850	AT4G12300	Cytochrome P450, family 706, subfamily A, polypeptide 4, CYP706A4	1259	6.61	6.01	17.48	16.29	1.42	2.37E-03	4.99E-02
JcLf4rcHa10528	AT3G26210	Cytochrome P450, family 71, subfamily B, polypeptide 23, CYP71B23	1610	10.98	13.02	31.88	41.56	1.61	2.04E-04	7.13E-03
JcLf4rcHa20002	AT3G14620	Cytochrome P450, family 72, subfamily A, polypeptide 8, CYP72A8	246	6.83	11.45	28.34	43.14	1.97	2.32E-03	4.92E-02
JcLf4rcHa25131	AT2G29090	Cytochrome P450, family 707, subfamily A, polypeptide 2, CYP707A2	822	1.69	0.77	5.63	5.80	2.21	3.75E-04	1.18E-02
JcLf4rcHa14617	AT5G57260	Cytochrome P450, family 71, subfamily B, polypeptide 10, CYP71B10	202	20.05	31.44	136.89	124.83	2.35	5.95E-05	2.70E-03
JcLf4rcHa17274	AT2G42250	Cytochrome P450, family 712, subfamily A, polypeptide 1, CYP712A1	437	3.16	1.53	9.37	16.60	2.47	2.22E-03	4.74E-02
JcLf4rcHa06504	AT3G26300	Cytochrome P450, family 71, subfamily B, polypeptide 34, CYP71B34	1785	0.58	0.95	5.44	4.50	2.70	4.11E-07	4.07E-05
JcLf4rcHa15681	AT3G14680	Cytochrome P450, family 72, subfamily A, polypeptide 14, CYP72A14	460	4.23	3.02	17.51	31.33	2.75	7.05E-04	1.94E-02
JcLf4rcHa10271	AT5G06900	Cytochrome P450, family 93, subfamily D, polypeptide 1, CYP93D1	1756	3.79	4.51	23.05	47.66	3.09	2.16E-03	4.66E-02
JcLf4rcHa23737	AT2G42250	Cytochrome P450, family 712, subfamily A, polypeptide 1, CYP712A1	338	2.55	1.31	14.68	22.02	3.25	9.08E-06	5.61E-04

(continued)

Table 6.2 (continued)

				RPKM[a]						
			Contig length	Toxic		Non-toxic		Fold Change		
Contig ID	*A. thaliana* Gene	Short description for *A. thaliana* Gene	(bp)	NBF-1_1	NBF-1_2	NBF-7_1	NBF-7_2	(log2)	*p* value	*q* value
JcLf4rcHa07441	AT1G13110	Cytochrome P450, family 71 subfamily B, polypeptide 7, CYP71B7	1782	8.54	35.15	188.28	233.26	3.27	2.83E-13	9.35E-11
JcLf4rcHa03436	AT4G31940	Cytochrome P450, family 82, subfamily C, polypeptide 4, CYP82C4	1098	1.75	1.07	16.18	14.53	3.44	1.63E-10	3.56E-08
JcLf4rcHa17350	AT4G31970	Cytochrome P450, family 82, subfamily C, polypeptide 2, CYP82C2	638	2.10	0.14	16.11	13.69	3.73	3.44E-09	5.64E-07
JcLf4rcHa11564	AT4G31940	Cytochrome P450, family 82, subfamily C, polypeptide 4, CYP82C4	1721	0.75	0.95	16.26	8.25	3.86	2.51E-04	8.42E-03
JcLf4rcHa11958	AT4G37370	Cytochrome P450, family 81, subfamily D, polypeptide 8, CYP81D8	1680	1.06	0.57	24.57	16.21	4.64	5.45E-11	1.28E-08
JcLf4rcHa03678	AT3G14630	Cytochrome P450, family 72, subfamily A, polypeptide 9, CYP72A9	1098	0.24	0.13	14.95	14.12	6.27	2.00E-20	2.24E-17
JcLf4rcHa16442	AT2G26710	Cytochrome P450 superfamily protein, CYP734A1	439	0.00	0.00	36.85	33.81	–	1.36E-21	1.98E-18
BAHD/BAHD-like genes										
JcLf4rcHa09206	AT3G26040	HXXXD-type acyl-transferase family protein	1113	186.88	40.19	0	0	–	1.08E-04	4.38E-03
JcLf4rcHa17446	AT5G48930	Hydroxycinnamoyl-CoA shikimate/quinate hydroxycinnamoyl transferase	1249	2.04	0.52	7.07	9.58	2.70	9.07E-07	7.98E-05
JcLf4rcHa09187	AT1G65450	HXXXD-type acyl-transferase family protein	1516	68.38	95.61	36.76	22.43	−1.47	1.35E-03	3.22E-02

[a]Reads Per Kilobase of exon per Million mapped reads

studies on the seeds are required to clarify the relationship between the toxicity and the involved genes.

6.4 Conclusion and Perspective

Research into the toxic substances of Jatropha is a growing area of research. Acceleration of a Jatropha molecular breeding program should help to generate novel elite plants with desirable characteristics such as high productivity, high stress tolerance, and low toxicity. Moreover, a combination of a traditional genetic approach and advanced sequencing technology will help to further identify the Jatropha genes involved in the biosynthesis of toxic substances, including PEs. Since Thymelaeaceae and Euphorbiaceae families are known to produce characteristic secondary metabolites, the results from studies of Jatropha should lead to unique biotechnological tools and potential drug candidates in addition to the development of sustainable bioenergy crops. Some toxic substances in Jatropha are now considered to be applicable to therapeutic use as mentioned in Sect. 6.2 (Thomas et al. 2008; Sabandar et al. 2013). The poison is a seed of the medicine; toxic Jatropha can bear rich fruits with many possibilities.

Acknowledgements We are thankful to Ms. Nobuko Shizawa and Ms. Shoko Mizobuchi (NAIST) for their technical assistance and to Dr. Toru Usami and Mr. Kazuo Sambongi (Nippon Biodiesel Fuel Co., Ltd.) for providing the Jatropha leaf samples. The Roche 454 GL FLX+ sequencing data were obtained from the sequencing service at Life Science Accelerator (LSA) Technical Service in RIKEN GeNAS. This work was supported in part by the Kanagawa Prefectural Government Program for Human Resources Utilization by Bio-venture Companies and by Grants-in-Aid from the MEXT NC-CARP project.

References

Adam SEI (1974) Toxic effects of *Jatropha curcas* in mice. Toxicology 2:67–76

Adam SEI, Magzoub M (1975) Toxicity of *Jatropha curcas* for goats. Toxicology 4:347–354

Adolf W, Opferkuch HJ, Hecker E (1984) Irritant phorbol derivatives from four Jatropha species. Phytochemistry 23:129–132

Aregheore EM, Makkar HPS, Becker K (1998) Assessment of lectin activity in a toxic and a non-toxic variety of *Jatropha curcas* using latex agglutination and haemagglutination methods and inactivation of lectin by heat treatments. J Sci Food Agri 77:349–352

Arulpandi I, Sangeetha R (2012) Antibacterial activity of fistulin: a protease inhibitor purified from the leaves of *Cassia fistula*. ISRN Pharm 2012:584073

Barbieri L, Battelli M, Stirpe F (1993) Ribosome-inactivating protein from plants. Biochim Biophy Acta 1154:237–282

Basha SD, Francis G, Makkar HPS, Becker K, Sujatha M (2009) A comparative study of biochemical traits and molecular markers for assessment of genetic relationships between *Jatropha curcas* L. germplasm from different countries. Plant Sci 176:812–823

Bourin MC, Delescluse C, Furstenberger G, Marks F, Schweizer J et al (1982) Effect of phorbol esters on guinea pig skin *in vivo*. Carcinogenesis 3:671–676

Cebrian D, Tapia A, Real A, Morcillo MA (2007) Inositol hexaphosphate: A potential chelating agent for uranium. Radiat Prot Dosim 127:479–1477

Chen F, Tholl D, Bohlmann J, Pichersky E (2011) The family of terpene synthases in plants: a mid-size family of genes for specialized metabolism that is highly diversified throughout the kingdom. Plant J 66:212–229

Colón-González F, Kazanietz MG (2006) C1 domain exposed: from diacylglycerol binding to protein-protein interactions. Biochem Biophys Acta 1761:827–837

Costa HPS, Cardoso KC, Del Bem LEV, Lima AC, Cunha MA et al (2010) Transcriptome analysis of the oil-rich seed of the bioenergy crop *Jatropha curcas* L. BMC Genom 11:462

Costa HPS, Oliveira JTA, Sousa DOB, Morais JKS, Moreno FB et al (2014) JcTI-I: a novel trypsin inhibitor from *Jatropha curcas* seed cake with potential for bacterial infection treatment. Front Microbiol 5:5

Coulibaly A, Kouakou B, Chen J (2011) Phytic acid in cereal grains: structure, healthy or harmful ways to reduce phytic acid in cereal grains and their effects on nutritional quality. Am J Plant Nutr Fertil Technol 1:1–22

D'Auria JC (2006) Acyltransferases in plants: a good time to be BAHD. Curr Opin Plant Biol 9:331–340

El Diwani G, Attia NK, Hawash SI (2009) Development and evaluation of biodiesel fuel and byproducts from jatropha oil. Int J Environ Sci Technol 6:219–224

Endo Y, Tsurugi K (1987) RNA *N*-glycosidase activity of ricin A-chain: mechanism of action of the toxic lectin ricin on eukaryotic ribosomes. J Biol Chem 263:8735–8739

Felke J (1914) The poisonous principles of the seeds of *Jatropha curcas* Linn. Landw Versuchsw 82:427–430

Francis G, Kerem Z, Makkar HPS, Becker K (2002) The biological action of saponins in animal systems: a review. Br J Nutr 88:587–605

Gandhi VM, Cherian KM, Mulky MJ (1995) Toxicological studies on Ratanjyot oil. Food Chem Toxicol 33:39–42

Goel G, Makkar HPS, Francis G, Becker K (2007) Phorbol esters: structure, biological activity, and toxicity in animals. Int J Toxicol 26:279–288

Grabherr MG, Haas BJ, Yassour M, Levin JZ, Thompson DA et al (2011) Full-length transcriptome assembly from RNA-Seq data without a reference genome. Nat Biotechnol 29:644–652

Haas W, Sterk H, Mittelbach M (2002) Novel 12-deoxy-16-hydroxyphorbol diesters isolated from the seed oil of *Jatropha curcas*. J Nat Prod 65:1434–1440

Hirakawa H, Tsuchimoto S, Sakai H, Nakayama S, Fujishiro T et al (2012) Upgraded genomic information of *Jatropha curcas* L. Plant Biotechnol 29:123–130

Hirota M, Suttajit M, Suguri H, Endo Y, Shudo K et al (1988) A new tumor promoter from the seed oil of *Jatropha curcas* L., an intramolecular diesterof 12-deoxy-16-hydroxyphorbol. Cancer Res 48:5800–5804

Huang K, Huang Q, Scott AL (1998) Overexpression, single-step purification, and site-directed mutagenetic analysis of casbene synthase. Arch Biochem Biophys 352:144–152

Islam AKMA, Yaakob Z, Anuar N (2011) Jatropha: A multipurpose plant with considerable potential for the tropics. Sci Res Essays 6:2597–2605

Jiang H, Wu P, Zhang S, Song C, Chen Y et al (2012) Global analysis of gene expression profiles in developing physic nut (*Jatropha curcas* L.) seeds. PLoS One 7:e36522

Jing L, Fang Y, Ying X, Wenxing H, Meng X et al (2005) Toxic impact of ingested Jatropherol-I on selected enzymatic activities and the ultrastructure of midgut cells in silkworm, *Bombyx mori* L. J Appl Entymol 129:98–104

Kim D, Pertea G, Trapnell C, Pimentel H, Kelley R et al (2013) TopHat2: accurate alignment of transcriptomes in the presence of insertions, deletions and gene fusions. Genome Biol 14:R36

King AJ, Li Y, Graham IA (2011) Profiling the developing *Jatropha curcas* L. seed transcriptome by pyrosequencing. BioEnergy Res 4:211–221

King AJ, Montes LR, Clarke JG, Affleck J, Li Y et al (2013) Linkage mapping in the oilseed crop *Jatropha curcas* L. reveals a locus controlling the biosynthesis of phorbol esters which cause seed toxicity. Plant Biotechnol J 11:986–996

King AJ, Brown GD, Larson Gilday AD, Graham TRIA (2014) Production of bioactive diterpenoids in the Euphorbiaceae depends on evolutionarily conserved gene clusters. Plant Cell 26:3286–3298

Kirby J, Nishimoto M, Withers Park JG, Nowroozi STF et al (2010) Cloning of casbene and neocembrene synthases from Euphorbiaceae plants and expression in *Saccharomyces cerevisiae*. Phytochemistry 71:1466–1473

Lima TB, Silva ON, Migliolo L, Souza-Filho CR, Gonçalves EG et al (2011) A Kunitz proteinase inhibitor from corms of *Xanthosoma blandum* with bactericidal activity. J Nat Prod 74:969–975

Lin J, Chen Y, Xu Y, Yan F, Tang L, Chen F (2003) Cloning and expression of Curcin, a ribosome-inactivating protein from the seeds of *Jatropha curcas*. Acta Bot Sin 45:858–863

Liu SY, Sporer F, Wink M, Jourdane J, Henning R et al (1997) Anthraquinones in *Rheum palmatum* and *Rumex dentatus* (Polygonaceae) and phorbolesters from *Jatropha curcas* (Euphorbiaceae) with molluscicidal activity against the schistosomias vector snails *Oncomelania*, *Biomphalaria* and *Bulinus*. Trop Med Int Health 2:179–188

Maere S, Heymans K, Kuiper M (2005) BiNGO: a cytoscape plugin to assess overrepresentation of gene Ontology categories in biological networks. Bioinformatics 21:3448–3449

Makkar HPS, Becker K (1997) Potential of *Jatropha* seed meal as a protein supplement to livestock feed and constraint to its utilization. In: Proceedings of *Jatropha* "97 international symposium biofuel and industrial products from *Jatropha curcas* and other tropical oil seed plants, Nicaragua, Mexico, 23–27 February 1997

Makkar HPS, Becker K (1999) Nutritional studies on rats and fish (carp *Cyprinus carpio*) fed diets containing unheated and heated *Jatropha curcas* meal of a non-toxic provenance. Plant Foods Hum Nutr 53:183–192

Makkar HPS, Becker K (2009) Challenges and opportunities for using byproducts from the production of biodiesel from *Jatropha* oil as livestock feed. Proceedings of Animal Nutrition Association World Conference, New Delhi, India, 14–17 February 2009, pp. 68–70

Makkar HPS, Aderibigde AO, Becker K (1998) Comparative evaluation of non-toxic and toxic varieties of *Jatropha curcas* for chemical composition, digestibility, protein degradability and toxic factors. Food Chem 62:207–215

Makkar HPS, Kumar V, Oyeleye OO, Akinkeye AO, Angulo-Escalante MA, Becker K (2011) *Jatropha platyphylla*, a new non-toxic Jatropha species: physical properties and chemical constituents including toxic and antinutritional factors of seeds. Food Chem 125:63–71

Martínez-Herrera J, Siddhuraju P, Francis G, Dávila-Ortíz G, Becker K (2006) Chemical composition, toxic/antimetabolite constituents, and effects of different treatments on their levels, in four provinces of *Jatropha curcas* L. from Mexico. Food Chem 96:80–89

Mau CJ, West CA (1994) Cloning of casbene synthase cDNA: evidence for conserved structural features among terpenoid cyclases in plants. Proc Natl Acad Sci USA 91:8501–9497

Molina DK, Kosma DK (2015) Role of HXXXD-motif/BAHD acyltransferases in the biosynthesis of extracellular lipids. Plant Cell Rep 34:587–601

Nakano Y, Ohtani M, Polsri W, Usami T, Sambongi K, Demura T (2012) Characterization of the casbene synthase homolog from Jatropha (*Jatropha curcas* L.). Plant Biotechnol 29:185–189

Natarajan P, Kanagasabapathy D, Gunadayalan G, Panchalingam J, Shree N et al (2010) Gene discovery from *Jatropha curcas* by sequencing of ESTs from normalized and full-length enriched cDNA library from developing seeds. BMC Genom 11:606

Ohtani M, Nakano Y, Usami T, Demura T (2012) Comparative metabolome analysis of seed kernels in phorbol ester-containing and phorbol ester-free accessions of *Jatropha curcas* L. Plant Biotechnol 29:171–174

Podolak I, Galanty A, Sobolewska D (2010) Saponins as cytotoxic agents: a review. Phytochem Rev 9:425–474

Prisic S, Xu J, Coates RM, Peters RJ (2007) Probing the role of the DXDD motif in Class II diterpene cyclases. Chembiochemistry 8:869–874

Qin W, Ming-Xing H, Ying X, Xin-Shen Z, Fang C (2005) Expression of a ribosome inactivating protein (curcin 2) in *Jatropha curcas* is induced by stress. J Biosci 30:351–357

Qin X, Zheng X, Shao C, Gao J, Jiang L et al (2009) Stress-induced curcin-L promoter in leaves of *Jatropha curcas* L. and characterization in transgenic tobacco. Planta 230:387–395

Qin X, Shao C, Hou P, Gao J, Lei N et al (2010) Different functions and expression profiles of curcin and curcin-L in *Jatropha curcas* L. Z Naturforsch C 65:355–362

Raja SA, Smart DR, Lee CLR (2011) Biodiesel production from jatropha oil and its characterization. Res J Chem Sci 1:81–87

Reddy NR, Pierson MD (1994) Reduction in antinutritional and toxic components in plant foods by fermentation. Food Res Int 27:281–290

Sabandar CW, Ahmat N, Jaafar FM, Sahidin I (2013) Medicinal property, phytochemistry and pharmacology of several *Jatropha* species (Euphorbiaceae): a review. Phytochemistry 85:7–29

Sato S, Hirakawa H, Isobe S, Fukai E, Watanabe A et al (2011) Sequence analysis of the genome of an oil-bearing tree, *Jatropha curcas* L. DNA Res 18:65–76

Schmidt RJ (1987) The biosynthesis of tigliane and related diterpenoids; an intriguing problem. Bot J Linn Soc 94:221–230

Shmidt ÉN, Kashtanova NK, Pentegova VA (1970) Neocembrene—a new diterpene hydrocarbon from *Picea obovata* and *Pinus koraensis*. Chem Nat Compd 6:705–707

Singh RK, Padhi SK (2009) Characterization of jatropha oil for the preparation of biodiesel. Nat Prod Rad 8:127–132

Sudheer Pamidimarri DVN, Singh S, Mastan SG, Patel J, Reddy MP (2009) Molecular characterization and identification of markers for toxic and non-toxic varieties of *Jatropha curcas* L. using RAPD, AFLP and SSR markers. Mol Biol Rep 36:1357–1364

Sujatha M, Makkar HPS, Becker K (2005) Shoot bud proliferation from axillary nodes and leaf sections of non-toxic *Jatropha curcas* L. Plant Growth Regul 47:83–90

Sun J, Nishiyama T, Shimizu K, Kadota K (2013) TCC: an R package for comparing tag count data with robust normalization strategies. BMC Bioinform 14:219

Thomas R, Sah NK, Sharma PB (2008) Therapeutic biology of *Jatropha curcas*: a mini review. Curr Pharm Biotechnol 9:315–324

Trapnell C, Hendrickson DG, Sauvageau M, Goff L, Rinn JL, Pachter L (2013) Differential analysis of gene regulation at transcript resolution with RNA-seq. Nat Biotechnol 31:46–53

Wang H, Zou A, Wang S, Gong M (2013) Global analysis of transcriptome responses and gene expression profiles to cold stress of *Jatropha curcas* L. PLoS One 8: e82817

Wink M, Koschmieder C, Sauerwein M, Sporer F (1997) Phorbol esters of *Jatropha curcas*—biological activities and potential applications. Gübitz GM, Mittelbach M

Wise A, Gilbert DJ (1981) Binding of cadmium and lead to the calcium-phytate complex *in vitro*. Toxicol Lett 9:45–50

Zerbe P, Hamberger B, Yuen M, Chiang A, Sandhu HK et al (2013) Gene discovery of modular diterpene metabolism in nonmodel system. Plant Physiol 162:1073–1091

Zheng Q, Xiong YL, Su ZJ, Zhang QH, Dai XY et al (2013) Expression of curcin-transferrin receptor binding peptide fusion protein and its anti-tumor activity. Protein Expr Purif 89:181–188

Lipid Biosynthesis and Regulation in Jatropha, an Emerging Model for Woody Energy Plants

7

Yonghuan Ma, Zhongcao Yin and Jian Ye

Abstract

Jatropha, *Jatropha curcas* L., a small seed-propagated Euphorbiaceae plant, is recognized as a promising new woody energy crop due to the high content of oil in its seeds, tolerance to drought, and ability to thrive in poor soil. Over one decade of research on this species has enhanced our understanding of the biological basis not only of reproduction, genetic diversity, stress resistance, genetic transformation, oil biosynthesis, and metabolism but also of transcriptional control. Several advantages such as easy genetic modification, short generation duration, and relative small plant size implement Jatropha as an emerging model for woody energy crops. Based on our research work on the functional genomics and transgenic analysis on Jatropha, here we review our understanding on the seed developmental regulation, structure genes in lipid biosynthesis, lipid turnover, and also a set of epigenetic or transcriptional regulators on lipid biosynthesis and seed development in Jatropha. The knowledge we generated from Jatropha will be also useful for improving agronomic traits in similar energy crops and oil crops.

7.1 Introduction

Plant seeds accumulate lipids (triacylglycerides, TAGs) as their major storage reservoir components for future consumption when the next round of life cycle starts. Plant TAGs are highly concentrated stores of renewable metabolic energy form. The seeds with high energy density of storage reservoir confer selective advantages when seeds counter unexpected environmental conditions. TAGs are biochemically esterified from three fatty acid molecules and a glycerol

Y. Ma · J. Ye (✉)
State Key Laboratory of Plant Genomics, Institute of Microbiology, Chinese Academy of Sciences, Beijing 100101, China
e-mail: jianye@im.ac.cn

Z. Yin
Temasek Life Sciences Laboratory, Research Link 1, National University of Singapore, Singapore 117604, Singapore

S. Tsuchimoto (ed.), *The Jatropha Genome*, Compendium of Plant Genomes,
DOI 10.1007/978-3-319-49653-5_7

backbone. The complete oxidation of 1-g fatty acids can generate about 9 kcal (around 38 kJ), in contrast with about 4 kcal from 1 g of either carbohydrates or proteins (Durrett et al. 2008). More importantly, the hyperhydrophobicity allows TAGs stored in a nearly anhydrous of small size in seed.

Lipid biosynthesis in plants especially in biofuel plant, e.g., Jatropha (*Jatropha curcas*) is of great interest due to the soaring renewable energy need to counter global warming and climate change. Biodiesel produced from Jatropha seed oil as renewable feedstock is receiving an increased interest worldwide as an alternative diesel fuel (Ye et al. 2012). This renewable biodiesel can be used unmixed or blended at various percentages with conventional fossil diesel fuel in diesel engines or as a heating fuel. This biodegradable and non-toxic biofuel is composed of simple alkyl esters of fatty acids obtained by transesterification of triglycerides with low-molecular weight alcohols, such as methanol and ethanol.

Lipid de novo biosynthesis and further storage in seed can divide into three closely linked processes: synthesis of fatty acids, assembly into triacylglycerols, and subsequent packaging into oil bodies. The detailed biochemical process has been mostly investigated in model plant such as *Arabidopsis thaliana* (Durrett et al. 2008). These processes are highly regulated, involving spatial compartmentalization between different organelles and the exquisite control of several biosynthetic steps by one or more of a variety of biochemical mechanisms. These biochemical processes have been reviewed elsewhere. Manipulation of a single biochemical step, even one of these bottleneck enzymes, has been proven to be of limited efficacy to tune up the whole pathway and often leads to weak or even no effect on lipid yield increasing. Meanwhile, mRNA levels of many genes involved in seed oil biosynthesis are well coordinated during seed development through tightly tissue and developmental orchestra by several transcription factors. Altering the levels of certain transcription factors can affect the expression of multiple lipid biosynthetic genes, resulting in altered seed oil content. The important role of genetic regulation and epigenetic regulation toward seed oil accumulation has recently emerged. Another major aspect of lipid metabolism is catabolism which involved with fatty acid oxidation to produce energy. Beside of TAGs biosynthesis, plant seeds TAGs are prone to contend with high level of free fatty acid (FFA) due to improper storage and non-optimized process procedure. As a reverse and essential biology step, actually, the degradation of TAGs in storage seed is vital for its germination. However, this high fatty acid level nature for plant oil is not optimized for developing them as feedstock of biodiesel.

Jatropha, *Jatropha curcas* L., is one of the major plants with non-edible oil-rich (27–40%) seeds using for biodiesel feedstock. However, Jatropha oil contains up to 14% free fatty acid (FFA), which is far beyond the limit of 1% FFA level that can be converted into biodiesel by transesterification. These degradations and oxidation processes are also enzyme catalyzing and highly regulated to secure plant germination properly. Hereafter, we provide our understanding on how epigenetic and transcriptional regulation on lipid biosynthesis and turnover in plant seed, beside of the strategies to optimize biofuel quality and seed oil productivity. Understanding in the lipid biosynthesis/turnover and then spatiotemporal manipulation of the process in Jatropha are essential for improvement of its seed productivity and quality as biofuels feedstock. Based on our research work on the functional genomic and transgenic analysis on Jatropha, we also review our understanding on the seed developmental regulation, lipid biosynthesis, and regulation in Jatropha, an emerging model for woody energy plants.

7.2 Structure and Development of Fruit and Seed of *Jatropha curcas*

The species *Jatropha*, family Euphorbiaceae, native to Central America, was spread by Portuguese seafarers via Cabo Verde and Guinea

Bissau to other countries in Africa and Asia. Today, this subtropical plant is widespread in different agricultural systems as hedges, wind protection systems, erosion barriers, or as a source of firewood. *Jatropha* is a large genus which includes about 180 species distributed mostly in America, Africa, and Asia countries such as India and Indonesia. The low range of agronomic traits variation in fatty acid composition and seed oil yield may also be associated with low level of genetic diversity in non-Central American *J. curcas*. *J. curcas* has a relatively small genome (321 Mb), organized in 22 chromosomes (2*n*), and contains 27,172 putative protein-coding genes (Sato et al. 2011; Wu et al. 2015), which makes it an attractive model for woody energy plant. Castor bean (*Ricinus communis*), another oilseed crop that belongs to the spurge (Euphorbiaceae) family, has a similar genome size but 4000 more protein-coding genes than that of *J. curcas* (Chan et al. 2010). Expressed sequence tag (EST) and proteomic analysis provide a convenient and efficient method for identification of genes/proteins expressed in specific tissues and cells, as well as allow the characterization of transcript/protein levels, high-throughput and cost-effective sequencing, and proteomic methods have been developed that can accelerate the process of breeding in the past decade. Several sequences from cDNA libraries of different seed developmental stages of *J. curcas* have been published (Natarajan and Parani 2011; Gu et al. 2012; Jiang et al. 2012). Several systematic proteomic analyses of different parts from developing seeds of *Jatropha curcas* were taken and used to further explore the protein machinery responsible for generating the lipids in seeds (Liu et al. 2009; Yang et al. 2009; Liu et al. 2011, 2013; Pinheiro et al. 2013; Liu et al. 2015).

Many of studies have been done to understand the structure of ovules and seeds in Jatropha and related genera of Crotonoideae (Gu et al. 2012). When compared morphological difference in development stages between two related species *J. curcas* and *J. integerrima*, 3–5 weeks after fertilization (WAF) was identified as the top vital stage to determinate the final size of seed and fruit (Fig. 7.1a–c). Beside of TAGs storage, the breakdown of TAGs into FAs is initiated by lipases, many of which have been identified in eukaryotes including yeasts, animals, and plants. In order to understand the process of storage lipid accumulation, especially the oil storage format–oil body development in endosperm cells of Jatropha seeds, developing endosperms at 5 and 8 WAF was subjected to transmission electron microscopy (TEM) observations. The TEM studies demonstrated that biosynthesis and accumulation of storage lipids in endosperm cells of Jatropha seeds were developmentally regulated (Gu et al. 2012). They showed that genes with similar functions were expressed differentially during endosperm development, and the majority of FA and lipid biosynthetic genes are highly consistent with the development of oil bodies and endosperm in Jatropha seeds. Temporal expression profiles of 21 lipid genes involved in FA and TAG synthesis pathways revealed that the expression of 17 genes was increased in developing Jatropha seeds compared to leaves. Only two diacylglycerol acyltransferase genes (*DGAT1* and *DGAT2*), representing rate-limiting enzymes in plant lipid accumulation, were specifically associated with the biosynthesis of TAG (Gu et al. 2012).

Jatropha seeds contained an increased number of oil bodies and positively correlated more peroxisomes in late stage than other stages (compare Fig. 7.1e with d). Intriguingly, much more peroxisomes were observed in *J. curcas* than that of *J. integerrima* (compare Fig. 7.1e with f), consistent with the fact that easier germination of *J. curcas*. TAG lipases initiate TAG hydrolysis into glycerol and FFAs, and the latter are metabolized through the β-oxidation pathway to release carbon sources for early growth of young seedling.

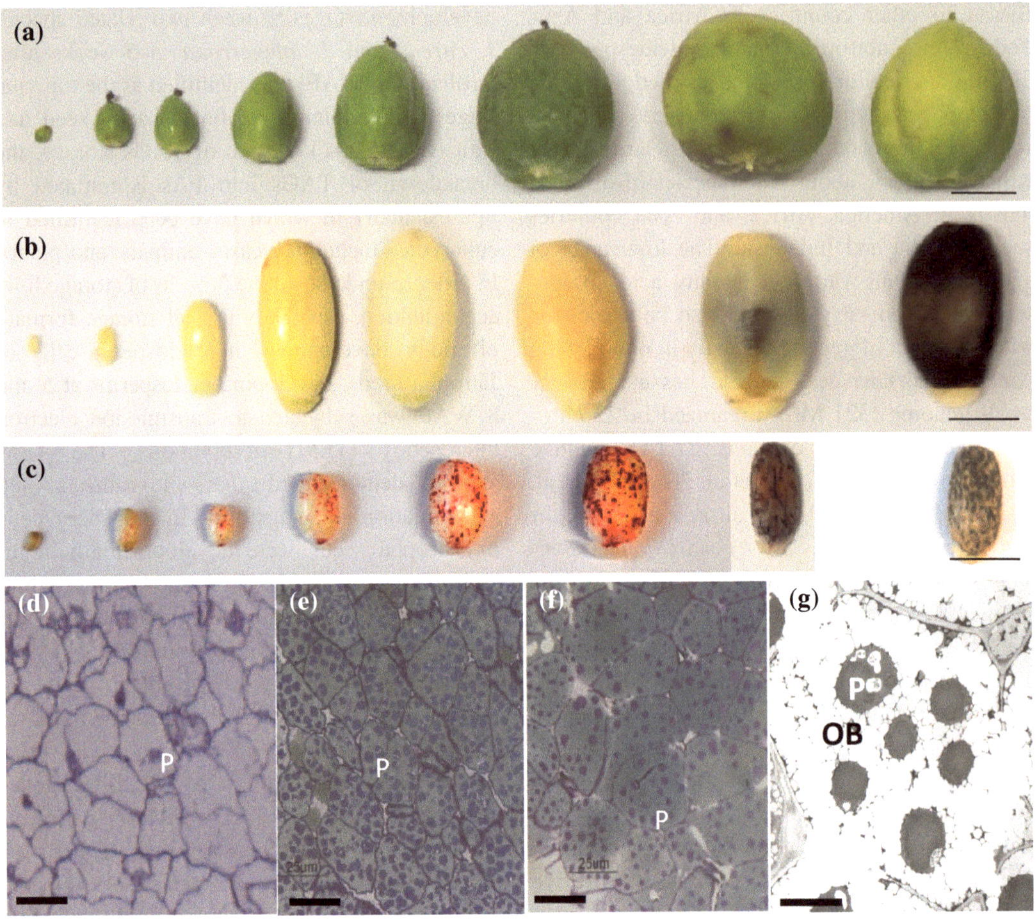

Fig. 7.1 Fruits and seeds developmental Biology for *Jatropha curcas* and *Jatropha integerrima*. **a** The fruit developmental stages of *J. curcas*. From *left* to *right*, each fruit represents individual stage of 1–8 weeks post fertilization. *Size bar* 10 mm. **b** The seed developmental stages of *J. curcas*. From *left* to *right*, each seed represents individual stage of 1–8 weeks post fertilization. *Size bar* 10 mm. **c** The seed developmental stages of *J. integerrima*. From *left* to *right*, each fruit represents individual stage of 1–8 weeks post fertilization. *Size bar* 10 mm. **d–f** Microscopic image of immature endosperm (**d**, 5 WAF) and mature endosperm (**e**, 7 WAF) of *J. curcas* and *J. integerrima* (**f**). *Size bar* 25 μm. *P* peroxisome. Noted the reduced peroxisome density in endosperm of *J. integerrima* compared with that of *J. curcas*. **g** Transmission electron microscopy of mature endosperm (7 WAF) of *J. curcas*. *P* peroxisome, *OB* oil body. *Size bar* 5 μm

7.3 Glycolysis

Glycolytic pathway (glycolytic pathway) is the way that glucose is converted to pyruvic acid and accompanied by ATP generated through a series of reactions. All the organisms use the way of glucose degradation to provide energy to creatures. Glycolysis enables the upstream sucrose importing and cleavage, and the contribution of the cytosolic and plastidial parallel routes for the oxidation of hexoses to pyruvate at the onset of oil deposition (Fig. 7.2).

Glucose is the major energy source of most organisms, under a condition which is lack of oxygen, the whole pathway consists of ten steps. There are ten enzymes catalyzing the ten reactions of glycolysis including hexokinase (HXK), phosphoglucose isomerase (PGi), phosphofructokinase-1 (PFK-1), Fru-bisphosphate aldolase (FBA),

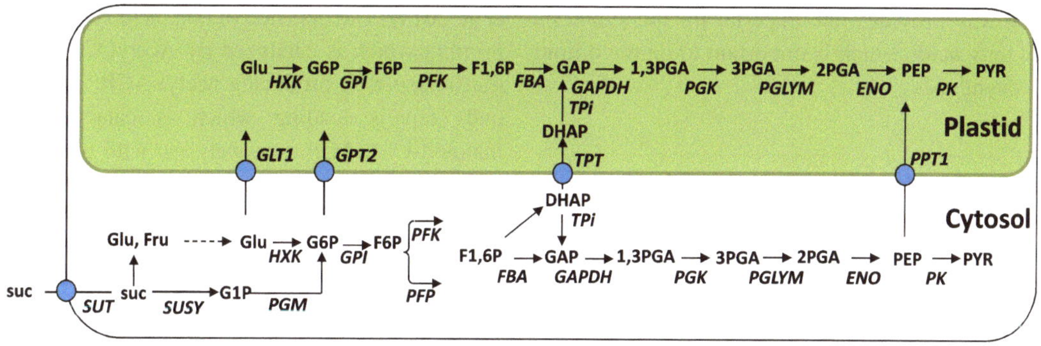

Fig. 7.2 Glycolysis pathway which mediates the conversion of Suc to pyruvate the precursor of Acetyl-CoA in Jatropha *Glu* glucose, *Fru* fructose, *Suc* sucrose, *DHAP* dihydroxyacetone-3-P, *FBA* Fru-bisphosphate aldolase, *FK* fructokinase, *F1.6P* Fru-1,6-bisphosphate, *F6P* Fru-6-P, *GAP* glyceraldehyde 3-P, *G1P* Glc-1-P, *G6P* Glc-6-P, *Glu* Glc, *GLT1* Glc transporter, *GAPDH* glyceraldehyde-3-phosphate dehydrogenase, *GPI* Glc-phosphate isomerase, *GPT2* Glc-6-phosphate/phosphate transporter, *HXK* hexokinase, *PEP* phosphoenolpyruvate, *PGK* 2-phosphoglycerate kinase, *PGM* phosphoglucomutase, *PFK* phosphofructokinase, *PFP* ppi-dependent phosphofructokinase, *2PGA* 2-phosphoglycerate, *1, 3 PGA* 3-phosphoglycerate, *PGYLM* 2,3-bisphosphoglycerate-independent phosphoglycerate mutase, *3PGA* 3-phosphoglycerate, *PGM* phosphoglucomutase, *PGI* phosphoglucose isomerase, *PK* pyruvate kinase, *PPT1* phosphoenolpyruvate/phosphate translocator, *Pyr* pyruvate, *SUSY* Suc synthase, *SUT* Suc transporter, *TPI* triose-P isomerase, *TPT* triose phosphate/phosphate translocator

triose phosphate isomerase (TPi), glyceraldehyde 3-phosphate dehydrogenase (GAPDH), phosphoglycerate kinase (PGK), phosphoglycerate mutase (PGM), enolase (ENO), and pyruvate kinase (PK). Glycolysis starts with glucose under the condition lacking or without oxygen, glucose would be catalyzed by hexokinase, and glucose would be transformed in glucose-6-phosphate, and this process is ATP-dependent in a same amount with glucose. Then glucose-6-phosphate would be catalyzed by PGi without consuming any ATP and was transformed into fructose-6-phosphate. PFK-1 is the main enzyme in the next step of transforming fructose-6-phosphate into fructose-1,6-bisphosphate, with consuming ATP as same amount as the first step in glycolysis. Then fructose-1,6-bisphosphate would be catalyzed by FBA, glyceraldehydes-3-phosphate, and dihydroxyacetone-phosphate in same amount would be produced. Only glyceraldehydes-3-phosphate can take part in the next steps of glycolysis in animals and human's cell; so dihydroxyacetone-phosphate would be transformed into glyceraldehydes-3-phosphate catalyzed by TPi. So 1 mol glucose would be transformed into 2 mol glyceraldehydes-3-phosphate. This is the first stage of glycolysis which consumes 2 mol ATPs without producing any energy. But the next stage will be an ATP-producing stage. In the second stage of glycolysis, glyceraldehydes-3-phosphate would be catalyzed by GAPDH and it would be transformed into 1,3-biophosphoglycerate, and then PGK would catalyze it into 3-phosphoglycerate and the same amount of ADP would be transformed into ATP. 3-phosphoglycerate would firstly transformed into 2-phosphoglycerate by PGM without consuming or producing any ATP. 2-Phosphoglycerate then transformed into phosphoenolpyruvate catalyse by enolase. In the last step of glycolysis, phosphoenolpyruvate would be catalyzed by pyruvate kinase and was transformed into pyruvate with same amount ATP was produced.

The whole reaction can be written as:

$$\text{Glucose} + 2\text{NAD}^+ + 2\text{ADP} + 2\text{Pi} \xrightarrow{\text{enzymes}} 2\text{pyruvate} + 2\text{NADH} + 2\text{ATP} + 2\text{H}^+ + 2\text{H}_2\text{O}.$$

In a word, 1 mol glucose is converted to 2 mol pyruvate and yields 2 mol ATP. All the steps can be showed in detail in Fig. 7.2. Glycolytic pathway provides raw material for the synthesis of fatty acids which is important to de novo lipid biosynthesis.

7.4 *De Novo*De Novo Lipid Biosynthesis

Plant lipids are synthesized as triacylglycerols (TAGs) via a complex series of pathways in which many FA biosynthetic enzymes are involved. Plastids are the main field where fatty acid is synthesized from the raw material in breeding seeds (Fig. 7.3). Sucrose is the original carbon (C) source of plant lipid synthesis. The sucrose of leaves transport to the developing seed and will be shaped into pyruvate (Pyr) through a series of biochemical reaction. Oxidative decarboxylation of pyruvate to acety-CoA by pyruvate dehydrogenase (PDH) is the first committing step in fatty acid biosynthesis. PDH is a four-subunit enzyme complex having pyruvate decarboxylase (α and β of E1 subunits), dihydrolipoyl acetyltransferase (E2 subunit), and dihydrolipoamide dehydrogenase (E3 subunit) enzymes. Fatty acid biosynthesis is initiated by the condensation of acetyl-CoA to malonyl-CoA which is catalyzed by heteromeric acetyl CoA carboxylase (ACCase). It consists of three nuclear encoded subunits (biotin carboxyl carrier protein-BCCP, biotin carboxylase-BC, and carboxyl transferase α-subunit-CTα) and a plastid-encoded subunit (carboxyl transferase β-subunit-CTβ). Transcription of these four subunits is coordinately regulated so that a constant molar stoichiometric ratio is expected. However, the plastid-encoded subunit-CTβ seems to be crucial for the accumulation of heteromeric ACCase and for increasing the oil content. Plants also have a homomeric ACCase localized in cytosol that does not play any major role in fatty acid biosynthesis that takes place almost exclusively in plastids. However, by targeting this enzyme to plastids, oil content was increased fivefold in potato tubers and 5% in rapeseed. Full-length contigs for all the four subunits of heteromeric ACCase (Gu et al. 2012) and a 5′ partial contig (3711 bps) for homomeric ACCase have been identified (Grover et al. 2014). The first step of fatty acid synthesis is priming which is catalyzed by Acetyl-CoA:ACP transferase with producing acetyl-ACP. The second step is loading which is catalyzed by malonyl-CoA:ACP transacetylase with the malonic acid acyl coenzyme A produced. The third step is condensation which is catalyzed by the β-ketoacyl:ACP synthase. The fourth step is reduction which is catalyzed by β-ketoacyl:ACP reductase and α, β-hydroxyl butyryl:ACP is generated. The next reaction will be dehydration which is catalyzed by β-hydrdoxyacyl:ACP dehydrase and the product is α,β- butene acyl: ACP. The last step of fatty acid synthesis is reduction which generates 4 carbon fatty acid. In every cycle of fatty acid synthesis, fatty acid chain extension of the two carbon atoms, condensation of malonyl-CoA and acyl carrier protein (ACP) to form 3-carbon malonyl-ACP is catalyzed by malony CoA acyl transferase. Polymerization of the fatty acid carbon chains starts with the formation of a 4-carbon butyryl-ACP by the addition of a carbon from acetyl-CoA to the 3-carbon malonyl-ACP by the action of keto-acyl ACP synthase III (KASIII). Further chain elongation up to 16-carbon palmitoyl-ACP is carried out by keto-acyl ACP synthase I (KASI) in six consecutive steps by adding two carbons in each step. Finally, keto-acyl ACP synthase II (KASII) converts the 16-carbon palmitoyl-ACP to 18-carbon stearoyl-ACP. While KASI is essential for seed oil content, KASII is useful for manipulating the chain length of the fatty acids. Longer the chain length of fatty acids more is the viscosity of the oil as well as the biodiesel derived from it. Viscosity affects atomization of the biodiesel and lowers the performance of the diesel engines due to engine deposits. Viscosity of Jatropha oil can be lowered by reducing its 18-carbon fatty acid content which is reported to be as high as 84.7%. Previous reports suggest that this can be achieved by silencing the activity of *KASII* that converts the palmitoyl-ACP to stearoyl-ACP. Alternatively, the activity of palmitoyl-ACP thioesterase (encoded by FATA1) could be enhanced to accelerate the cleavage of palmitic acid from the palmitoyl-ACP before it is converted to stearoyl-ACP.

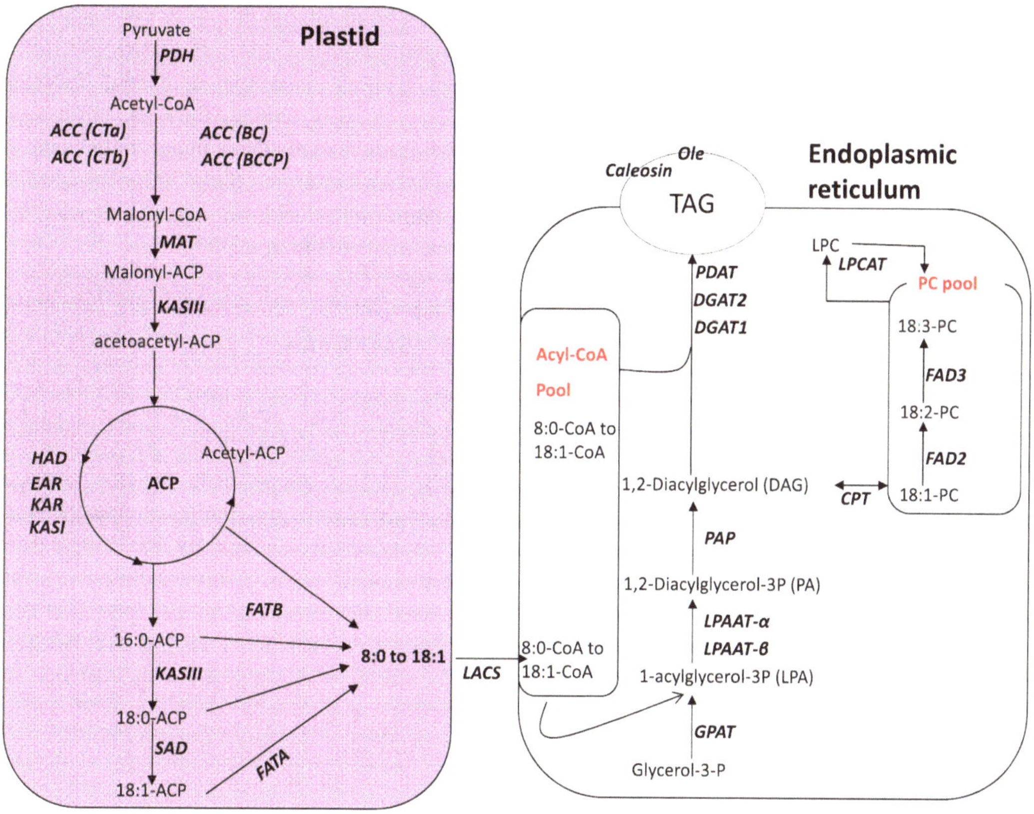

Fig. 7.3 De novo fatty acid biosynthesis and assembly pathway in Jatropha. *ACP* acyl carrier protein, *ACC* (*BC*) biotin carboxylase subunit of heteromeric Acetyl-CoA carboxylase (ACCase), *ACC* (*BCCP*) biotin carboxyl carrier protein of heteromeric ACCase, *ACC* (*CTα*) carboxyltransferase α-subunit of heteromeric ACCase, *ACC* (*CTβ*) carboxyltransferase β-subunit of heteromeric ACCase, *EAR* enoyl-ACP reductase, *KASI*, *II*, *III* β-ketoacyl-[acyl carrier protein] synthase I, II, III, *KAR* ketoacyl-ACP reductase, *MAT* malonyl-CoA:ACP malonyltransferase, *FAD2* ω-6 desaturase, *FAD3* ω-3 desaturase, *FATA* acyl-ACP thioesterase A, *FATB* acyl-ACP thioesterase B, *HAD* hydroxyacyl-ACP dehydrase, *PAP* phosphatidate phosphatase, *PDH* (*E1α*) subunit α of E1 component of PDH complex, *PDH* (*E1β*) subunit β of E1 component of PDH complex, *PDH* (*E2*) E2 component of PDH complex, *PDH* (*E3*) E3 component of PDH complex, *SAD* stearic acid desaturase, *LCAS* long-chain acyl-CoA synthases, *LPAAT* lysophosphatidic acid acyltransferase, *PAP* phosphatidic acid phosphatase, *DGAT* diacylglycerol acyltransferase-2, *CPT* CDP-choline:diacylglycerol cholinephosphotransferase, *GPAT* glycerol-3-phosphate acyltransferase, *LPCAT* Acyl-CoA: lysophosphatidylcholine Acyltransferases

Free fatty acids that are released from acyl-ACPs are converted to respective acyl-CoAs by long-chain acyl-CoA synthase (LCACS). The activated fatty acids are bound by acyl-CoA-binding proteins (ACBPs) to protect them from acyl-CoA hydrolases and to transport them to endoplasmic reticulum (ER).

With the advance of genome sequencing in *J. curcas*, the expression of genes involved in the biosynthetic pathways has become more feasible, which should be beneficial for selective breeding of *J. curcas*, especially for oil quality and yield (Sato et al. 2011; Wu et al. 2015). Here we summarize some key genes which have been functionally characterized in Jatropha.

7.4.1 *KAS* Genes

Three *KASI*, *KASII*, and *KASIII* encoding the β-ketoacyl-acyl carrier protein synthases which are the condensing enzymes for the elongation of the carbon chain from C4 to C18. KASI participates from the 2nd to the 6th carbon-chain extension cycles. KASII participates in the 7th carbon-chain extension cycle. KASIII is the initiation enzyme of FA chain elongation which is responsible for the condensation reaction of malonyl-ACP and acetyl-ACP. The expression profiles of *KASII* and *KASIII* are similar with each other but different from the abundance profile of *KASI*.

KASII encoding the β-ketoacyl-acyl carrier protein synthase II is responsible for the elongation of 16:0-ACP to 18:0-ACP (Carlsson et al. 2002). Virus-induced gene silencing of *KASII* in Jatropha suggested that *KASII* is one key gene controlling the ratio of C16 FAs to C18 FAs in plants (Ye et al. 2009). Palmitic acid (16:0) consists of 14.1–18% of total FAs in Jatropha oil. The overexpression of *KASII* cDNA under the CaMV 35S promoter in Arabidopsis resulted in decreasing C16 FAs and increasing C18 FAs in leaves and seeds (Wei et al. 2012).

7.4.2 *FATB* and *FATA* Genes

Fatty acid biosynthesis in higher plants involves the condensation of malonyl-ACP with an acetyl-CoA primer or with previously synthesized acyl-ACP derivatives, resulting in the successive elongation of the acyl-ACP chain by two carbon units. Fatty acid synthesis is terminated by acyl-ACP thioesterases that hydrolyze the thioester bond existing between the acyl moiety and the ACP. Acyl-ACP thioesterases are plastid-targeted soluble enzymes encoded by nuclear genes. The specificity of plant acyl-acyl carrier protein (ACP) thioesterases is the major determinant of the chain length and level of saturated fatty acids found in most plant tissues. FATB and FATA are two distinct thioesterase playing an essential role in the chain determination of fatty acid synthesis and in the channeling of carbon flux in higher plants.

FATB encodes thioesterases preferring acyl-ACPs having saturated acyl groups, and *FATA* encodes the 18:1-ACP thioesterase. In the seed development of *J. curcas*, the expression of *FATA* occurred later than *FATB*, consistent with the presumption of the ubiquitous 18:l-ACP thioesterase is a derivative of a 16:0 thioesterase (Ye et al. 2009), and the fatty acid component analysis indicated that the major fatty acids in *J. curcas* seed were the oleic, linoleic, palmitic, and stearic fatty acid, in agreement with the previous report (Liu et al. 2013).

7.4.3 *SAD* Gene

There are at least seven stearoyl-acyl carrier protein (SAD) genes in *J. curcas* genome (Ye et al. 2009). In Jatropha seeds, the component of C16:0 reached their maximum abundance at 25 DAF and decreased gradually at the late seed development. C16:0 is firstly synthesized at the early seed development as the precursor for elongation to C18:1 and C18:2. SADs catalyze the first desaturation step in FA biosynthesis by converting stearoyl-ACP to oleoyl-ACP. Therefore, SADs play important roles in determining FA chain length and the ratio of saturated to unsaturated Fas (Shanklin and Somerville 1991). At least three putative Jatropha *SAD* genes in a Jatropha seed cDNA library were identified. *SAD1* is highly expressed in seeds at 5–7 weeks after fertilization, when Jatropha triglycerides mainly accumulate in the endosperm. All data so far support that SAD1 is the key enzyme responsible for converting not only 18:0-ACP, but also 16:0-ACP to oleic acid. Therefore, it should be possible to generate transgenic Jatropha with even higher oleic acid (>75%) and lower saturated FA (C16:0 and C18) contents in seeds by specifically up-regulating the expression of SAD1 in seeds.

7.4.4 *FAD2* Gene

FAD2, the enzyme 1-acyl-2-oleoyl-sn-glycero-3-phosphocholine delta 12-desaturase, is the key enzyme responsible for the production of linoleic acid with oleic acid as the substrate in plants. We identified three putative delta-12 fatty acid desaturase genes in Jatropha (*JcFAD2s*) through genome-wide analysis (Ye et al. 2009; Qu et al. 2012) and the function of several key genes from Jatropha that regulate FA chain length and saturation levels (Ye et al. 2009, 2012). Among these *FAD2* genes, *JcFAD2-1* is of particular interest as it mediates the conversion of oleic acid to linoleic acid. The expression of *FAD2-1* is high in seeds and weaker in vegetative tissues, while *FAD2-2* is highly expressed in seeds and not detectable in leaves. The expression pattern of these two *FAD2* genes in Jatropha is very similar with those of other members of Euphorbiaceae, i.e., the genes *FAD2* and *FAH12* (fatty acid hydroxylase) in castor bean (Shanklin and Somerville 1991), *FAD2* and *FADX* in tung tree (Dyer et al. 2002). Combining the *SAD1* up-regulation together with down-regulation of *FAD2-1* by RNAi, both oleic acid level and unsaturated FA content will be further increased; therefore, flow properties at low temperatures and oxidative stability will be further improved.

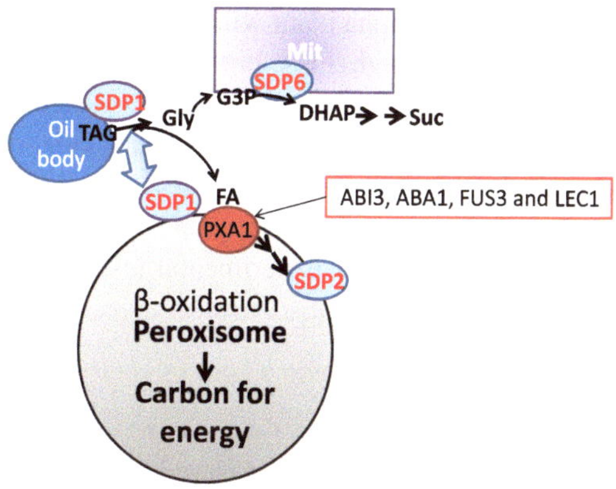

Fig. 7.4 Sugar-dependent (SDP) lipases-mediated TAG degradation pathway. *Gly* glycerol, *DHAP* dihydroxyacetone-3-P, *Mit* mitochondrial, *Suc* sucrose

7.5 Lipid Turnover

During Jatropha seed germination, TAGs are hydrolyzed into FAs and glycerol and this reaction is catalyzed by TAG lipases, which are widely distributed in plants but also found in animals and microorganisms (Figs. 7.1, 7.4). Among the known lipases are the unorthodox patatin-like TAG lipases (PTLs), which are oil body-associated enzymes that play a major role in the initiation of TAG degradation in yeast, mammals, and insects. During seed germination, TAG lipases initiate TAG hydrolysis into glycerol and FFAs, and the latter are metabolized through the β-oxidation pathway to release carbon sources for early seedling growth.

7.5.1 *SDP1* Gene

TAGs consist of three fatty acid chains (usually C16 or C18) covalently linked to glycerol. *Sugar-dependent 1*(*SDP1*) gene encodes a patatin-like acyl hydrolase domain (Eastmond 2006). The encoded protein, SDP1, is specifically responsible for the first step of TAG degradation during Arabidopsis and Jatropha seed germination (Eastmond 2006; Kim et al. 2014). Recently, SDP1 is found to being first localized to the peroxisome membrane at early stages of seedling growth and then possibly moving to the oil body surface through peroxisome tubulations. The timely transfer of SDP1 to the oil body membrane requires a functional retromer complex (Thazar-Poulot et al. 2015).

Using Arabidopsis as a model plant proves that a deficiency of *SDP1* led to higher TAG accumulation and a larger number of oil bodies in seeds compared with wild type (Columbia-0; Col-0). The function of *JcSDP1* was verified by complementation of the Arabidopsis *sdp1-5* mutant. Further functional analysis with RNA interference (RNAi) technology to generate *JcSDP1* deficiency transgenic Jatropha further confirmed its function in Jatropha. Jatropha *JcSDP1*-RNAi plants accumulated 13–30% higher total seed storage lipid, along with a 7% compensatory decrease in protein content, compared with control plants. Free fatty acid (FFA) content in seeds was reduced from 27% in control plants to 8.5% in *JcSDP1*-RNAi plants. Up to 14% FFA can be reached in Jatropha seed when it served as feedstock for biodiesel production. The significant reduction of FFA in

SDP1-RNAi Jatropha plant will be of great value to address this high FFA problem in Jatropha industry (Kim et al. 2014). However, whether the reduction in TAG degradations and oxidation processes can affect on energy recycling during seed germination to secure plant germination properly needs further experimental test.

7.5.2 *SDP2* Gene

Beside of SDP1, another conserved peroxisome-located enzyme monodehydroascorbate reductase (MDAR or SDP2) was identified in Arabidopsis and Jatropha (Eastmond 2006) (23) (Ye J. unpublished data) to be critical to protect oil bodies that are in close proximity to peroxisomes from incurring oxidative damage, which otherwise inactivates the triacylglycerol lipase SUGAR-DEPENDENT1 and cuts off the supply of carbon for seedling establishment. Following germination, Arabidopsis seeds rely on storage oil breakdown to supply carbon skeletons and energy for early seedling growth, and massive amounts of H_2O_2 are generated within the peroxisome as a by-product of fatty acid β-oxidation. Without SDP2, the escaping H_2O_2 inactivates the SDP1 and SDP1-like lipases on the oil body surface that are responsible for catalyzing the first step in storage oil breakdown.

7.5.3 *SDP6* Gene

Recently, mitochondrial FAD-dependent glycerol-3-phosphate dehydrogenase SDP6 has been identified to be involved in storage lipid catabolism and glycerol assimilation, and in glycerol-3-phosphate shuttle which transports reducing power from cytosol to mitochondrion (Quettier et al. 2008). We also identified SDP6 homolog from Jatropha (Ye J. unpublished data). Knockdown multiple genes (*SDP1*, *SDP2*, and *SDP6*) provide a good opportunity to generate much low seed FFA Jatropha.

7.6 Epigenetic and Transcriptional Cascade Regulation on Lipid Biosynthesis and Accumulation

On contrast of lots of functional data in oil biosynthesis process in Jatropha seed, no transcription factor has been functionally characterized. Transcription factors are sequence-specific DNA-binding proteins that help recruit the transcriptional machinery to gene promoters and sometimes activate or inactivate gene transcription. This typically involves conformational changes in chromatin structure to facilitate access to the DNA, although this aspect of gene regulation has not been well characterized for genes involved in seed oil synthesis. Plant genomes contain large numbers of genes capable of encoding transcription factors, which can be grouped into families based upon the structure of their DNA-binding domains. Based on extensive research works and knowledge from model plant especially Arabidopsis, some of the families with members implicated in regulating seed oil deposition have been identified from Jatropha and the TFs are listed in Table 1. TFs may control a whole metabolic biogenesis pathway by recognizing a specific short (6–10 bp) motif of DNA sequences in promoter region of various genes. That trait provides cells to coordinate themselves upon environmental or cellular programming and phytohormone stimulus such as auxin (Fig. 7.5).

The B3-domain family of transcription factors such as ABSCISIC ACID INSENSITIVE3 (ABI3), FUSCA3 (FUS3), LEAFY COTYLEDON1 (LEC1), and LEC2 are generally considered as master regulators of embryogenesis and seed maturation/oil content. These B3-domain TFs are believed to be situated at the top of transcriptional cascades. Overexpression of the master regulators has been shown to increase oil content but also often associated with negative side effects (Braybrook and Harada 2008).

Beside of these B3-family, TFs can regulate another master regulator WRINKLED1 (WRI1), a member of the large, plant-specific APETALA2

Table 1 Transcription factors and epigenetic regulators identified as key factor to control seed size in Arabidopsis

Family	Transcription factor name	Summary of role in seed oil deposition	Ref
AP2	WRINKED1 (WRI1)	Downstream master regulators having more specific role towards seed oil biosynthesis; control the energy and carbohydrate distribution by regulation of carbohydrate and lipid metabolism genes, particularly plastidial fatty acid synthesis	Shen et al. (2006), Shanklin and Somerville (1991), Thazar-Poulot et al. (2015)
	APETALA2 (AP2)	Mutant seed size increased by uncharacterised mechanism	Ye et al. (2014a)
	SMALL ORGAN SIZE (SMOS1)	Auxin-induced TF, acts as an auxin-dependent regulator for cell expansion during organ size control, and that its function is conserved among land plants via microtubule orientation	Zang et al. (2012)
ARF	auxin response factor (ARF)	Auxin-induced TF, acts as auxin-dependent regulator for cell expansion during organ size control via cytoskeleton regulatory mechanism	Ye et al. (2009), Sun et al. (2017)
B3 domain;	HIGH-LEVEL EXPRESSION OF SUCROSE INDUCIBLE GENE2 (HS12) and its like parologs HSIL1 and HSL2	Act redundantly to repress AFL Clade genes and other positive regulators of seed maturation during germination and in seedlings; possible chromatin remodeling activities	Veerappan et al. (2014), Wang et al. (2007)
	ABSCISIC ACID INSENTIVE3 (ABI3), LEAFY COTYLEDON2 (LEC2), FUSCA3(FUS3)	Mastrer regulators of embryogenesis and seed maturation; mutation/overexpression often associated with pleiotropic effects; direct and indirect regulation of suites of genes involved in carbohydrate and lipid metabolism, including fatty acid synthesis, triacylglycerol assembly and packaging	Sato et al. (2011)
CHD3	PICKLE (PKL)	Chromatin remodeling factor, regulates the repression chromatin mark H3K27me3 on key genes in seed and embryo development	Wu et al. (2015), Yang et al. (2009)
Dof	Soybean Dof4, Dof11	Overexpression plant yields higher seed oil levels by direct upregulation of lipid metabolism genes;	Tsukagoshi et al. (2007)
HAP3	LEAFY COTYLEDON1 (LEC1)	Subunit of CCAAT binding proteins in various plants including monocot and dicot by direct regulation of lipid genes or via other transcription factor such as WRI1; key regulators of embryo development and seed maturation	Sato et al. (2011)
HD-ZIP	GLABRA2 (GL2)	Mutant seed oil content increased due to partition more carbohydrate into lipid biosynthesis	Zhang et al. (2008)
PRC2	FERTILIZATION INDEPENDENT ENDOSPERM (FIE), SWINGER (SWN), EMBRONIC FLOWER2 (EMF2)	Components of Polycomb Repressive Complex 2 (PRC2) that catalyze deposition of H3K27me3; repressors of seed maturation genes in vegenative tissues	Yang et al. (2009)

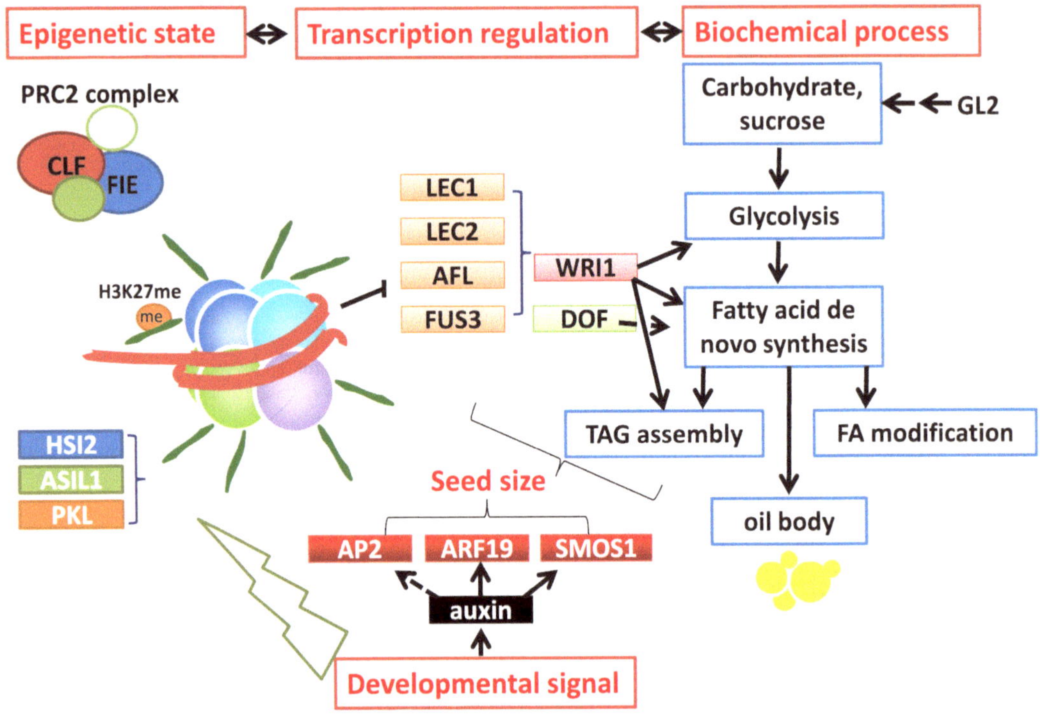

Fig. 7.5 Epigenetic regulation and transcriptional cascade of lipid biosynthesis and TAG storage

(AP2) family of transcription factors in controlling the seed lipid biosynthesis. WRI1 has been firstly identified from Arabidopsis (Cernac and Benning 2004) and later from Jatropha (Ye J. unpublished data). The expression of *WRI1* is up-regulated by FUS3/LEC1/LEC2, and the gene is a direct target of LEC2, and possibly of FUS3 (Baud et al. 2007). Overexpression of *Jatropha curcas WRI1* (*JcWRI1*) increased plant oil levels in seeds in Arabidopsis and Jatropha (Ye J. unpublished data). Increased seed oil in *WRI1* transgene does not appear to be generally correlated with altered fatty acid composition, indicating a separation between oil yield and oil composition.

There is a consensus DNA-binding motif for WRI1, the AW box [(CnTnG)(n)7[CG]] in the WRI1-regulated genes' promoter (Maeo et al. 2009), especially fatty acid biosynthesis genes in Jatropha (Ye J. unpublished data). WRI1 has been shown to bind directly to the AW box of several fatty acid synthesis genes in vitro, and mutation of the AW box abolishes WRI-mediated transcriptional activation in transient protoplast assays (Maeo et al. 2009). There is another Dof TF family (soybean Dof 1 and Dof4) acting as positive regulator in fatty acid biosynthesis. Overexpression of Dof TFs makes plant yielding higher seed oil levels by direct up-regulation of lipid-related metabolism genes (Wang et al. 2007).

Beside of positive regulator, there is another clade of B3-domain transcription factors, HIGH-LEVEL EXPRESSION OF SUCROSE INDUCIBLE GENE2 (HSI2)/VIVIPAROUS1/ABI3-LIKE1 (VAL1) and its HSI2-like (HSL) parologs in lipid biosynthesis and storage (Tsukagoshi et al. 2007). They act redundantly to suppress seed maturation programs in vegetative tissues. Double mutants in two of these factors produce seedlings expressing B3-domain genes and large numbers of embryo-specific genes containing Sph/RY elements in the promoters, accumulate seed storage compounds, and exhibit ectopic embryo formation, shared with phenotypes with mutants of chromatin re-modeling factors such as

PICKLE (PKL) (Tsukagoshi et al. 2007). It may be explained by HSI2/VAL1 family proteins structure with homeodomain (PHD)-like domain, proteins contain domains associated with chromatin re-modeling factors and transcriptional repressors, and HSI2 has been shown to be a potent transcriptional repressor in vitro. PHD-like domain of HSI2 is required for the deposition of H3K27me3, a typical repressive histone methylation mark to repress gene expression in vivo (Veerappan et al. 2014).

In addition to regulation of TFs, the chromatin structure modifications by epigenetic regulator are also important for proper transcription regulation in tissue- or organ-specific gene expression and lipid biosynthesis. The repression H3K27m3 mark is one of the most important epigenetic codes to confer developmental regulation and storage lipid biosynthesis and storage exclusively in seeds. In plant, the H3K27m3 mark is deposited by Polycomb Repressor Complex 2 (PRC2) which contributes to repressing the master seed regulators in vegetative plant tissues, as mutation of several of its components result in the derepression of master regulator genes and the production of seed oil in seedlings (Pu and Sung 2015). The first suppressor to be characterized was *PICKLE* (*PKL*), which encodes a putative chromatin re-modeling factor belonging to the chromodomain/helicase/DNA-binding (CHD3) family (Ogas et al. 1999; Zhang et al. 2008). CHD3 proteins are believed to act as transcriptional repressors possibly mediated by histone de-acetylase enzyme complex. However, chromatin of *pkl* mutants shows no differences in histone acetylation but is depleted in H3K27m3, a modification widely associated with gene silencing. In germinating wild-type seeds, PKL is physically associated with promoters enriched in H3K27m3, including *LEC1*, *LEC2*, and *FUS3*, suggesting that it acts to directly promote this repressive chromatin mark on the genes encoding master regulators of seed maturation at the transition from seed to vegetative growth (Zhang et al. 2012). PKL function appears to be transient and is no longer required two weeks following germination in Arabidopsis (Zhang et al. 2012).

7.7 Future Perspective

Genome sequencing and functional analysis by virus-induced gene silencing and transgenic technology have revolutionized plant functional genomics research and development (Ye et al. 2009; Qu et al. 2012). New genome editing technology such as transcription activator-like effector (TALE) and CRISPR/CAS9 confers broader horizon of genetic manipulation and biotechnological improving on woody energy plants (Hsu et al. 2014). Once expression has been altered, mRNA, protein, and/or metabolite levels are quantified through various profiling approaches. Knowledge about the pattern of gene expression in plant tissues under variable culture conditions will help to increase production efficiency. Our information on epigenetic and transcriptional regulation on oil content and oil quality could provide the necessary data for breeding and genetic engineering to increase oil content or optimize FA composition in Jatropha seeds. Manipulation of other aspects including flowering (Ye et al. 2014a), disease resistance (Ye et al. 2014c), seed developmental regulation (Ohto et al. 2005; Shen et al. 2006; Aya et al. 2014; Ye et al. 2014b), and many others are useful for improving Jatropha yield level.

Acknowledgments The work is supported by State key laboratory of Plant genomics, Institute of Microbiology, Chinese Academy of Sciences. It is also a part of the project-Genetic Improvement of Jatropha in Singapore initiated and coordinated by Prof. Nam-Hai Chua (Rockefeller University, USA). We would like to thank him for his invaluable suggestions during the course of this work.

References

Aya K, Hobo T, Sato-Izawa K, Ueguchi-Tanaka M, Kitano H, Matsuoka M (2014) A novel AP2-type transcription factor, SMALL ORGAN SIZE1, controls organ size downstream of an auxin signaling pathway. Plant Cell Physiol 55:897–912

Baud S, Mendoza MS, To A, Harscoet E, Lepiniec L, Dubreucq B (2007) WRINKLED1 specifies the regulatory action of LEAFY COTYLEDON2 towards fatty acid metabolism during seed maturation in Arabidopsis. Plant J 50:825–838

Braybrook SA, Harada JJ (2008) LECs go crazy in embryo development. Trends Plant Sci 13:624–630
Carlsson AS, LaBrie ST, Kinney AJ, von Wettstein-Knowles P, Browse J (2002) A KAS2 cDNA complements the phenotypes of the Arabidopsis fab1 mutant that differs in a single residue bordering the substrate binding pocket. Plant J 29:761–770
Cernac A, Benning C (2004) WRINKLED1 encodes an AP2/EREB domain protein involved in the control of storage compound biosynthesis in Arabidopsis. Plant J 40:575–585
Chan AP, Crabtree J, Zhao Q, Lorenzi H, Orvis J, Puiu D, Melake-Berhan A, Jones KM, Redman J, Chen G, Cahoon EB, Gedil M, Stanke M, Haas BJ, Wortman JR, Fraser-Liggett CM, Ravel J, Rabinowicz PD (2010) Draft genome sequence of the oilseed species *Ricinus communis*. Nat Biotechnol 28:951–956
Durrett TP, Benning C, Ohlrogge J (2008) Plant triacylglycerols as feedstocks for the production of biofuels. Plant J 54:593–607
Dyer JM, Chapital DC, Kuan JC, Mullen RT, Turner C, McKeon TA, Pepperman AB (2002) Molecular analysis of a bifunctional fatty acid conjugase/desaturase from tung. Implications for the evolution of plant fatty acid diversity. Plant Physiol 130:2027–2038
Eastmond PJ (2006) SUGAR-DEPENDENT1 encodes a patatin domain triacylglycerol lipase that initiates storage oil breakdown in germinating Arabidopsis seeds. Plant Cell 18:665–675
Grover A, Kumari M, Singh S, Rathode SS, Gupta SM, Pandey P, Gilotra S, Kumar D, Arif M, Ahmed Z (2014) Analysis of *Jatropha curcas* transcriptome for oil enhancement and genic markers. Physiol Mol Biol Plants 20:139–142
Gu K, Yi C, Tian D, Sangha JS, Hong Y, Yin Z (2012) Expression of fatty acid and lipid biosynthetic genes in developing endosperm of *Jatropha curcas*. Biotechnol Biofuels 5:47
Hsu PD, Lander ES, Zhang F (2014) Development and applications of CRISPR-Cas9 for genome engineering. Cell 157:1262–1278
Jiang H, Wu P, Zhang S, Song C, Chen Y, Li M, Jia Y, Fang X, Chen F, Wu G (2012) Global analysis of gene expression profiles in developing physic nut (*Jatropha curcas* L.) seeds. PLoS One 7:e36522
Kim MJ, Yang SW, Mao HZ, Veena SP, Yin JL, Chua NH (2014) Gene silencing of Sugar-dependent 1 (JcSDP1), encoding a patatin-domain triacylglycerol lipase, enhances seed oil accumulation in *Jatropha curcas*. Biotechnol Biofuels 7:36
Liu H, Liu YJ, Yang MF, Shen SH (2009) A comparative analysis of embryo and endosperm proteome from seeds of *Jatropha curcas*. J Integr Plant Biol 51:850–857
Liu H, Yang Z, Yang M, Shen S (2011) The differential proteome of endosperm and embryo from mature seed of *Jatropha curcas*. Plant Sci 181:660–666
Liu H, Wang C, Komatsu S, He M, Liu G, Shen S (2013) Proteomic analysis of the seed development in Jatropha curcas: from carbon flux to the lipid accumulation. J Proteom 91:23–40
Liu H, Wang C, Chen F, Shen S (2015) Proteomic analysis of oil bodies in mature Jatropha curcas seeds with different lipid content. J Proteom 113:403–414
Maeo K, Tokuda T, Ayame A, Mitsui N, Kawai T, Tsukagoshi H, Ishiguro S, Nakamura K (2009) An AP2-type transcription factor, WRINKLED1, of Arabidopsis thaliana binds to the AW-box sequence conserved among proximal upstream regions of genes involved in fatty acid synthesis. Plant l 60:476–487
Natarajan P, Parani M (2011) De novo assembly and transcriptome analysis of five major tissues of Jatropha curcas L. using GS FLX titanium platform of 454 pyrosequencing. BMC Genom 12:191
Ogas J, Kaufmann S, Henderson J, Somerville C (1999) PICKLE is a CHD3 chromatin-remodeling factor that regulates the transition from embryonic to vegetative development in Arabidopsis. Proc Natl Acad Sci USA 96:13839–13844
Ohto MA, Fischer RL, Goldberg RB, Nakamura K, Harada JJ (2005) Control of seed mass by APETALA2. Proc Natl Acad Sci USA 102:3123–3128
Pinheiro CB, Shah M, Soares EL, Nogueira FC, Carvalho PC, Junqueira M, Araujo GD, Soares AA, Domont GB, Campos FA (2013) Proteome analysis of plastids from developing seeds of *Jatropha curcas* L. J Proteome Res 12:5137–5145
Pu L, Sung ZR (2015) PcG and trxG in plants—friends or foes. Trends Genet 31(5):252–262
Qu J, Mao HZ, Chen W, Gao SQ, Bai YN, Sun YW, Geng YF, Ye J (2012) Development of marker-free transgenic Jatropha plants with increased levels of seed oleic acid. Biotechnol Biofuels 5:10
Quettier AL, Shaw E, Eastmond PJ (2008) SUGAR-DEPENDENT6 encodes a mitochondrial flavin adenine dinucleotide-dependent glycerol-3-p dehydrogenase, which is required for glycerol catabolism and post germinative seedling growth in Arabidopsis. Plant Physiol 148:519–528
Sato S, Hirakawa H, Isobe S, Fukai E, Watanabe A, Kato M, Kawashima K, Minami C, Muraki A, Nakazaki N, Takahashi C, Nakayama S, Kishida Y, Kohara M, Yamada M, Tsuruoka H, Sasamoto S, Tabata S, Aizu T, Toyoda A, Shin-i T, Minakuchi Y, Kohara Y, Fujiyama A, Tsuchimoto S, Kajiyama S, Makigano E, Ohmido N, Shibagaki N, Cartagena JA, Wada N, Kohinata T, Atefeh A, Yuasa S, Matsunaga S, Fukui K (2011) Sequence analysis of the genome of an oil-bearing tree, *Jatropha curcas* L. DNA Res 18:65–76
Shen B, Sinkevicius KW, Selinger DA, Tarczynski MC (2006) The homeobox gene GLABRA2 affects seed oil content in Arabidopsis. Plant Mol Biol 60:377–387
Shanklin J, Somerville C (1991) Stearoyl-acyl-carrier-protein desaturase from higher plants is structurally unrelated to the animal and fungal homologs. Proc Natl Acad Sci USA 88:2510–2514
Sun YW, Wang CM, Wang N, Jiang XY, Mao HZ, Zhu CX, Wen FJ, Wang XH, Lu ZJ, Yue GH, Xu ZF, Ye J (2017) Manipulation of Auxin Response Factor

19 affects seed size in the woody perennial *Jatropha curcas*. Sci Rep 7:40844

Thazar-Poulot N, Miquel M, Fobis-Loisy I, Gaude T (2015) Peroxisome extensions deliver the Arabidopsis SDP1 lipase to oil bodies. Proc Natl Acad Sci USA 112:4158–4163

Tsukagoshi H, Morikami A, Nakamura K (2007) Two B3 domain transcriptional repressors prevent sugar-inducible expression of seed maturation genes in Arabidopsis seedlings. Proc Natl Acad Sci USA 104:2543–2547

Veerappan V, Chen N, Reichert AI, Allen RD (2014) HSI2/VAL1 PHD-like domain promotes H3K27 trimethylation to repress the expression of seed maturation genes and complex transgenes in Arabidopsis seedlings. BMC Plant Biol 14:293

Wang HW, Zhang B, Hao YJ, Huang J, Tian AG, Liao Y, Zhang JS, Chen SY (2007) The soybean Dof-type transcription factor genes, GmDof4 and GmDof11, enhance lipid content in the seeds of transgenic Arabidopsis plants. Plant J 52:716–729

Wei Q, Li J, Zhang L, Wu P, Chen Y, Li M, Jiang H, Wu G (2012) Cloning and characterization of a beta-ketoacyl-acyl carrier protein synthase II from *Jatropha curcas*. J Plant Physiol 169:816–824

Wu P, Zhou C, Cheng S, Wu Z, Lu W, Han J, Chen Y, Chen Y, Ni P, Wang Y, Xu X, Huang Y, Song C, Wang Z, Shi N, Zhang X, Fang X, Yang Q, Jiang H, Chen Y, Li M, Wang Y, Chen F, Wang J, Wu G (2015) Integrated genome sequence and linkage map of physic nut (*Jatropha curcas* L.), a biodiesel plant. Plant J 81:810–821

Yang MF, Liu YJ, Liu Y, Chen H, Chen F, Shen SH (2009) Proteomic analysis of oil mobilization in seed germination and postgermination development of *Jatropha curcas*. J Proteome Res 8:1441–1451

Ye J, Geng Y, Zhang B, Mao H, Qu J, Chua N (2014a) The Jatropha FT ortholog is a systemic signal regulating growth and flowering time. Biotechnol Biofuels 7:19

Ye J, Hong Y, Qu J, Wang C (2012) Improvement of Jatropha oil by genetic transformation. Jatropha, challenges for a new energy crop-genetic improvement and biotechnology, vol 2. Springer Science Publishers, New York

Ye J, Liu P, Zhu C, Qu J, Wang X, Sun Y, Sun F, Jiang Y, Yue G, Wang C (2014b) Identification of candidate genes JcARF19 and JcIAA9 associated with seed size traits in Jatropha. Funct Integr Genomics 14:757–766

Ye J, Qu J, Bui HT, Chua NH (2009) Rapid analysis of *Jatropha curcas* gene functions by virus-induced gene silencing. Plant Biotechnol J 7:964–976

Ye J, Qu J, Mao HZ, Ma ZG, Rahman NE, Bai C, Chen W, Jiang SY, Ramachandran S, Chua NH (2014c) Engineering geminivirus resistance in *Jatropha curcus*. Biotechnol Biofuels 7:149

Zhang H, Bishop B, Ringenberg W, Muir WM, Ogas J (2012) The CHD3 remodeler PICKLE associates with genes enriched for trimethylation of histone H3 lysine 27. Plant Physiol 159:418–432

Zhang H, Rider SD Jr, Henderson JT, Fountain M, Chuang K, Kandachar V, Simons A, Edenberg HJ, Romero-Severson J, Muir WM, Ogas J (2008) The CHD3 remodeler PICKLE promotes trimethylation of histone H3 lysine 27. J Biol Chem 283:22637–22648

Part III

Genetics

Forward and Reverse Genetics for the Improvement of *Jatropha*

8

Fatemeh Maghuly and Margit Laimer

Abstract

The domestication, introduction, and selection of crop plants for different uses require knowledge of genetic variability as well as gene function, which helps to understand and improve the efficiency of exploitation of genes and to identify better breeding strategies. The present chapter covers experimental research in *Jatropha* from sequences, markers, and chromosomes to functions and provides an overview of the major forward and reverse genetic approaches used to improve the performance of this biofuel crop. The key elements are characterization, identification, localization, profiling as well as determination of the expression or function of different genes of interest.

8.1 Introduction

With the advancement of whole-genome sequencing in plants, the determination of gene functions is now the most challenging goal. Basically, there exist two approaches to link function and sequence of a specific gene: forward and reverse genetics (Peters et al. 2003).

Sequencing entire genomes and expressed sequence tags (ESTs) has greatly facilitated the genetic analysis from "phenotype to gene," the so-called forward genetics approaches. Traditionally, mutagens were used to identify genes responsible for specific phenotype by high-density genetic maps, physical maps, and positional cloning of almost any gene. The strategies of map-based or positional cloning depends on the availability of following factors: (1) a crossing population accurately segregating for the target trait; (2) suitable genomic libraries and/or nucleotide sequence information corresponding to the candidate genomic region; and (3) an efficient transformation and evaluation system of transgenic plants. Advances in sequencing, available molecular markers, and methods have facilitated fast mapping of genes to detect DNA polymorphisms (Peters et al. 2003). However, forward genetics is not suitable for genome-wide gene analyses, since it requires

F. Maghuly (✉) · M. Laimer
Plant Biotechnology Unit (PBU), Department of Biotechnology, BOKU VIBT, Muthgasse 18, 1190 Vienna, Austria
e-mail: margit.laimer@boku.ac.at

S. Tsuchimoto (ed.), *The Jatropha Genome*, Compendium of Plant Genomes,
DOI 10.1007/978-3-319-49653-5_8

more labor and time effort to find affected genes from newly generated mutants (Alonso and Ecker 2006).

One of the most important tools of functional genomics is reverse genetics, which in contrast to forward genetics can identify phenotypes for specific genes much faster. Several reverse genetic approaches have been developed in order to understand gene function including homologous recombination (Hanin and Paszkowski 2003), antisense or RNA interference (RNAi) suppression (Chuang and Meyerowitz 2000), insertional mutagenesis (Feldmann 1991), target gene inactivation (Belhaj et al. 2013), and target-induced local lesions in genomes (TILLING) (Peters et al. 2003; Maghuly et al. 2013a). Among these methods, random insertional mutagenesis by transposons or T-DNA has been most widely used for large-scale analyses. Since the T-DNA sequence is known, it can be used as a bait to characterize the flanking sequences (Gambino et al. 2009). If the entire genome sequence is available, the exact location of the insertion is easily determined through homology searches (i.e., BLAST) against a whole-genome database.

Jatropha curcas L. is a perennial, monoecious shrub that belongs to the Euphorbiaceae family (Wen et al. 2010). The species is native to America but is distributed widely in the tropical and subtropical areas (Wen et al. 2010). Semi-cultivated *Jatropha* can grow well under all unfavorable climatic and soil conditions (Maghuly et al. 2013b). In addition, *Jatropha* has a relatively small genome (2C DNA content of 0.850 ± 0.006 pg or C DNA content of 0.416 ± 109 bp) compared with other members of the Euphorbiaceae (Carvalho et al. 2008), which make it attractive for genomic analysis. Further, the biochemical composition of *Jatropha* is of great interests for several reasons (Sabandar et al. 2013). Different parts of *Jatropha* are reported to be used in traditional medicine (Soomro and Memon 2007), and for their antimicrobial properties (Thomas 1989). Oil from the seeds can be used as a biofuel, for producing soap (Vollmann and Laimer 2013), and as fertilizer (Sherchan et al. 1989) because of high protein content compared to soybean in the press cake. Earlier chemical analyses have revealed high concentrations of several interesting compounds (Sabandar et al. 2013). Currently, *Jatropha* seeds contain 30–45% oil, comprising 14.1–15.3% palmitic acid (16:0), 3.7–9.8% stearic acid (18:0), 34.3–45.8% oleic acid (18:1) and 29–44.2% linoleic acid (18:2) (Maghuly et al. 2013b).

Although the knowledge about *Jatropha* increases, there are still a number of challenges that should be overcome; it has not really been domesticated yet; it is susceptible to many pathogens; it contains a range of toxins and antinutritional compounds that make the seed-cake and oil unsuitable for use as animal feed and/or for human consumption (Sabandar et al. 2013); and its fruiting is continuous, which causes harvesting difficult (Brittaine and Lutaladio 2010). Therefore, before starting any improvement program in order to develop elite *Jatropha* cultivars, understanding gene function to decipher important metabolic pathways would allow to address these challenges (Fig. 8.1).

8.2 Physical and Chemical Mutagenesis

Physical mutagens have been used widely in plant breeding for crop improvement, with gamma irradiation accounting for more than 60% of known mutant cultivars (Matijevic et al. 2013). A number of physical mutagens have been used to induce variability in plant genome: (1) ionizing atomic particle irradiation induces mutagenesis by using subatomic particles such as electrons, protons, neutrons, deuterons, alpha, and beta; and (2) ionizing electromagnetic radiation induces mutagenesis by energy in the form of electromagnetic waves. Fast neutron and ion beams as ionizing atomic particles and X-rays, γ-rays, and cosmic rays as ionizing electromagnetic irradiation are the most commonly used for mutation breeding (Mba et al. 2010). γ-ray mutagenesis using different facilities such as gamma phytotron, gamma house, gamma field, and gamma cell irradiator with cobalt-60 or cesium-137 as radioactive source is the most

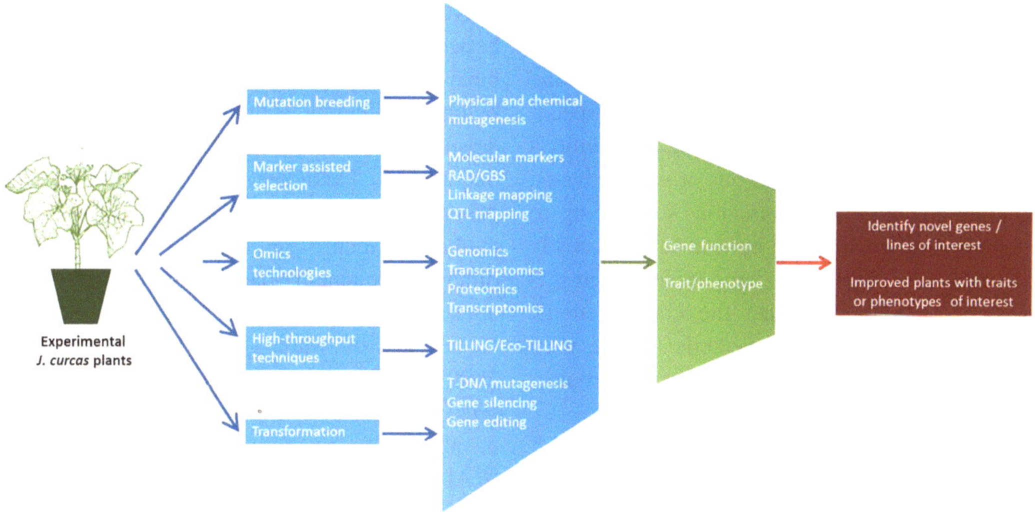

Fig. 8.1 An overview of the major forward and reverse genetic approaches used to improve the performance of *Jatropha curcas*

commonly available worldwide. Protocols for plant mutagenesis using γ-rays (Kodym and Afza 2003; Mba et al. 2010) and especially for *J. curcas* (Maghuly et al. 2015) are available. In addition, Dhakshanamoorthy et al. (2011) and Dhillon et al. (2014) showed that different doses of γ-rays (5–50 Kr) in *J. curcas* changed various morphological traits like flowering and germination percentage; however, these preliminary data were recorded in M_1 populations, and no reverse genetic work was carried out to validate gene functions.

X-ray mutation breeding using various X-ray irradiators adapted to plant mutagenesis such as Faxitron 650 (Faxitron bioptics LLC, Tuscon, AZ, USA), RS-2400 self-contained low-energy irradiator (Rad-Source), and Hitachi (http://www-naweb.iaea.org/nafa/pbg/public/Protocol-X-raymutagenesis-plants.pdf) can be observed since some decades.

A number of chemical mutagens have been used to alter the genome sequences of plants to induce variability in the species such as ethyl methanesulfonate (EMS) and sodium azide (NaN_3). EMS is a popular and commonly used chemical mutagen, with the formula $CH_3SO_3C_2H_5$. It produces random mutations by nucleotide substitution, particularly by guanine alkylation, and typically produces only point mutations. Mutations are induced at a rate of 5×10^{-4} to 5×10^{-2} per gene. DNA polymerases frequently replace thymine with cytosine, which in the following subsequent rounds of replication replaces G/C base pairs with A/T pairs, while sodium azide mutagenesis preferentially generates A/T to G/C transitions.

Crop plant improvement by physical and chemical mutagenesis techniques has been described by Kodym and Afza (2003) and Mba et al. (2010) and has been adapted to seeds, in vivo and in vitro cuttings to induce variability in *Jatropha* spp. (Maghuly et al. 2015). DNA polymorphisms induced by different concentrations of EMS (1–4%) were studied in *J. curcas* (Dhakshanamoorthy et al. 2011) using random amplified polymorphic DNA (RAPD) markers.

However, physical and chemical mutagenesis methods are time and cost intensive to reduce chimeras, since they require large populations (5000–10,000 individuals) of M_2 or higher. In addition, the entire mutated populations should be screened for selection of phenotype of interest (forward genetics). In fact, cost-effective and high-throughput methods for screening mutation at genetic level (reverse genetics) are described in Sects. 8.3 and 8.7.

8.3 Marker Systems

DNA polymorphisms are generated over the course of evolution by various molecular mechanisms producing either nucleotide substitutions or insertions or deletions of one or several bases. Despite advances in automation, sequencing to visualize sequence polymorphism remains costly and time-consuming on a large scale. When it is not necessary to visualize the entire genome sequence, some indirect methods can be used which are much faster and cheaper, i.e., molecular markers (Semagn et al. 2006).

Molecular markers can be classified into different groups based on: (a) method of analysis, i.e., hybridization-based or polymerase chain reaction (PCR)-based markers; (b) mode of transmission, i.e., biparental nuclear inheritance, maternal nuclear inheritance, and maternal or paternal organelle inheritance; and (c) mode of gene action, i.e., dominant or codominant markers (Falque and Santoni 2007).

PCR-based molecular markers, such as RAPDs (Dhillon et al. 2014; Jiang and Ramachandran 2013), or simple sequence repeats (SSRs, also known as microsatellites) as well as intersimple sequence repeats (ISSRs; Maghuly et al. 2015), amplified fragment length polymorphisms (AFLPs; Maghuly et al. 2015), and simple nucleotide polymorphisms (SNPs) were used for studying genetic variation of *Jatropha*. Among them, SSRs and SNPs are the most preferred and useful markers for linkage mapping, breeding, and genotyping and provide information concerning gene structure, linkage disequilibrium, and population structure, due to their high-abundance, high-throughput, and cost-effective scoring (Yue et al. 2013). Detection of SSRs is easy in genome sequences (Sun et al. 2008; Metzker 2009; Sudheer et al. 2010; Wang et al. 2011; Yadav et al. 2011); thus, they were already identified in draft genome sequences in *J. curcas* (Sato et al. 2011).

Although a number of SNPs were identified in *J. curcas* (Silva-Junior et al. 2011; Wang et al. 2011; Gupta et al. 2012), they were found only in low frequency (Silva-Junior et al. 2011; Gupta et al. 2012; Yue et al. 2013; Maghuly et al. 2015). Therefore, still more SNPs are necessary that can be accessed by high-throughput and cost-effective methods. To provide sufficient SNP markers, a number of different techniques have been recently developed, but all have their limitations. Sequencing candidate genes from multiple genotypes is the most accurate alternative to these methods (Gilchrist et al. 2006). Additionally, different sequencing-based approaches to SNP allele calling have been developed including sequencing of whole genomes (Hillier et al. 2008), exome capture (Ng et al. 2009), RNA-seq (Hansey et al. 2012), methylated DNA (Brunner et al. 2009), restriction-site associated DNA (RAD-seq) (Baird et al. 2008), and genotyping-by-sequencing (GBS) (Elshire et al. 2011), the last two being based on restriction enzyme digestion (reviewed by Davey et al. 2011). In contrast to whole-genome sequencing, the other approaches represent efficient and cost-effective methods to produce genetic information (a) by reducing the representation of the genome and (b) by applying efficient pooling strategies (Beissinger et al. 2013; Schlötterer et al. 2014).

8.4 Genetic Mapping

In different studies of crop plant genomes, attempts to improve the growth and quality of fruits and resistance of trees to pathogens have produced valuable linkage maps (Vilanova et al. 2003, Abbott et al. 2007, Lambert et al. 2007). However, many agriculturally important traits such as plant height, seed yield, and oil content are polygenically controlled and greatly depend on genetic $\times$ environmental (G $\times$ E) interactions (Abdurakhmonov and Abdukarimov 2008). Genetic mapping can facilitate the detection and understanding of the number of neutrally inherited markers in close proximity to the genetic causatives or genes controlling the complex quantitative traits (Abdurakhmonov and Abdukarimov 2008). Linkage analyses require experimental populations such as F_2 with a large number of individuals from the crossing of two parental genotypes with an alternative trait of interest as well as polymorphic DNA markers.

These marker positions along the linkage map are statistically correlated with phenotypic characteristics of individuals of the mapping populations (Abdurakhmonov and Abdukarimov 2008).

Although genetic diversity of *J. curcas* showed a low-to-moderate level of variability (Sun et al. 2008; Sudheer et al. 2010; Shen et al. 2012; Maghuly et al. 2015), clear phenotypic variation among *Jatropha* collections is well documented (Yi et al. 2010). Moreover, some studies reported a higher variation among toxic populations in Mexico (Ovando-Medina et al. (2011), however, nontoxic accessions, which could be only located in Mexico, show a high level of genetic similarity. The relatively small genome size of *J. curcas* and its close taxonomic distances to important Euphorbiaceae species, such as *Ricinus* or *Manihot*, allow the comparison of diverging orthologs, since partial genomes of these species are available (King et al. 2011, Sato et al. 2011). Since there is no biological barrier to cross-pollination between *J. curcas* and *J. integerrima*, the first mapping populations were produced by interspecific crosses (Wang et al. 2011). Moreover, the high level of microsynteny between *J. curcas* and *Ricinus* allows to conclude the relative position of candidate genes for finer mapping (King et al. 2013). Based on the present synteny, Wen et al. (2010) transferred 187 EST-SSR and 54 genomic (G)-SSR markers from cassava to *J. curcas*.

This information provides useful hints for linkage maps. In comparison with the map from a cross between *J. curcas* and *J. integerrima* (Wang et al. 2011), King et al. (2013) obtained higher density over 11 linkage groups, corresponding to 11 chromosome pairs of the *J. curcas* genome (Carvalho et al. 2008). Analysis of F_2 plants and quantitative trait loci (QTL) revealed that a locus on linkage group 8 is responsible for PE biosynthesis (King et al. 2013). Wang et al. (2011) detected 18 QTLs for oil traits and 28 QTLs for tree growth and seed traits. They found that the overlapping QTL qC18:1-1 and QTL qC18:2-1 is located on the end of linkage group 1 and responsible for the oleic and linoleic acid content. The developed SNP markers in three *J. curcas* oleosin genes (Popluechai et al. 2011) showed that *OleI* and *OleII* controlling oleic and stearic acids were mapped to linkage group 5. The QTL *qOil*C-4 influencing the total oil content was mapped to linkage group 4, and two QTLs *qTSW*-5 and *qTSW*-7, controlling seed yield were located on linkage group 5 and 7, respectively. Using the same mapping population, Ye et al. (2014) found the interaction of two genes *ARF19* and *IAA9* involved in auxin signal transduction and play important roles in seed size.

Although QTL-mapping is an important tool in gene tagging and marker-assisted selection, still many important traits are not studied in *J. curcas* such as pathogen resistance, tree height, flowering time, female and male flower ratio, seed yield, and number of branches. In fact, it is very costly, labor-intensive, and low genetic variation in *J. curcas* hampering generation of informative reference families for mapping (Yue et al. 2013).

8.5 Microscopy and Associated Bioimaging

Genetic maps and genomic sequences tell relatively little about the physical organization of the chromosomes (Schmidt and Heslop-Harrison 1998; Maghuly et al. 2009). Chromosomes can differ in size and morphology between and within species and can be identified on the basis of their physical characteristics (length, position of centromere, arm ratio, secondary construction). Very often, chromosomes are numerous, small, and poorly differentiated in morphology (Maghuly et al. 2009). Cytogenetics provides a powerful system for looking at the organization of DNA repeat motifs on a chromosome using in situ hybridization. Probes for DNA repeats, e.g., ribosomal, microsatellite, and telomeric regions, have become powerful tools for discerning chromosomal organization, evolutionary history, and phylogenetic relationships (Maghuly et al. 2009).

Among the highly repetitive sequence motifs, some are highly conserved between species, and some are easily recognized: (1) tandemly repeated sequences; (2) retroelements, which amplify through an RNA intermediate (acting as a template for protein translation as well as reverse transcription into DNA) before reinsertion into the genome; and (3) special classes such as telomeric sequences or rDNA units.

The 45S rDNA loci consist of tandem arrays of repeat units, each comprising the highly conserved 18S-5.8S-26S rRNA genes and variable non-transcribed spacer regions. Hundreds or thousands of copies of the repeat units may be present, together representing up to ~10% of the genome (8% in *Arabidopsis*; Heslop-Harrison 2000). The units, along with the nuclear genes encoding 5S rRNA genes (occurring as tandem repeats independent of the 45S rDNA), are usually physically separated along the chromosome of angiosperms. In plants, the 5S rDNA genes are arrayed independently, while the 18S, 5.8S, and 26S rRNAs are produced together from a 45S rDNA precursor gene. The units themselves are highly conserved, and the chromosomal distribution of the rDNA sites varies among species (Castilho and Heslop-Harrison 1995; Taketa et al. 1999). Also, in plants with chromosomes of similar size and morphology, identification of homologous pairs can be difficult. Physical mapping of rDNA genes by fluorescence in situ hybridization (FISH) provides valuable landmarks that can be used for chromosome identification of non-homologous chromosomes with similar morphology (Doudrick et al. 1995). Using FISH in *J. curcas* showed that 2 loci of 45S (Witkowska et al. 2009) located at the terminal region of chromosomes 7 and 9 (Gong et al. 2013), one locus of 5S located on the same chromosome with 45S (Witkowska et al. 2009), and 5 *copia*-type retrotransposons are spread across the chromosomes but mainly in the distal region (Alipour et al. 2013). Additionally, Alipour et al. (2014) located *Gypsy*-type retrotransposon in the pericentromic region of *Jatropha* chromosomes.

Localizing important traits using genetic map can help to understand the genetic basis of plants, comparative mapping, and evaluation studies. In combination with FISH, this will also provide information on the conservation of the gene order.

8.6 Genome and Transcriptome Sequencing

Genome sequencing can be used for screening natural or mutagenized populations to identify single nucleotide polymorphisms, even in plants with complex genomes, especially when reference sequences are available. Several groups have sequenced the entire *J. curcas* genome (Life Technology Corporation, USA; Synthetic Genomics, Inc., USA; SG Biofuel, USA; Temasek Life Science Laboratory, Singapore; and Asiatic Centre for Genome Technology, Malaysia), but only a draft genome sequence obtained by combining conventional and new sequencing techniques (Sato et al. 2011) and upgraded later by additional new data (Hirakawa et al. 2012) is available for public use (http://www.kazusa.or.jp/jatropha/). As a result, a total of 19,454 independent tentative consensus sequences (TCs) collected from all available transcriptome data and update genome sequencing were used for the gene prediction, producing a total of 30,203 genes and 17,575 transposon-related genes as well as 2,124 putative pseudogenes. The authors could also assign 2,402 genes to the metabolic pathways in the KEGG database (Sato et al. 2013). In comparison, the genome assembled by Wu et al. (2015) covers a total length of 320.5 Mbp and contains 27,172 putative protein-coding genes.

In addition, the whole chloroplast genome sequence of *J. curcas* was completed by pyro- and Sanger sequencing, showing that the size and arrangement of genes located on chloroplasts in *J. curcas* are similar to other plants, with a particularly strong similarity to cassava, except for the loss of two genes (*rps16* and *infA*) (Asif et al. 2010).

Besides whole-genome sequencing, transcriptome profiling and analyses of its dynamics and regulation is important to understand the genome's function and to identify candidate gene for traits of interest and their underlying biological process (Yue et al. 2013). Various ESTs were generated from developing and germinating seeds of *J. curcas* as well as from endosperm (Gomes et al. 2010; Natarajan et al. 2010). Costa et al. (2010) found genes involved in seed toxicity, related to lipid synthesis and degradation and transposable element. Also next-generation sequencing (NGS) was used to analyze transcriptomes from mature leaves, callus, embryos, root, flower, and developing seed (King et al. 2011; Sato et al. 2011). King et al. (2011) found that storage proteins are the most abundant proteins, with ribosomal proteins and late embryogenesis abundant (LEA) proteins, but also curcin were highly expressed in developing seed of *J. curcas*. Using NGS, Purushothaman and Madasamy (2011) obtained transcripts related to oil biosynthesis as well as other biochemical pathways. Eswaran et al. (2010) isolated 32 full-length genes involved in biotic and abiotic stresses from a cDNA library constructed from salt-stressed roots of *J. curcas*. Tang et al. (2011) isolated the *JcERF* gene, encoding an AP2/EREBP domain containing transcriptional factor, which functioned effectively as trans-activator in yeast one hybrid (YOH) assay. In addition, overexpression of this gene in *Arabidopsis* induced salt and frost tolerance. Sriram et al. (2010) provided a method to identify and functionally characterized genes related to abiotic stress tolerance, which can be used to improve *J. curcas*. Gu et al. (2011) cloned and characterized four genes encoding subunits of heteromeric ACCase (*accA*, *accB1*, *accC*, and *accD*), which are expressed temporally and spatially in leaves and endosperm of *J. curcas*. Expression analyses of 68 genes that encode enzymes involved in fatty acid and lipid biosynthesis at different developmental stages of endosperm showed that the expression of majority of genes is comparable with the development of oil bodies and endosperm in *Jatropha* seeds; however, some genes with similar function may be differentially expressed (Gu et al. 2012). Gene expression profiles of developing seeds 14–45 days after pollination (DAP) showed that main period of storage synthesis reversion appears between 29–41 DAP (Jiang et al. 2012). RNA-seq analyses of 14-day-old seedlings of *J. curcas* under cold conditions (12 °C) at different time points revealed 4185 unigenes that might be associated with cold resistance (Wang et al. 2013). All together, 3178 genes were significantly upregulated, while 1244 were downregulated under cold stress. Transcriptome changes in roots and leaves of *Jatropha* plants exposed to 100 mM NaCl were investigated, and 1504 and 1115 genes were significantly up- or downregulated (Zhang et al. 2014). The differently regulated genes were related to trehalose synthesis and cell wall structure in roots, and to raffinose synthesis and reactive oxygen scavengers in leaves. Yang et al. (2010) identified one group of DNA-binding with one finger (Dof) transcription factor genes, which showed to have important role in the control of flowering time in *J. curcas*. Using YOH systems showed that JcDof1 was localized to the onion epidermal cell nucleus and was characterized by a circadian-clock oscillation under long day, short day, and continuous light conditions, suggesting as a transcriptional factor gene responding to light signals (Yang et al. 2010). Transcriptional levels of genes involved in response to phytohormones were studied by Pan et al. (2014), who identified novel genes involved in the biosynthesis, metabolism and signaling of cytokinin and other hormones, as well as flower development and cell division. Expression of ABCE model genes was mostly downregulated (*JcAP1-2*, *JcPI*, *JcAG*, and *JCSEP1-3*), while (*JcAP3*) was upregulated, which may show different mechanisms in floral organ development. Further several genes related to cell division (*JcCycD1;3*, *JcCycD3;2*, and *JcTSO1*) were up regulated, explaining the increased number of flowers after treatment. In fact, functional classification of unique ESTs by BLASTX revealed a broad range of cellular, molecular, and biological functions (Sato et al. 2011; Yadav et al. 2011). Although the available information and sequences are valuable, further efforts for completing the whole-genome sequence and a reference transcriptome, which contains all transcripts, as well as coding, noncoding, and large and small RNAs in *J. curcas*, are necessary.

8.7 TILLING and EcoTILLING

The ability of individuals to adapt to different environments resides in their genetic diversity, most commonly manifested as SNPs, providing information for the adaptive strategies and population histories, which are important for species evolution. In addition, natural DNA polymorphisms form the basis for the intraspecific variation that is of great relevance to breeders. TILLING is a general strategy for identifying point mutations that combine with PCR-based screening of gene regions of interest. It is a reverse genetic approach and can be applied to almost any organism that combines traditional mutagenesis with high-throughput mutation discovery. It is a low-cost, high-throughput reverse genetic method for haplotyping and SNP discovery (Comai et al. 2004) and can be used to identify point mutations in specific genes to study gene function and interaction studies. The method has also been adapted for the discovery of natural polymorphisms, called EcoTILLING (Henikoff and Comai 2003; Comai et al. 2004), able to detect small deletions, insertions, and microsatellite polymorphisms in addition to single-bp changes in the DNA sequence (Gilchrist and Haughn 2005). These approaches were shown to provide a robust, high-throughput, fast, and cost-effective method for *J. curcas* with a narrow genetic basis (Maghuly et al. 2013b, 2015). Further, functional analyses could discover new SNPs that can affect expression of selected genes in *J. curcas* involved in oil and toxin biosynthesis (Maghuly et al. 2015). Three important components are necessary for optimal results in TILLING and EcoTILLING: (1) a good pooling strategies, (2) a good gene model (exon/intron position), and (3) a good PCR primer pair. In this approach, first gene model and protein conservation model are used to choose the region of a gene with the highest number of possible deleterious changes or a best screening region using CODDLE (http://www.proweb.org/coddle/) to design primers. DNA samples are normalized and mixed to ensure that each individual will be equally represented in the pool. The PCR products are of approximately 1000–2000 bp with fluorescent labeling of site-specific primers (left primer 5′-labeled with IRD 700 and right primer with IRD 800). These are denatured and renatured to produce several heteroduplexes, which are digested using mismatch endonuclease CEL I, and the products are analyzed by gel electrophoresis. Two electronic image files are produced per gel run, containing data from the 700 and 800 nm channels, and are scored using the GelBuddy program (Zerr and Henikoff 2005). Identified SNPs are confirmed and evaluated for their potential effect on protein function by sequencing and SIFT programs (NG and Henikoff 2003).

8.8 T-DNA Mutagenesis

Genetic transformation is a key technology to determine the function of a specific gene or the potential of existing cultivars as well as to develop new cultivars with improved traits of economic value. This could answer questions related to gene expression, localization of specific proteins, altering phenotypes by overexpressing or knockingdown selected genes, and achieving a fine-tuned control about the spatial and temporal expression of gene coding for desired characteristics. Essential knowledge in plant stress physiology will lead to the identification of useful genes.

The construction of appropriate transformation vectors and experience in plant transformation and regeneration systems will allow the regeneration of transgenic plants. Different methods are available for plant transformation, with *Agrobacterium tumefaciens* and particle bombardment being the most widely used techniques to transfer DNA into plant tissues. Regardless of the delivery method, the development of an efficient regeneration system amenable to genetic transformation is a prerequisite for plant genetic engineering in *J. curcas* (Misra and Misra 2010; Mazumdar et al. 2010). First reports concerning *J. curcas* regeneration and subsequent transformation since the mid-1990s produced shoots from cotyledon, epicotyl, hypocotyl, and leaf tissue (Sujatha and Mukta 1996; da Câmara Machado et al. 1997; Lu et al. 2003; Wei et al. 2004; Sujatha et al. 2005; Rajore and Batra 2005; Jha et al. 2007; Kumar and Reddy 2010, 2012).

Agrobacterium tumefaciens-mediated transformation is the most widely used method to introduce genes in *J. curcas* (Li et al. 2008a, b; He et al. 2009; Kumar et al. 2010; Pan et al. 2010; Zong et al. 2010). It involves the processing of transferred (T)-DNA characterized by two imperfect 25 bp direct border repeats from the left and the right borders and its transfer to the plant genome (Gelvin 2003; Maghuly et al. 2006; Kim et al. 2007; Gambino et al. 2009).

The first successful transformation of cotyledons of *J. curcas* was carried out by Li et al. (2006) using *A. tumefaciens* strain LBA4404 carrying marker genes. Further, Mazumdar et al. (2010) and Kumar et al. (2010) developed an efficient protocol for genetic transformation of *Jatropha* using *Agrobacterium-mediated* transformation of leaf explants. Subroto et al. (2014) used plasmid CAMBIA 1303 containing the *glucuronidase* gene, which was successfully transferred into *J. curcas* cotyledon explants and could be expressed by the plant.

Since biodiesel with high monounsaturated fatty acid content (oleic acid) and low percentage of saturated fatty acids are desirable, future breeding goals for *J. curcas* are to increase the total amount of oil and improve the quality by increasing monounsaturated and decreasing saturated fatty acids (Maghuly et al. 2013b). To improve the quality of the biodiesel, the first important step is to identify the key genes and enzymes involved in fatty acid metabolisms, oil composition, and lipid accumulation. Ye et al. (2009) developed a virus-induced gene silencing (VIGS) strategy for rapid analysis of *J. curcas* gene function. The method produced robust and reliable gene silencing in more than 13 genes related to fatty acid and toxic biosynthesis as well as developmental regulation of *J. curcas*. Functional analysis showed that *FAD2-1*, *FAD2-2*, and *FAD6* are involved in oleate desaturase and silencing of *SAD1* could generate transgenic *J. curcas* with high stearic acid content, while suppressing *FATB* expression led to higher production of 18:0-ACP (Yeh et al. 2009) Marker-free transgenic *J. curcas* with increased oil quality were reported by down regulating FAD2-1 using RNAi technology, showing a high increase of oleic acid (>78%) and a clear reduction in polyunsaturated fatty acids (<3%) (Qu et al. 2012). Further, overexpression of *Jatropha KASII* gene (beta-ketoacyl-acyl carrier protein synthase II) in *Arabidopsis* decreased 16-carbon fatty acids and increased 18-carbon fatty acids in leaves and seeds, showing its important role in accumulation of 18-carbon fatty acids (Wei et al. 2012).

Jatropha contains different components, but the main toxicity of *J. curcas* seeds is related to the presence of ribosome-inactivating proteins (RIPs; curcin), diterpenes (phorbol esters; PEs), saponins, trypsin inhibitor, protease inhibitors, and many more, which make its seedcake unsuitable for use as animal feed (Maghuly et al. 2013b). Therefore, the reduction or even the loss of toxicity of *J. curcas* seed would increase safety for fieldworkers, when handling the plant, and would also make the press cake available for animal feed. Currently, it is used as organic fertilizer, exhibiting a nitrogen content similar to chicken manure, ranging from 3.2 to 3.8% (Heller 1996).

According to Sato et al. (2011), two curcin and three curcin-like genes were identified in *J. curcas*, which were addressed for curcin reduction in transgenic *Jatropha* (Ye et al. 2009; Yin et al. 2010) using RNAi technology. Yin et al. (2010) used an endosperm-specific promotor to generate curcin-deficient transgenic *Jatropha*.

On the other hand, PEs (12-hydroxy-16-deoxyphorbol) are the major toxic components of *Jatropha* seeds. Without doubt, a better understanding of PE biosynthesis is expected to elucidate an effective strategy for the utilization of *Jatropha* plants (Nakano et al. 2012). So far, the complete biosynthetic pathway of PEs is unknown, but it is supposed that geranylgeranyl pyrophosphate (GGPP) is converted by casbane synthase (CS) to casbane (a monocyclic diterpene), which is the first cyclization step toward tigliane (a tetracyclic diterpene) (Nakano et al. 2012). Accordingly, altering the level of the expression of genes and enzymes involved in CS and GGPPS biosynthesis could help to clarify the influence of these enzymes in the regulation of diterpene biosynthesis. Lin et al. (2010) confirmed the function of *Jc- GGPPS* as general precursor for diterpens and carotenoid biosynthesis.

Although *J. curcas* is well adapted to grow in arid to semiarid area, slightly tolerant to salt and short mild frosts, its yield is limited by cold, drought, as well as abiotic stresses (Li et al. 2008a, b). Thus, Tsuchimoto et al. (2012) developed transgenic *Jatropha* that can produce high amount of glycine obtained by overexpression of glycine sarcosine methyltransferase (GSMT) and dimethylglycine methyltransferase (DMT) genes from *Synechococcus* under the control of a CaMV 35S promoter, the *PPAT* genes, catalyzing the CoA biosynthesis, and *NF-YB*, which encodes a subunit of NF-Y transcription factor. In fact, GSMT and DMT catalyze the production of the glycine betaine, suggested to maintain high osmotic pressure in the cells and thus enabling efficient water absorption even from saline conditions, resulting in salt and drought resistance. Recent studies by Li et al. (2014) showed that the expression of *JcR1MYB1* gene, encoding a R1-MYM transcription factor, is higher in roots than in stems and also increased by stresses such as NaCl, cold, abscisic acid, jasmonic acid, and ethylen treatment. In addition, overexpression of *JcR1MYB1* gene in transgenic tobacco showed its important role in salt stress tolerance. Although the results showed an increased expression of the introduced genes, further studies are needed to confirm the effective improvement of drought tolerance in *Jatropha*.

Only a few studies used microprojectile bombardment for transformation in *J. curcas* (Pukayastha et al. 2010; Joshi et al. 2011; Jha et al. 2013). Jha et al. (2013) demonstrated inheritance and stable transformation of the desirable traits in the transgenic *J. curcas* plants with the *sbNHX1* gene encoding an active vacuolar Na^+/H^+ antiporter from an extreme halophyte *Salicornia brachiata*, showing enhanced tolerance at 200 mM NaCl compared to wild type.

8.9 Gene Silencing

Fire et al. (1998) demonstrated that dsRNA induced RNA silencing, a process called as RNAi. In the first plant case study of gene silencing via dsRNA-triggered RNAi technology, inverted repeats were used to overexpress dsRNAs that triggered highly efficient silencing of flower identity genes (Chuang and Meyerowitz 2000). RNAi can be initiated by sense, antisense, or inverted repeat transgenes or by viruses, since they all can produce dsRNAs in plants (Maghuly et al. 2006; Brosnan et al. 2008). When a sense or antisense transgene is expressed above threshold level, it may be recognized as foreign or aberrant RNA (abRNA) and amplified into dsRNA.

Utilizing developed protocols for *J. curcas* transformation and regeneration, it was possible to employ RNAi technology to produce transgenic *Jatropha* plants with enhanced characteristics. Patade et al. (2014) designed a RNAi construct for the curcin precursor gene to reduce its expression using a sense and antisense construct separated by an intron and cloned into the binary vector pRI101 AN. Results showed more than 98% reduction to undetectable level of curcin expression in transformed seedlings. Further, Ye et al. (2014) used the RNAi strategy to produce virus-resistant transgenic *J. curcas* by expressing a hairpin ds RNA targeting 5 genes of the DNA-A genome of a geminivirus. With 94% nucleotide identity, the transgenic plant showed broad resistance to geminiviruses.

To improve FA composition by increasing the content of monosaturated fatty acids (oleate) and decreasing linoleic acid, Qu et al. (2012) used RNAi for the enzyme 1-acyl-2-oleoyl-sn-glycero-3-phosphocholine delta 12-desaturase (FAD2) of *J. curcas* to suppress its expression. The generated marker-free transgenic *Jatropha* plants showed more than 78% oleic acid, an important strategy for improvement of *Jatropha* traits.

RNA silencing also involves RNA-dependent DNA methylation (RdDM), when DNA homologous to a dsRNA that causes gene silencing is methylated de novo (Gambino et al. 2009, 2010). SiRNAs homologous to the coding region of the target gene induce PTGS, which involves mRNA-specific degradation in the cytoplasm and, in some cases, methylation of the coding sequences (Baulcombe 2004; Gambino et al. 2010), which are involved in gene regulation and developmental processes in plant.

Yi et al. (2010) found that epigenetics might also play an important role in the development of *J. curcas*. They found epigenetic bands ranging from 22.42 to 25.58% in the studied populations and proved that phenotypes generated by DNA methylation are heritable and follow Mendelian segregation patterns, which also may influence gene expression. Further, salinity and methylation were studied in *J. curcas* plants, and results showed different methylation patterns compared to control plants (Mastan et al. 2012; Kanchanaketu et al. 2012). The portion of the genome immediately methylated by salt stress is increased in roots compared to leaf tissues (Mastan et al. 2012). In plants, DNA methylation occurs at both symmetric sites (CpG and CpNpG, where N is A, T, or C) and asymmetric sites (CpNpN), with several distinct RNA silencing pathways, causing silencing that is associated with siRNAs and/or DNA methylation (Baulcombe 2004; Gambino et al. 2010). In *J. curcas*, CpG methylation was found to be the most frequent cysteine methylation, which was followed by double cysteine methylation (Yi et al. 2010).

Besides the regulatory role in plant development, RNA silencing also functions as a natural antiviral defense mechanism, a process also known as VIGS (Kumar and Sarin 2013). Host RNA silencing machinery targets and processes the virus-derived dsRNA, resulting into virus-derived small RNAs (vsRNAs) (Duan et al. 2012). Although the vsRNAs are generally considered to be siRNAs, many of these molecules may be microRNAs (miRNAs) because their hairpins have great similarity to miRNA precursors (Dunoyer and Voinnet 2005).

The method for VIGS involves a partial cDNA sequence of a candidate gene cloned to a DNA or RNA plant virus. The recombinant virus can be transmitted systemically from cell to cell over a short distance, or through the phloem for longer distances, and alternate gene expression and phenotypes. VIGS is also an attractive strategy to investigate gene functions, especially for plant with long life cycle like *J. curcas*. Ye et al. (2009) infected *Jatropha* plant with recombinant RNA virus using *Tobacco rattle virus* vector with constructs carrying an insertion of partial sequence of different candidate genes. These results showed that VIGS can be used as a rapid and high-throughput reverse genetic system that has been successfully used to investigate gene functions in *J. curcas*. The effect of phenotypic alteration and gene expression changes in 13 *Jatropha* genes involved in FA and toxic biosynthesis and regulatory networks in leaf and fruit patterning were assayed 1–3 week after infiltration. The effect of two plasma membrane genes (*JcPIP1* and *JcPIP2*) in the recovery of salt and drought stresses in *J. curcas* was examined using VIGS-mediated gene knockdown (Jang et al. 2013). Although both genes have a positive role in response to both stresses, *JcPIP2* plays a more important role than *JcPIP1* in seedlings.

8.9.1 MicroRNA and Artificial MicroRNA

MiRNAs are small noncoding RNA molecules of approximately 21–24 nt in length that are key regulators of gene expression (Bartel 2004; Baulcombe 2004). After the discovery of the first members of the miRNA family, lin-4 and let-7 from *Caenorhabditis elegans* (Reinhart et al. 2000), several miRNAs have been identified in plants, animals, viruses, and green algae (*Chlamydomonas reinhardtii*).

In plants, miRNA biogenesis is confined to the nucleus, and only mature miRNA is transported to the cytoplasm (Kurihara and Watanabe 2004). The miRNA-loaded RNA-induced silencing complex (RISC) binds to the target mRNA in a sequence-specific manner that cleaves the target or prevents the translation inhibition (Selbach et al. 2008). Perfect or near perfect complementarity results in cleavage, whereas imperfect complementarity results in impaired translation (Hutvagner and Zamore 2002).

MiRNAs in plants have been found to regulate genes involved in plant growth and development, and biotic and abiotic responses (Jones-Rhoades et al. 2006; Bazzini et al. 2007). The majority of the early identified miRNAs are abundantly expressed in plants, and approximately 50% of their validated targets are transcription factors involved in leaf, shoot, root vascular and flower development, floral identity, flowering time, and hormone signaling (Mallory et al. 2004; Nikovics et al. 2006). Although some studies identified microRNA in *J. curcas* (Zeng et al. 2010; Wang et al. 2012; Gu et al. 2012; Vishwakarma and Jadeja 2013; Galli et al. 2014; Maghuly et al. 2014), the number of identified miRNA should be essentially increased. Wang et al. (2012) isolated 52 miRNAs by sequencing 2000 clones of leaves and seeds library of *J. curcas*, among them 46 miRNAs were novel; further, they analyzed the expression pattern by QPCR of miRNAs in different tissues. Results also showed that ten miRNAs could be involved in seed development or fatty acid biosynthesis because of their high expression. BLASTX analyses showed 16 target genes that were associated with three major gene ontology categories, out of them four targets were identified for JcmiR004. Expression of these four target genes was increased, and oil composition was modified by silencing the primary miRNA of JcumiR004. Vishwakarma and Jadeja (2013) sequence 46862 EST and searched 1596 genome survey sequence (GSS) for identifying 24 new potential miRNA in *J. curcas*. They predicted 78 potential target genes, which are related to 3 miRNA families, and encoding transcription factors regulating cell growth and development, as well as signaling and metabolism. Galli et al. (2014) used computational analysis for identification of 180 conserved miRNA involving in cellular structure, nuclear function, translation, transport, hormone synthesis, defense, and lipid biosynthesis. Bioinformatics predictions show that miR156, miR159, miR166, miR160, and miR395 have potential targets in the virus genomes (Pérez-Quintero et al. 2010). In silico analysis could also reveal that several tomato miR/miR* sequences exhibit propensity to bind to *Tomato leaf curl New Delhi* virus (ToLCNDV) associated (Naqvi et al. 2011).

Plant miRNAs can also accumulate to a higher extent in virus-infected plants (Tagami et al. 2007). For example, begomovirus infection leads to a decreased translation of genes involved in the development of plants (Amin et al. 2011). Further, identification of conserved and highly expressed families of plant miRs/miRs* was shown to have potential targets in the genome of both begomoviruses, important for antiviral defense in plant like *J. curcas* (Maghuly et al. 2014). *Nicotiana benthamiana* plants infected by begomoviruses ACMV, *Cabbage leaf curl virus*, *Tomato yellow leaf curl* virus, and *Cotton leaf curl Multan* virus showed an increase in the level of miR159, miR164, miR165/166, miR167, and miR168 (Amin et al. 2011).

The identification of miRNA and its target play an important role for understanding miRNA function. Further, because of miRNA ability to silence a target gene of interest, artificial miRNA-based technology can be used for reverse genetic to suppress a gene activity in plant. Research using genetically modified viruses and plants has been able to show that complementarities between a plant miRNA and the virus genome are enough for antiviral activity (Pérez-Quintero et al. 2010). This has made it possible to modify miRNA sequences to create artificial miRNAs (amiRNA) directed against any gene of interest resulting in its posttranscriptional silencing (Niu et al. 2006). It is important to select target sites of amiRNA that have optimal accessibility in order to obtain high resistance (Duan et al. 2008). Although off-target is less problematic in plants compared to animals, it is possible to avoid off-target in plant like

J. curcas where a draft of genome exists when planning experiments. Artificial pre-miR171a cleaves the 2b (silencing suppressor of CMV) and inhibits its gene expression conferring resistance to CMV (Qu et al. 2007). Multiple virus resistance by introducing amiRNA targeting more than one virus family is also possible.

8.10 Perspectives

Powerful forward and reverse genetic approaches facilitated the study of gene function in *J. curcas*; however, the choice of method to use largely depends on questions addressed and available facilities.

Targeted genome engineering or genome editing can be broadly defined as the ability to make directed changes to a genomic sequence, by removing or integrating a piece of DNA to produce a specific alleles of a gene in target plant. This was possible to achieve in certain model organisms by homologous recombination for many years, but reliable systems for higher plants have been difficult to implement (Belhaj et al. 2013). The development of synthetic, site-specific DNA double-strand breaks was restricted to zinc-finger nuclease (ZFN) launched genome editing and are changing the way geneticists study gene function. Furthermore, ZFNs are time and cost intensive. Transcription activator-like effectors (TALEs) address this problem; however, in the same manner like ZFNs, TALENs require the time-consuming assembly of small building blocks to create synthetic DNA-binding proteins (Lozano-Juste and Cutler 2014). The latest incarnation of various methods for targeted genome editing is the CRISPR/Cas9 system. The CRISPR/Cas system allows targeted cleavage of genomic DNA guided by a customizable small noncoding RNA, resulting in gene modifications by both, non-homologous end joining (NHEJ) and homology-directed repair (HDR) mechanisms (Belhaj et al. 2013). Achieving a fine-tuned control about the spatial and temporal expression of gene coding for desired characters would allow to carry out marker-assisted breeding to produce elite genotypes of *J. curcas*.

References

Abbott AG, Arús P, Scorza R (2007) Peach. In: Kole C (ed) Genome mapping and molecular breeding in plants, vol 4., Fruits and nutsSpringer, Berlin, Heidelberg, pp 137–156

Abdurakhmonov IY, Abdukarimov A (2008) Application of association mapping to understanding the genetic diversity of plant germplasm resources. Int J Plant Genom 2008:574927

Alonso JM, Ecker JR (2006) Moving forward in reverse: genetic technologies to enable genome-wide phenomic screens in *Arabidopsis*. Nat Rev Genet 7:524–536

Alipour A, Tsuchimoto S, Sakai H, Ohmido N, Fukui K (2013) Structural characterization of copia-type retrotransposons leads to insights into the marker development in a biofuel crop, *Jatropha curcas* L. Biotechnol Biofuels 6:129

Alipour A, Cartagen JA, Tsuchimoto S, Sakai H, Ohmido N et al (2014) Identification and characterization of novel Gypsy-type retrotransposons in a biodiesel crop, *Jatropha curcas* L. Plant Mol Biol Rep 32:923–930

Amin I, Patil BL, Briddon RW, Mansoor S, Fauquet CM (2011) A common set of developmental miRNAs are upregulated in *Nicotiana benthamiana* by diverse begomoviruses. Virol J 8:143

Asif MH, Mantri SS, Sharma A, Srivastava A, Trievedi I et al (2010) Complete sequence and organization of the *Jatropha curcas* (Euphorbiaceae) chloroplast genome. Tree Genet Genom 6:941–952

Baird NA, Etter PD, Atwood TS, Currey MC, Shiver AL et al (2008) Rapid SNP discovery and genetic mapping using sequenced RAD markers. PLoS One 3:e3376

Bartel DP (2004) MicroRNAs: genomics, biogenesis, mechanism, and function. Cell 116:281–297

Baulcombe D (2004) RNA silencing in plants. Nature 431:356–363

Bazzini AA, Hopp HE, Beachy RN, Asurmendi S (2007) Infection and coaccumulation of *Tobacco mosaic virus* proteins alter microRNA levels, correlating with symptom and plant development. Proc Natl Acad Sci USA 104:12157–12162

Beissinger TM, Hirsch CN, Sekhon RS, Foerster JM, Johnson JM et al (2013) Marker density and read depth for genotyping populations using genotyping-by-sequencing. Genetics 193:1073–1081

Belhaj K, Chaparro-Garcia A, Kamoun S, Nekrasov V (2013) Plant genome editing made easy: targeted mutagenesis in model and crop plants using the CRISPR/Cas system. Plant Methods 9:39

Brittaine R, Lutaladio N (2010) Jatropha: a smallholder bioenergy crop, the potential for propoor development. Integrated Crop Management, vol 8. FAO, Rome

Brosnan CA, McCallum EJ, Botella JR, Bernard JC (2008) Gene silencing, mutation analysis and

functional genomics. In: Kahl G (ed) The handbook of plant functional genomics. Meksem K. Wiley-VCH verlag GmbH & Co.KGaA, Weinheim
Brunner AL, Johnson DS, Kim SW, Valouev A, Reddy TE et al (2009) Distinct DNA methylation patterns characterize differentiated human embryonic stem cells and developing human fetal liver. Genome Res 19:1044–1056
Carvalho CR, Clarindo WR, Praça MM, Araújo FS, Carels N (2008) Genome size, base composition and karyotype of *Jatropha curcas* L., an important biofuel plant. Plant Sci 174:613–617
Castilho A, Heslop-Harrison JS (1995) Physical mapping of 5S and 18S–25S rDNA and repetitive DNA sequences in *Aegilops umbellulata*. Genome 38:91–96
Chuang CF, Meyerowitz EM (2000) Specific and heritable genetic interference by doublestranded RNA in *Arabidopsis thaliana*. Proc Natl Acad Sci USA 97:4985–4990
Comai L, Young K, Till BJ, Reynolds SH, Greene EA et al (2004) Efficient discovery of DNA polymorphisms in natural populations by EcoTILLING. Plant J 37:778–786
Costa GGL, Cardoso KC, Del Bem LEV, Lima AC, Cunha MAS et al (2010) Transcriptome analysis of the oil-rich seed of the bioenergy crop *Jatropha curcas* L. BMC Genom 11:462
da Camara Machado A, Frick NS, Kremen R, Katinger H, Laimer da Camara Machado M (1997) Biotechnological approaches to the improvement of *Jatropha curcas*. In: Proc of Jatropha 97, 23–27 February, Managua, Nicaragua
Davey JW, Hohenlohe PA, Etter PD, Boone JQ, Catchen JM et al (2011) Genome-wide genetic marker discovery and genotyping using next-generation sequencing. Nat Rev Genet 12:499–510
Dhakshanamoorthy D, Selvaraj R, Chidambaram ALA (2011) Induced mutagenesis in *Jatropha curcas* L. using gamma rays and detection of DNA polymorphism through RAPD marker. Crit Rev Biol 334: 24–30
Dhillon RS, Saharan RP, Jattan M, Rani T, Sheokand RN et al (2014) Molecular characterization of induced mutagenesis through gamma radiation using RAPD markers in *Jatropha curcas* L. Afr J Biotechnol 13:806–813
Doudrick RL, Heslop-Harrison JS, Nelson CD, Schmidt T, Nance WL et al (1995) Karyotype of Slash Pine (*Pinus elliottii* var. elliottii) using patterns of fluorescence in situ hybridization and fluorochrome banding. J Hered 86:289–296
Duan CG, Wang CH, Guo HS (2012) Application of RNA silencing to plant disease resistance. Silence 3:5
Duan CG, Wang CH, Fang RX, Guo HS (2008) Artificial MicroRNAs highly accessible to targets confer efficient virus resistance in plants. J Virol 82:11084–11095
Dunoyer P, Voinnet O (2005) The complex interplay between plant viruses and host RNA silencing pathways. Curr Opin Plant Biol 8:415–423
Elshire RJ, Glaubitz JC, Sun Q, Poland JA, Kawamoto K et al (2011) A robust, simple genotyping-by-sequencing (GBS) approach for high diversity species. PLoS ONE 6:e19379
Eswaran N, Parameswaran S, Balaji S, Anatharaman B, Sudhakar JT (2010) Nucleic acids of *Jatropha curcas* and its applications. WO2010095150 A3
Falque M, Santoni S (2007) Molecular markers and high-throughput genotype analysis. In: Morot-Gaudry JF, Lea P, Briat JF (eds) Functional plant genomics. Science, Enfield, p 714
Feldmann KA (1991) T-DNA insertion mutagenesis in *Arabidopsis*: mutational spectrum. Plant J 1:70–82
Fire A, Xu S, Montgomery MK, Kostas SA, Driver SE et al (1998) Potent and specific genetic interference by double-stranded RNA in *Caenorhabditis elegans*. Nature 391:806–811
Galli V, Guzman F, de Oliveira LFV, Loss-Morais G, Körbes AP et al (2014) Identifying MicroRNAs and transcript targets in *Jatropha*Seeds. PLoS One 9: e83727
Gambino G, Chitarra W, Maghuly F, Laimer M, Boccacci P et al (2009) Characterization of T-DNA insertions in transgenic grapevines obtained by *Agrobacterium*-mediated transformation. Mol Breed 24:305–320
Gambino G, Perrone I, Carra A, Chitarra W, Boccacci P et al (2010) Transgene silencing in grapevines transformed with GFLV resistance genes: analysis of variable expression of transgene, siRNA production and cytosine methylation. Transgen Res 19:17–27
Gelvin SB (2003) *Agrobacterium*-Mediated plant transformation: the biology behind the "Gene-Jockeying" tool. Microbiol Mol Biol R 67:16–37
Gilchrist EJ, Haughn GW (2005) TILLING without a plough: a new method with applications for reverse genetics. Curr Opin Plant Biol 8:211–215
Gilchrist EJ, Haughn GW, Ying CC, Otto SP, Zhuang J et al (2006) Use of EcoTILLING as an efficient SNP discovery tool to survey genetic variation in wild populations of *Populus trichocarpa*. Mol Ecol 15:1367–1378
Gomes KA, Almeida TC, Gesteira AS, Lôbo IP, Guimarães ACR et al (2010) ESTs from seeds to assist the selective breeding of *Jatropha curcas* L. for oil and active compounds. Genom Insights 3:29–56
Gong Z, Xue C, Zhang M, Guo R, Zhou Y et al (2013) Physical Localization and DNA Methylation of 45S rRNA Gene Loci in *Jatropha curcas* L. PLoS One 8: e84284
Gu KY, Chiam H, Tian DS, Yin ZC (2011) Molecular cloning and expression of heteromeric ACCase subunit genes from *Jatropha curcas*. Plant Sci 180:642–649
Gu K, Yi C, Tian D, Sangha JS, Hong Y et al (2012) Expression of fatty acid and lipid biosynthetic genes in developing endosperm of *Jatropha curcas*. Biotechnol Biofuels 5:1–15
Gupta P, Idris A, Mantri S, Asif MH, Yadav HK et al (2012) Discovery and use of single nucleotide

polymorphic (SNP) markers in *Jatropha curcas* L. Mol Breed 3:1325–1335

Hanin M, Paszkowski J (2003) Plant genome modification by homologous recombination. Curr Opin Plant Biol 6:157–162

Hansey CN, Vaillancourt B, Sekhon RS, de Leon N, Kaeppler SM et al (2012) Maize (*Zea mays* L.) genome diversity as revealed by RNA-sequencing. PLoS One 7:e33071

He Y, Pasapula V, Li X, Lu R, Niu B et al (2009) *Agrobacterium tumefaciens*-mediated transformation of *Jatropha curcas*: factors affecting transient transformation efficiency and morphology analysis of transgenic calli. Silvae Genet 58:123–128

Heller J (1996) Physic nut. *Jatropha curcas* L. promoting the conservation and use of underutilized and neglected crops. IPGRI, Gatersleben/International Plant Genetic Resource Institute, Rome, Italy

Henikoff S, Comai L (2003) Single-nucleotide mutations for plant functional genomics. Annu Rev Plant Physiol Plant Mol Biol 54:375–401

Heslop-Harrison JS (2000) Comparative genome organization in plants: From sequence and markers to chromatin and chromosomes. Plant Cell 12:617–635

Hillier LW, Marth GT, Quinlan AR, Dooling D, Gewell G et al (2008) Whole-genome sequencing and variant discovery in *C. elegans*. Nat Methods 5:183–188

Hirakawa H, Tsuchimoto S, Sakai H, Nakayama S, Fujishiro T et al (2012) Upgraded genomic information of *Jatropha curcas* L. Plant Biotechnol 29:123–130

Hutvagner G, Zamore PD (2002) A microRNA in a multiple-turnover RNAi enzyme complex. Science 297:2056–2060

Jang HY, Yang SW, Carlson JE, Ku YG, Ahn SJ. (2013) Two aquaporins of *Jatropha* are regulated differentially during drought stress and subsequent recovery. J Plant Physiol 170:1028–1038

Jha T, Mukherjee P, Datta MM (2007) Somatic embryogenesis in *Jatropha curcas* L. an important biofuel plant. Plant Biotechnol Rep 1:135–140

Jha B, Mishra A, Jha A, Joshi M (2013) Correction: developing transgenic *Jatropha* using the *SbNHX1* gene from an extreme halophyte for cultivation in saline wasteland. PLoS ONE 8:e71136

Jiang HW, Wu PZ, Zhang S, Song C, Chen YP et al (2012) Global analysis of gene expression profiles in developing physic nut (*Jatropha curcas* L.) seeds. PLoS One 7:e36522

Jones-Rhoades MW, Bartel DP, Bartel B (2006) MicroRNAS and their regulatory roles in plants. Annu Rev Plant Biol 57:19–53

Joshi M, Mishra A, Jha B (2011) Efficient genetic transformation of *Jatropha curcas* L. by microprojectile bombardment using embryo axes. Ind Crop Prod 33:67–77

Kanchanaketu T, Sangduen N, Toojinda T, Hongtrakul V (2012) Genetic diversity analysis of *Jatropha curcas* L. (Euphorbiaceae) based on methylation-sensitive amplification polymorphism. Genet Mol Res 11:944–955

Kim S, Veena Gelvin SB (2007) Genome-wide analysis of *Agrobacterium* T-DNA integration sites in the *Arabidopsis* genome generated under non-selective conditions. Plant J 51:779–791

King AJ, Li Y, Graham IA (2011) Profiling the developing *Jatropha curcas* L. seed transcriptome by pyrosequencing. Bioenerg Res 4:211–221

King AJ, Montes LR, Clarke JG, Affleck J, Li Y et al (2013) Linkage mapping in the oilseed crop *Jatropha curcas* L. reveals a locus controlling the biosynthesis of phorbol esters, which cause seed toxicity. Plant Biotechnol J 11:986–996

Kodym A, Afza R (2003) Physical and chemical mutagenesis. Methods Mol Biol 236:189–204

Kumar A, Sarin NB (2013) RNAi: a promising approach to develop transgenic plants against geminiviruses and insects. J Plant Physiol Pathol 1:1

Kumar N, Anand KGV, Pamidimarri DVNS, Sarkar T, Reddy MP et al (2010) Stable genetic transformation of *Jatropha curcas* via *Agrobacterium tumefaciens*-mediated gene transfer using leaf explants. Ind Crops Prod 32:41–47

Kumar N, Reddy MP (2010) Plant regeneration through the direct induction of shoot buds from petiole explants of *Jatropha curcas*: a biofuel plant. Ann Appl Biol 156:367–375

Kumar N, Reddy MP (2012) Thidiazuron (TDZ) induced plant regeneration from cotyledonary petiole explants of elite genotypes of *Jatropha curcas*: a candidate biodiesel plant. Ind Crops Prod 39:62–68

Kurihara Y, Watanabe Y (2004) Arabidopsis micro-RNA biogenesis through Dicer-like 1 protein functions. Proc Natl Acad Sci USA 101:1278–12753

Lambert P, Dicenta F, Rubio M, Audergon JM (2007) QTL analysis of resistance to sharka disease in the apricot (*Prunus armeniaca* L.) 'Polonais' Å∼ 'Stark Early Orange' F1 progeny. Tree Genet Genom 3:299–309

Li MR, Li HQ, Wu GJ (2006) Study on factors influencing *Agrobacterium*-mediated transformation of *Jatropha curcas*. J Mol Cell Biol 39:83–89

Li J, Li MR, Wu PZ, Tian CE, Jiang HW et al (2008a) Molecular cloning and expression analysis of a gene encoding a putative β-ketoacyl-acyl carrier protein (ACP) synthase III (*KAS III*) from *Jatropha curcas*. Tree Physiol 28:921–927

Li MR, Li HQ, Jiang HW, Pan XP, Wu GJ (2008b) Establishment of an *Agrobacterium*-mediated cotyledon disc transformation method for *Jatropha curcas*. Plant Cell Tissue Organ Cult 92:173–181

Li H-L, Guo D, Peng S-Q (2014) Molecular characterization of the *Jatropha curcas* JcR1MYB1 gene encoding a putative R1-MYB transcription factor. Genet Mol Biol 37:549–555

Lin J, Jin Y, Zhou X, Wang JY (2010) Molecular cloning and functional analysis of the gene encoding geranylgeranyl diphos-phate synthase from *Jatropha curcas*. Afr J Biotechnol 9:3342–3351

Lozano-Juste J, Cutler SR (2014) Plant genome engineering in full bloom. Trends Plant Sci 19:284–287

Lu WD, Wei Q, Tang L, Yan F, Chen F (2003) Induction of callus from *Jatropha curcas* and rapid propagation. Chin J Appl Environ Biol 9:127–130

Maghuly F, St Leopold, da Câmara Machado A, Borroto Fernandez E, Khan MA et al (2006) Molecular characterization of grapevine plants with GFLV resistance genes: II. Plant Cell Rep 25:546–553

Maghuly F, Schmöllerl B, Temsch E, Laimer M (2009) Genome size, Karyotype and FISH physical mapping of 45S and 5S genes in *Prunus subhirtella*. J Biotechnol 149:88–94

Maghuly F, Jankowicz J, Till B, Laimer M (2013a) The use of EcoTILLING for the genetic improvement of *Jatropha curcas* L. In: Bahadur B, Sujatha M, Carels N (eds) Jatropha, challenges for a new energy crop, vol 2. Springer, New York, pp 335–350

Maghuly F, Laimer M (2013b) *Jatropha curcas*, a biofuel crop: functional genomics for understanding metabolic pathways and genetic improvement. Biotechnol J 8:1172–1182

Maghuly F, Ramkat RC, Laimer M (2014) Virus versus host plant MicroRNAs: who determines the outcome of the interaction? PLoS ONE 9:e98263

Maghuly F, Jankowicz J, Pabinger F, Till B, Laimer M (2015) Geographic origin is not supported by the genetic variability found in a large living collection of *Jatropha curcas* with accessions from three continents. Biotechnol J 10:536–551

Mallory AC, Dugas DV, Bartel DP, Bartel B (2004) MicroRNA regulation of NAC-domain targets is required for proper formation and separation of adjacent embryonic, vegetative, and floral organs. Curr Biol 14:1035–1046

Mastan SG, Rathore MS, Bhatt VD, Yadav P, Chikara J (2012) Assessment of changes in DNA methylation by methylation-sensitive amplification polymorphism in *Jatropha curcas* L. subjected to salinity stress. Gene 508:125–129

Matijevic M, Bado S, Lagoda PJL, Forster BP (2013) Impact of induced mutations in plant breeding. In: Plant Genetics and Breeding Technologies, Plant Diseases and Resistance Mechanisms: Proceedings, February 18–20, Vienna, Austria. Medimond, Monduzzi Editore international Proceedings Division, Pianoro, Italy, pp 45–47

Mazumdar P, Basu A, Paul A, Mahanta C, Sahoo L (2010) Age and orientation of the cotyledonary leaf explants determine the efficiency of de novo plant regeneration and *Agrobacterium tumefaciens*-mediated transformation in *Jatropha curcas* L. S Afr J Bot 76:337–344

Mba C, Afza R, Bado S, Jain SM (2010) Induced mutagenesis in plants using physical and chemical agents. In: Davey MR, Anthony P (eds) Plant cell culture, essential methods. Wiley, Chichester, pp 111–130

Metzker ML (2009) Sequencing technologies the next generation. Nat Rev Genet 11:31–46

Misra M, Misra AN (2010) *Jatropha*: the biodiesel plant biology, tissue culture and genetic transformation—a review. Int J Pure Appl Sci Technol 1:11–24

Nakano Y, Ohtani M, Polsri W, Usami T, Sambongi K et al (2012) Characterization of the casbene synthase homolog from *Jatropha* (*Jatropha curcas* L.). Plant Biotechnol 29:185–189

Naqvi AR, Choudhury NR, Mukherjee SK, Haq QM (2011) *In silico* analysis reveals that several tomato microRNA/microRNA* sequences exhibit propensity to bind to *Tomato leaf curl virus* (ToLCV) associated genomes and most of their encoded open reading frames (ORFs). Plant Physiol Biochem 49:13–17

Natarajan P, Kanagasabapathy D, Gunadayalan G, Panchalingam J, Shree N et al (2010) Gene discovery from *Jatropha curcas* by sequencing of ESTs from normalized and full-length enriched cDNA library from developing seeds. BMC Genom 11:606

Ng PC, Henikoff S (2003) SIFT: predicting amino acid changes that affect protein function. Nucleic Acid Res 31:3812–3814

Ng SB, Turner EH, Robertson PD, Flygare SD, Bigham AW et al (2009) Targeted capture and massively parallel sequencing of 12 human exomes. Nature 461:272–276

Nikovics K, Blein T, Peaucelle A, Ishida T, Morin H et al (2006) The balance between the MIR164A and CUC2 genes controls leaf margin serration in Arabidopsis. Plant Cell 18:2929–2945

Niu QW, Lin SS, Reyes JL, Chen KC, Wu HW et al (2006) Expression of artificial microRNAs in transgenic *Arabidopsis thaliana* confers virus resistance. Nat Biotechnol 24:1420–1428

Ovando-Medina I, Sánchez-Gutiérrez A, Adriano-Anaya L, Espinosa-García F, Núñez-Farfán J et al (2011) Genetic diversity in *Jatropha curcas* populations in the State of Chiapas, Mexico. Diversity 3:641–659

Pan B-Z, Chen M-S, Ni J, Xu Z-F (2014) Transcriptome of the inflorescence meristems of the biofuel plant *Jatropha curcas* treated with cytokinin. BMC Genom 15:974

Pan J, Fu Q, Xu Z-F (2010) *Agrobacterium tumefaciens*-mediated transformation of biofuel plant *Jatropha curcas* using kanamycin selection. Afr J Biotechnol 9:6477–6481

Patade V, Khatri D, Kumar K, Grover A, Kumari M et al (2014) RNAi Mediated curcin precursor gene silencing in *Jatropha* (*Jatropha curcas* L.). Mol Biol Rep 41:4305–4312

Pérez-Quintero AL, Neme R, Zapata A, López C (2010) Plant microRNAs and their role in defense against viruses: a bioinformatics approach. BMC Plant Biol 10:138

Peters JL, Cnudde F, Gerats T (2003) Forward genetics and map-based cloning approaches. Trends Plant Sci 8:484–491

Popluechai S, Froissard M, Jolivet P, Breviario D, Gatehouse AMR et al (2011) *Jatropha curcas* oil body proteome and oleosins: L-form *JcOle3* as a potential phylogenetic marker. Plant Physiol Biochem 49:352–356

Purkayastha J, Sugla T, Paul A, Solleti S, Mazumdar P et al (2010) Efficient in vitro plant regeneration from shoot apices and gene transfer by particle bombardment in *Jatropha curcas*. Biol Plant 54:13–20

Purushothaman N, Madasamy P (2011) De novo assembly and transcriptome analysis of five major tissues of *Jatropha curcas* L. using GS FLX titanium platform of 454 pyrosequencing. BMC Genom 12:191

Qu J, Ye J, Fang R (2007) Artificial microRNA-mediated virus resistance in plants. J Virol 81:6690–6699

Qu J, Mao HZ, Chen W, Gao SQ, Bai NY et al (2012) Development of marker-free transgenic *Jatropha* plants with increased levels of seed oleic acid. Biotechnol Biofuels 5:10

Rajore S, Batra A (2005) Efficient plant regeneration via shoot tip explant in *Jatropha curcas*. J Plant Biochem Biotechnol 14:73–75

Reinhart BJ, Slack FJ, Basson M, Pasquinelli AE, Bettinger JC et al (2000) The 21-nucleotide let-7 RNA regulates developmental timing in *Caenorhabditis elegans*. Nature 403:901–906

Sabandar CW, Ahmat N, Jaafar FM, Sahidin I (2013) Medicinal property, phytochemistry and pharmacology of several *Jatropha* species (Euphorbiaceae): a review. Phytochemistry 85:7–29

Sato S, Hirakawa H, Isobe S, Fukai E, Kato M et al (2011) Sequence analysis of the genome of an oil-bearing tree, *Jatropha curcas* L. DNA Res 18:65–76

Sato S, Hirakawa H, Tsuchimoto S, Sakai H, Shibagaki N (2013) Genome structure of *Jatropha curcas* L. In: Bahadur B, Sujatha M, Carels N (eds) Jatropha, challenges for a new energy crop, vol 2. Springer, New York, pp 563–576

Schlötterer C, Tobler R, Kofler R, Nolte V (2014) Sequencing pools of individuals—mining genome-wide polymorphism data without big funding. Nat Rev Genet 15:749–763

Schmidt T, Heslop-Harrison JS (1998) Genomes, genes and junk: the large scale organization of plant chromosomes. Trends Plant Sci 3:195–199

Selbach M, Schwanhäusser B, Thierfelder N, Fang Z, Khanin R et al (2008) Widespread changes in protein synthesis induced by microRNAs. Nature 455:58–63

Semagn K, Bjornstad A, Ndjiondjop MN (2006) An overview of molecular marker methods for plants. Afr J Biotechnol 5:2540–2568

Shen J, Pinyopusarerk K, Bush D, Chen X (2012) AFLP-based molecular characterization of 63 populations of *Jatropha curcas* L. grown in provenance trials in China and Vietnam. Biomass Bioenergy 37:265–274

Sherchan DP, Thapa YB, Khadka JT, Tiwari TP (1989) Effect of green manure on rice production. In: PAC Occasional Paper—Pakhribas Agricultural Centre, Dhankuta Koshi Zone, Nepal, Italy, p 12

Silva-Junior O, Rosado T, Laviola B, Pappas M, Pappas G, Grattapaglia D (2011) Genome-wide SNP discovery from a pooled sample of accessions of the biofuel plant *Jatropha curcas* based on whole-transcriptome Illumina resequencing. BMC Proc 5:P57

Soomro R, Memon RA (2007) Establishment of callus and suspension culture in *Jatropha curcas*. Pak J Bot 39:2431–2441

Sriram P, Nalini E, Balaji S, Bhagyam A, Tangirala SJ. (2010) Identification of genes related to abiotic stress tolerance in *Jatropha curcas*. WO2010058428

Subroto AP, Utomo C, Darmawan C, Hendroko R, Liwang T (2014) Tissue culture media optimization and genetic transformation of *Jatropha curcas* genotype Jatromas cotyledon explants. Energy Procedia 47:15–20

Sudheer P, Rahman H, Mastan SG, Reddy MP (2010) Isolation of novel microsatellites using FIASCO by dual probe enrichment from *Jatropha curcas* L. and study on genetic equilibrium and diversity of Indian population revealed by isolated microsatellites. Mol Biol Rep 37:3785–3793

Sujatha M, Mukta N (1996) Morphogenesis and plant regeneration from tissue cultures of *Jatropha curcas*. Plant Cell Tiss Org Cult 44:135–141

Sujatha M, Makkar HPS, Becker K (2005) Shoot bud proliferation from axillary nodes and leaf sections of nontoxic *Jatropha curcas* L. Plant Growth Regul 47:83–90

Sun Q-B, Li L-F, Li Y, Wu G-J, Ge X-J (2008) SSR and AFLP markers reveal low genetic diversity in the biofuel plant *Jatropha curcas* in China. Crop Sci 48:1865–1871

Tagami Y, Inaba N, Kutsuna N, Kurihara Y, Watanabe Y (2007) Specific enrichment of miRNAs in *Arabidopsis thaliana* infected with *Tobacco mosaic virus*. DNA Res 14:227–233

Taketa S, Harrison G, Heslop-Harrison JS (1999) Comparative physical mapping of the 5S and 18S–25S rDNA in nine wild *Hordeum* species and cytotypes. Theor Appl Genet 98:1–9

Tang M, Liu X, Deng H, Shen S (2011) Over-expression of JcDREB, a putative AP2/EREBP domain-containing transcription factor gene in woody biodiesel plant *Jatropha curcas*, enhances salt and freezing tolerance in transgenic *Arabidopsis thaliana*. Plant Sci 181:623–631

Thomas OO (1989) Re-examination of the antimicrobial activities of *Xylopia aethiopica*, *Carica papaya*, *Ocimum gratissimum* and *Jatropha curcas*. Fitoterapia 60:147–155

Tsuchimoto S, Cartagena J, Khemkladngoen N, Singkaravanit S, Kohinata T et al (2012) Development of transgenic plants in *Jatropha* with drought tolerance. Plant Biotechnol 29:137–143

Vilanova S, Romero C, Abbott AG, Llácer G, Badenes ML (2003) An apricot (*Prunus armeniaca* L.) F2 progeny linkage map based on SSR and AFLP markers, mapping *plum pox* virus resistance and self-incompatibility traits. Theor Appl Genet 107:239–247

Vishwakarma NP, Jadeja VJ (2013) Identification of miRNA encoded by *Jatropha curcas* from EST and GSS. Plant Signal Behav 8:e23152

Vollmann J, Laimer M (2013) Novel and traditional oil crops and their biorefinery potential. In: Yang ST, El Enshasy HA, Thongchul N, Lo YM (eds) Bioprocessing technologies in integrated biorefinery for production of biofuels, biochemicals, and biopolymers from biomass. Wiley, Hoboken, pp 47–60

Wang CM, Liu P, Yi CX, Gu KY, Sun F et al (2011) A first generation microsatellite- and SNPbased linkage map of Jatropha. PLoS One 6:e3632

Wang CM, Liu P, Sun F, Li L, Liu P et al (2012) Isolation and identification of miRNAs in *Jatropha curcas*. Int J Biol Sci 8:418–429

Wang H, Zou Z, Wang S, Gong M (2013) Global analysis of transcriptome responses and gene expression profiles to cold stress of *Jatropha curcas* L. PLoS One 8: e82817

Wei Q, Lu WD, Liao Y, Pan SL, Xu Y et al (2004) Plant regeneration from epicotyl explants of *Jatropha curcas*. J Plant Physiol Mol Biol 30:475–478

Wei Q, Li J, Zhang L, Wu P, Chen Y et al (2012) Cloning and characterization of a β-ketoacyl-acyl carrier protein synthase II from *Jatropha curcas*. J Plant Physiol 169:816–824

Wen M, Wang H, Xia Zou M, Lu C et al (2010) Developmenrt of EST-SSR and genomic-SSR markers to assess genetic diversity in *Jatropha curcas* L. BMC Res Notes 3:42

Witkowska M, Ohmido N, Cartagena J, Shibagaki N, Kajiyama S, Fukui K (2009) Physical mapping of ribosomal DNA genes on *Jatropha curcas* chromosomes by multicolor FISH. Cytologia 74:133–139

Wu P, Zhou C, Cheng S, Wu Z, et al (2015) Integrated genome sequence and linkage map of physic nut (*Jatropha curcas* L.), a biodiesel plant. Plant J 81:810–821

Yadav HK, Ranjan A, Asif MH, Mantri S, Sawant SV et al (2011) EST-derived SSR markers in *Jatropha curcas* L.: development, characterization, polymorphism, and transferability across the species/genera. Tree Genet Genom 7:207–219

Yang J, Yang MF, Wang D, Chen F, Shen SH (2010) JcDof1, a Dof transcription factor gene, is associated with the light-mediated circadian clock in *Jatropha curcas*. Physiol Plant 139:324–334

Ye J, Qu J, Bui HT, Chua NH (2009) Rapid analysis of *Jatropha curcas* gene functions by virus induced gene silencing. Plant Biotechnol J 7:964–976

Ye J, Qu J, Mao H-Z, Ma Z-G, Rahman NEB et al (2014) Engineering geminivirus resistance in *Jatropha curcas*. Biotechnol Biofuels 7:149

Yi C, Zhang S, Liu X, Bui HT, Hong Y (2010) Does epigenetic polymorphism contribute to phenotypic variances in *Jatropha curcas* L.? BMC Plant Biol 10:259

Yin ZC, Wu LF, Mao HZ, Qiu CX (2010) *Jatropha curcas* curcin genes, tissue-specific promoters and generation of curcin-deficient transgenic *Jatropha* plants. WO2010140981

Yue GH, Sun F, Liu P (2013) Status of molecular breeding for improving *Jatropha curcas* and biodiesel. Renew Sustain Energy Rev 26:332–343

Zeng C, Wang W, Zheng Y, Chen X, Bo W et al (2010) Conservation and divergence of microRNAs and their functions in Euphorbiaceous plants. Nucleic Acids Res 38:981–995

Zerr T, Henikoff S (2005) Automated band mapping in electrophoretic gel images using background information. Nucleic Acids Res 33:2806–2812

Zhang L, Zhang C, Wu P, Chen Y, Li M et al (2014) Global analysis of gene expression profiles in physic nut (*Jatropha curcas* L.) seedlings exposed to salt stress. PLoS One 9:e97878

Zong H, Wang S, Ouyang C, Deng X, Li L et al (2010) Agrobacterium-mediated transformation of *Jatropha curcas* young leaf explants with lateral shoot-inducing factor (LIF). Int J Agric Biol 12:891–896

9 Flowering Genes and Homeotic Floral Gene Analysis in Jatropha

Nobuko Ohmido, Eri Makigano, Suguru Tsuchimoto and Kiichi Fukui

Abstract

Jatropha, *Jatropha curcas* L., is a monoecious species, which forms unisexual flowers. The unisexual flowers are produced on the same inflorescence with 10 to 30-folds higher ratio of male flowers to female flowers. The bias in the inflorescence limits Jatropha seed production. Flowering genes analysis would provide important knowledge about the multiple environmental and endogenous factors to inflorescence formation. To understand the flowering genes' characters determined by sequence and phylogenetic analysis could open the process of molecular breeding in future. In Jatropha genomic sequences, Jatropha orthologs of flowering-related genes consisting eight flowering regulators, including five flowering regulators: *CONSTANS* (*CO*), *FLOWERING LOCUS D* (*FD*), *FLOWERING LOCUS T* (*FT*), *LEAFY* (*LFY*), *SUPRESSOR OF OVEREXPRESSION OF CONSTANS 1* (*SOC1*), and three floral identity genes: *APETALA2* (*AP2*), *APETALA3* (*AP3*), *PISTILLATA* (*PI*), are well identified. Phylogenetic analysis demonstrated that Jatropha flowering-related genes are closely related to those of woody and herbaceous plants. The information of the flowering regulation generated in this study would be used for the qualitative and quantitative improvements in an oil material resource in Jatropha breeding.

9.1 Introduction

Jatropha, *Jatropha curcas* L., is a biomass plant belonging to the family Euphorbiaceae that is endemic to tropical America and originated in Mexico. It is grown commercially in tropical and subtropical Africa and Asia. Jatropha includes approximately 37% oil in the seeds. Furthermore, the quality of oil in the seeds is suitable for

N. Ohmido (✉) · E. Makigano
Graduate School of Human Development and Environment, Kobe University, Kobe 657-8501, Hyogo, Japan
e-mail: ohmido@kobe-u.ac.jp

S. Tsuchimoto · K. Fukui
Department of Biotechnology, Graduate School of Engineering,, Osaka University, Osaka, Japan

S. Tsuchimoto (ed.), *The Jatropha Genome*, Compendium of Plant Genomes,
DOI 10.1007/978-3-319-49653-5_9

production of biodiesel as they contain more than 75% unsaturated fatty acids (Sato et al. 2011). It has phorbol ester toxicity, so it does not compete with food crop and can be cultivated in the field unfavorable for cultivation. Oil can be extracted from Jatropha seeds, which is suitable for production of biodiesel fuel. However, because Jatropha has not been a major crop, it has been improved little by breeding. Especially, Jatropha flowering is an important factor to expect high seed production. In this chapter, we discuss about genes related to bud differentiation in male and female flower of Jatropha.

9.2 Genes Related to Flower Formation of Annual Plants

Plants flowering system is well studied in androgynous flowering plants including *Arabidopsis thaliana* (long-day plant) and *Oryza sativa* (short-day plant) as the distinct transition from vegetative to reproductive growth. Flower formation in plants is controlled by many factors such as individual's age and the nutritional condition, photoperiod, temperature, light wavelength, and environmental factors. Therefore, flower differentiation is controlled by a more complex signaling system, and interaction relation exists in the system. Primary concerning factors of flower differentiation from genetic analysis of the mutants are photoperiodism, gibberellin biosynthesis regulation, and autonomous control. Various genes react with each other to suppress and promote among them, and floral induction is clearly demonstrated in model plants, Arabidopsis and rice (see the reviews, Andrés and Coupland 2012). The genes related to the differentiation of the floral organs: petals, calyxes, stamens, and pistils , are also studied. These genes underlying formation of flower buds and floral organ also play important roles.

Representative flowering genes in the model plants, *FT* (*FLOWERING LOCUS T*), *FD* (*FLOWERING FLOCUS C*), and *CO* (*CONSTANS*), relates the photoperiodism. *SOC1* (*SUPRESSOR OF OVEREXPRESSION OF CONSTANS*) relates the gibberellin biosynthesis. *LFY* (*LEAFY*) and *AP2* (*APETALLA2*) in flower bud differentiation and *AP3* (*APETTALA3*) and *PI* (*PISTILLATA*) in flower organ differentiation are related, respectively (Thomas et al. 2006; Yu et al. 2004). The FT protein serves as a flowering inducing signal (Liu et al. 2013). *FT* is coded as a flowering integrator gene, florigen (Turck et al. 2008). By contrast, in rice, a short-day plant, the expression of the florigen gene *Hd3a* with *FT* ortholog is regulated by the induction under short-day conditions and repression under long-day conditions (see the review, Itoh and Izawa 2013.)

9.3 Phylogenic Analysis of Flowering and Homeotic Floral Genes in Jatropha

Jatropha is a monoecious species, which forms unisexual flowers, male and female flowers, separately in inflorescences. The unisexual flowers are produced on the same inflorescence, with the ratio of male flowers to female flowers ranging from 10:1 to 30:1 (Dehgan and Webber 1992). Therefore, the number of female flowers can significantly affect the yield of seed, and the functions of flowering-related genes including floral organ formation are important.

Many of the genes in Jatropha have been analyzed by the research group of Kazusa DNA Research Institute to perform genome analysis (Sato et al. 2011). The evolutionary relationships of proteins of MADS-box genes, *COL* genes, and flowering genes were analyzed using predicted amino acid sequences from different databases aligned with the program CLUSTALW (Ver.1.83, DDBJ) (Thompson et al. 1994). Gene sequence data decoded from the whole genome analysis is used for phylogenetic analysis in the comparison with the genes of euphorbiaceous and other phanerogamae plants described in Fig. 9.1. The use of phylogenetic analysis showing homology between genes has suggested the similarity of function. The application of this analysis method is able to identify many new genes in Jatropha.

To detect the interaction of the flowering genes of Jatropha with those of other plant

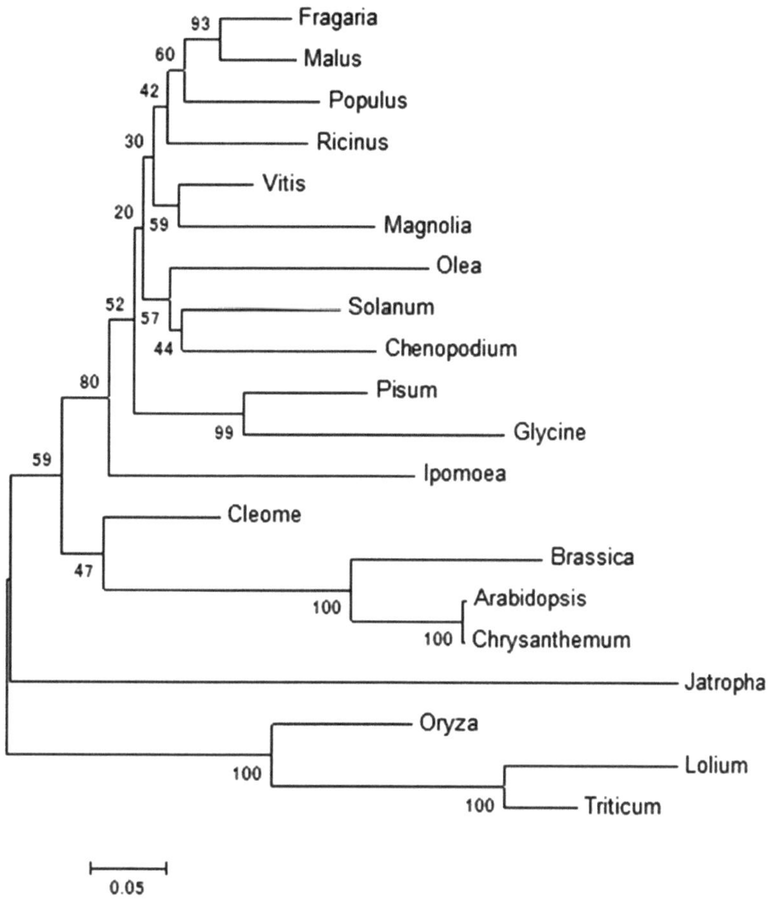

Fig. 9.1 Phylogenic analysis of flowering-related gene, *CONSTANS*

species, we investigated the phylogenetic analysis using open reading frame by 25% homology of the amino acid to the previously reported flowering genes by BLAST program (NCBI). In the results, five flowering regulators in Jatropha including *CONSTANS* (*CO*), *FLOWERING LOCUS D* (*FD*), *FLOWERING LOCUS T* (*FT*), *LEAFY* (*LFY*), and *SUPRESSOR OF OVER-EXPRESSION OF CONSTANS 1* (*SOC1*), and three floral identity genes including *APETALA2* (*AP2*), *APETALA3* (*AP3*), and *PISTILLATA* (*PI*) were investigated (Sato et al. 2011). The criteria to functionally identify genes demonstrated that high homology more than 80% sequences level to the functional genes queries are phylogenetic analyzed by neighbor-joining method (Saitou and Nei 1987). As the results, floral regulator *JcSOC1/AGL20* is nearly classified to the euphorbiaceous plant, genus *Hevea* and *Populus*. *JcFD* has highly homology to euphorbiaceous *Ricinus* and Arabidopsis clade that well functionally characterized in plants. *JcAP2* was nearly classified to *Ricinus*, *Vitis*, and *Populus*; *JcAP3* was nearly classified to *Cucumis* and legume. *JcPI* close to angiosperm plants, *Betula* and *Hydrandia* was classified to the woody plants clade. Floral meristem integrator *JcLFY* is closed to *Ricinus*, woody *Populus*, and *Salix*. Phylogenic analysis of *JcFT* gene is nearly classified to the euphorbiaceous, although the gene is evolutionary not close to Arabidopsis and rice *FT* genes. *FT* gene of the all plants used in this analysis might be similar function. Ye et al. (2014) demonstrated the identification of *FT* related and one *JcFT* genes. The results show 86% homology in the amino acid level in comparison with Arabidopsis *FT*. These data represents *Jatropha* gene sequences appear similar to the previously known flowering functional genes. Li et al. (2014) demonstrated that *JcFT* was expressed in all tissues and the highest expression level in female flowers, but young leaves. Moreover, Li et al. (2015) isolated six Jatropha genes that are highly similar to Arabidopsis *FT/TFL1* genes and investigated their expression patterns throughout the Jatropha developmental stages. The *FT/TFL1* gene family demonstrated the differential expression patterns in Jatropha developmental organs. The complex expression patterns of flowering-related genes in several organs promote to flowering patterns' diversification even in euphorbiaceous plants.

On the other hand, phylogenic gene analysis demonstrated *JcCO* does not belong to any of the clades—dicotyledonous plant and monocotyledonous (Fig. 9.1). *JcCO* gene might lose the function of original *CONSTANS*. Then the gene sequence was analyzed for phylogenic analysis in the comparison with *CONSTANS-LIKE* (*COL*) gene of Arabidopsis and rice. In the results of *COL* gene phylogenetic analysis, the gene is not belonging to the clade of flowering inducing genes or daylight response genes of Arabidopsis, rice, and other plants. However, the transcription factor (B-box and CCT motif) conserved in *CO* was also detected in *JcCO*. This suggests that *JcCO* is not directly involved in flowering regulation, although *JcCO* has all CO-conservative domains as a transcription factor including B-box and CCT motif. There were other CO homologues in the Jatropha genome suggests that different components participate in response to photoperiod.

Many plant species are manipulated by multiple MADS-box genes that belong to a range of functionally divergent subfamilies. They are transcription factors involving the gene sequences of a MADS-box domain, and it is mainly controlling a flower formation (Alvarez-Buylla et al. 2000). MADS-box gene mutant, a part of organ of a flower changes to another organ, and it is known that the double petals mutant from stamens to petals (Becker and Theissen 2003). Moreover, the MADS gene is conserved not only in plants, but in an animal and fungi, and is participating in organic formation (Alvarez-Buylla et al. 2000). K-box gene sequences known as a DNA binding domain, and MADS-box is called MIKC type II in the MADS gene cluster (Rijpkema et al. 2007). As homeotic regulating genes, potential MIKC type II MADS-box genes families were also investigated in Jatropha using amino acid sequences of *PI* as a query in *Arabidopsis*. A total of 28 candidate MADS-box genes (*JcMADS01–JcMADS28*) were identified (Fig. 9.2). The isolated 28 potential MIKC genes were compared with the similar functional genes of the other plants analyzed. The phylogenetic analysis classified these genes into nine subfamilies, *i.e*, *SEPALLATA* (*SEP*), *APETALA1*, *SQUAMOSA/AP1*(*SQUA*), *APETALA*, *DEFICIENS* (*AP/DEF*),*PISTILLATA*, *GLOBOSA* (*PI/GLO*), *AGAMOUS-LIKE 12* (*AGL12*), *AGAMOUS* (*AG*), *AGAMOUS-LIKE 15* (*AGL15*), *AGAMOUS-LIKE 62* (*AGL62*), and*SHORT VEGETATIVE PHASE* (*SVP*). The functions were estimated from the Jatropha sequence corresponding to as same as previously reported nine genes in other plants.

SVP controls negatively regulating flowering time as the expression of a floral integrator. One *SVP* gene in Arabidopsis and rice is identified in annual herbaceous plant (Hartmann et al. 2000; Lee et al. 2008). In poplar which is a model plant of the woody plant, eight paralogs *SVP* gene family are identified. Four and six paralogs of *SVP* are identified in peach and kiwi tree, respectively (Bielenberg et al. 2004, 2008; Wu et al. 2011). Eight paralogs of *SVP* copies have been found in *MIKC* type II MADS-box genes of *Populus trichocarta*. Therefore, the expression timing and the function of these genes are analyzed, and it has been reported that *SVP* paralogs also differ among them. Since, we identified five *SVP* paralogs, which are suggesting amplification and functional diversification of the *SVP* gene in woody plants.

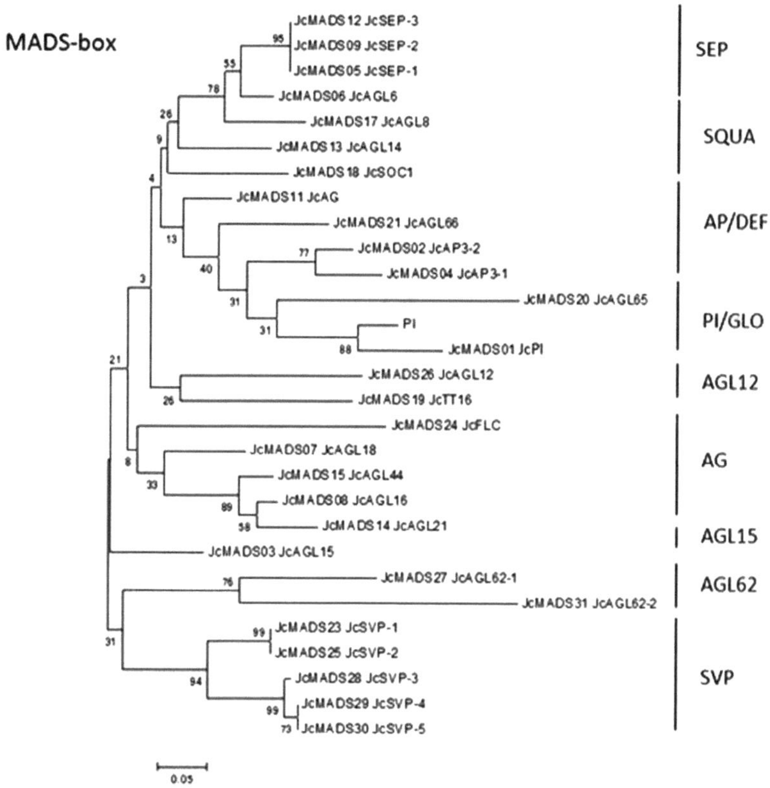

Fig. 9.2 Phylogenic analysis of MADS-box genes family

9.4 Interaction of Signal Transduction of Jatropha Flowering Systems

Plant flowering is controlled by environmental factor, such as day length, temperature, and wavelength, additionally plant physiological conditions, such as growth stages and the nutrient state. Therefore, flowering is controlled by two or more interaction of signal transduction systems, they are interacted with each other and it should be complicated (Fig. 9.3).

The genetic analysis of the mutant genes of which control flowering process is clearly promoted by three pathways in Arabidopsis. The first pathway is called a photoperiodism. It relates the function to induce the flowering formation on both of long-day plants, *e.g.*, Arabidopsis, and short-day plants *e.g.*, rice carried out under long- or short-day period plants, respectively. In this pathway, *CO*, *FT*, and *FD* have influenced. In both Arabidopsis and rice, *CO* recognizes photoperiod and promotes *FT* gene expression and forms a complex with *FD* in shoot apices to induce the flowering initiation. *ft* and *co* mutant in Arabidopsis and rice, respectively, even if they are under the considerable photoperiod condition for each flowering, it has flowering delayed. On the other hand, *FT* and *CO* overexpression lines initiate earlier flowering than wild type, regardless of the photoperiod (Kojima et al. 2002; Abe et al. 2005; Tamaki et al. 2007).

The second pathway is relating the plant hormone gibberellin. Arabidopsis flowering is promoted by gibberellin applied. The loss of a part of biosynthetic pathway of gibberellin mutant *ga-1* in Arabidopsis shows the flowering time delay (Nakajima et al. 2006).

The third pathway is the autonomous pathway controlling flowering time in plants. Arabidopsis have focused on daily light duration, or photoperiod, and transient exposure to winter-like

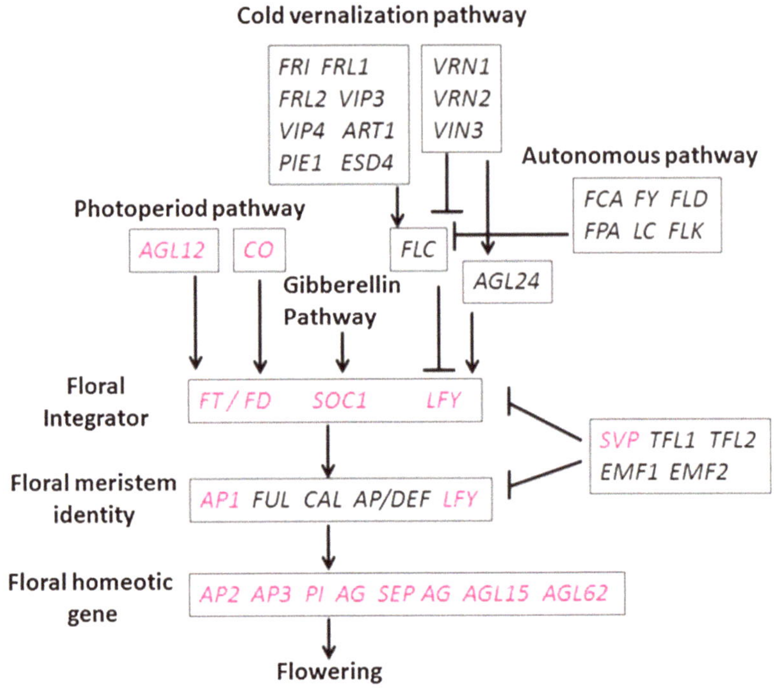

Fig. 9.3 Interaction and phylogenic analysis of flowering-related genes in Jatropha

temperatures, or vernalization (Blázquez et al. 2003). The thermosensory regulators of the floral repressor *FLC* are negatively regulated the induction of flowering by *FT* (Simpson and Dean 2002). The downstream of *FT/FD* gene are related the floral regulation genes *LFY* and *AP1*. Furthermore, *AP2*, *AP3*, and *PI* belonging to the MADS-box gene family on the downstream regulation of *LFY* and *AP1* induce the floral organs formation and flowering (Alvarez-Buylla et al. 2000).

9.5 The Internal Forms of the Male Flower and Meiosis Cell

Morphological analyses of the flowering formation are promoted in the immature male buds. Inflorescence of soaked flower buds is fixed into 4% paraformaldehyde solution. Embedded the bud samples in Technovit 7100 resin are sliced into 3 μm with the microtome. The sliced sections are stained with Toluindine Blue O. Meiosis stage in the immature anther is stained with 4′,6-diamidino-2-phenylindole (DAPI). A pistil, three carpels wraps, three ovules and five stamens of the combination of two anthers, and a filament are involved in a male bud.

Flower formation was a cyme inflorescence and a sympodial branching style. Female and male flowers are in a ratio of about 1:30, and the timing of the settlement of flower buds was greatly different in both flowers. Female flowers involving three ovules, which are composite pistil consisting of three carpels and five petals, and five calyxes. Male flowers involving 10 anthers, five filaments, five petals, and five calyxes. Stamens have two anthers and one filament.

Figure 9.4 demonstrated male flower bud development. Immature stamens growth was observed in buds of 1.06–1.40 mm in diameter (Fig. 9.4a, b). A shape of anther was clearly observed in buds of 1.61–1.68 mm in diameter (Fig. 9.4c, d). About the internal structure, male flower C developed to male flower D. The meiotic cell division in anther was observed in approx. 1.7–2.0 mm male flowers. Pollens

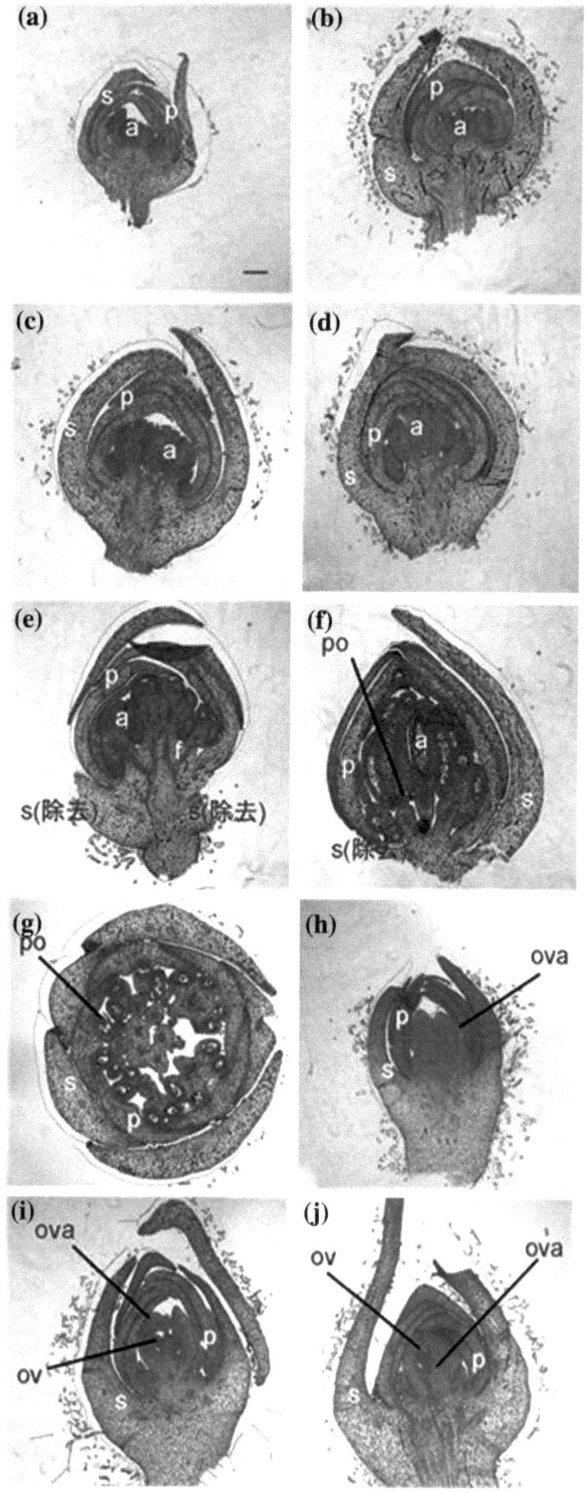

Fig. 9.4 Flower bud morphology in each development stage. *a* anther, *po* pollen, *f* filament, *ova* ovary, *ov* ovule, *p* petal, and *s* sepal. **a–g** Male flower. *Bar* 200 μm

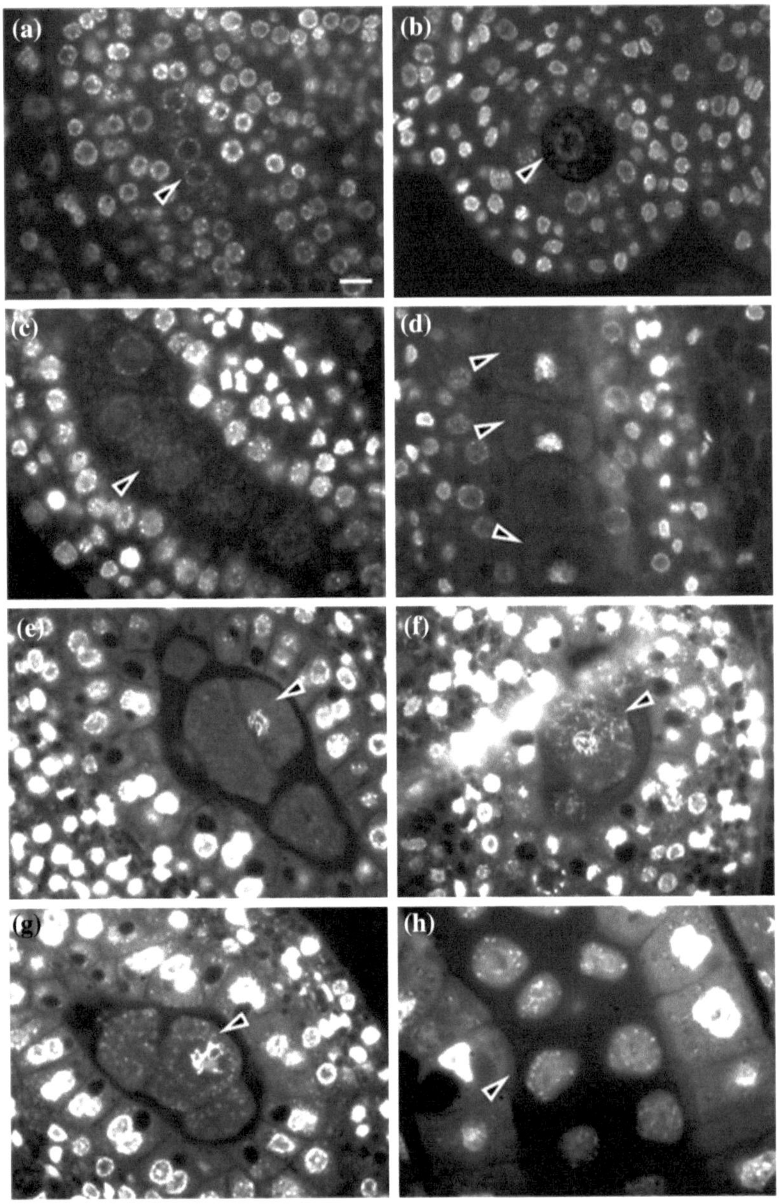

Fig. 9.5 Meiosis cell in anther of male flower. *Arrowhead* shows the meiosis cell. **a–c** Premature pollen mother cell, **d** leptotene, **e**, **f** pachytene, **g** diakinesis, and **h** pollen grains. *Bar* 10 μm

development starts from both ends to the center in the anther (Fig. 9.5a, b). In this stage, cytoplasm regions in pollen mother cell are enlarged (Fig. 9.5c). In male flower (Fig. 9.4e), the meiotic division begins in the cell inside, and it was observed leptotene (Fig. 9.5d) of the earlier period of meiosis, pachytene (Fig. 9.5e, f), diplotene, and diakinesis (Fig. 9.5g) stages. The pollen cells are matured (Fig. 9.5h). Female flower development was analyzed in Fig. 9.4h–j.

9.6 Conclusions

We have achieved the gene analysis of the flowering control and have evaluated the relationship with other plants relating to flowering system of Jatropha. Genome analysis could improve the agricultural characters related to the biomass production, e.g., genetically modified (GM) and/or marker-assisted selection. Ye et al.

(2014) to produce transgenic Jatropha introduced *FT* gene by transformation method using Agrobacterium transformation. Manipulation of Jatropha flowering time can lead to yield increase. A rapid breeding technology would be applied for Jatropha. These information and the poetical of flowering control would contribute the breeding of the biofuel plant for sustainable oil.

Acknowledgments This study was funded by the Joint Research Program of Arid Land Research Center, Tottori University to N.O.

References

Abe M, Kobayashi Y, Yamamoto S, Daimon Y, Yamaguchi A, Ikeda Y, Ichinoki H, Notaguchi M, Goto K, Araki T (2005) FD, a bZIP protein mediating signals from the floral pathway integrator FT at the shoot apex. Science 309:1052–1056

Alvarez-Buylla ER, Pelaz S, Liljegren SJ (2000) An ancestral MADS-box duplication occurred before the divergence of plants and animals. Proc Natl Acad Sci USA 97:5328–5333

Andrés F, Coupland G (2012) The genetic basis of flowering responses to seasonal cues. Nat Rev Gene 13:627–639

Becker A, Theissen G (2003) The major clades of MADS-box genes and their role in the development and evolution of flowering plants. Mol Phylogenet 29:464–489

Bielenberg DG, Wang Y, Fan S, Reighard GL, Scorza R, Abbott AG (2004) A deletion affecting several gene candidates is present in the Evergrowing peach mutant. J Heredity 95:436–444

Bielenberg DG, Wang Y, Li ZG, Zhebentyayeva T, Fan SH, Reighard GL, Scorza R, Abbott AG (2008) Sequencing and annotation of the evergrowing locus in peach [*Prunus persica* (L.) Batsch] reveals a cluster of six MADS-box transcription factors as candidate genes for regulation of terminal bud formation. Tree Genetics and Genomes 4:495–507

Blázquez MA, Ahn JH, Weigel D (2003) A thermosensory pathway controlling flowering time in *Arabidopsis thaliana*. Nat Genet 33:168–171

Dehgan B, Webster G (1992) Morphology and infrageneric relationships of the genus *J. curcas*. University of California Press, Berkeley, CA, USA

Hartmann U, Ho¨hmann S, Nettesheim K, Wisman E, Saedler H, Huijser P (2000) Molecular cloning of SVP: a negative regulator of the floral transition in Arabidopsis. Plant J 21:351–360

Itoh H, Izawa T (2013) The coincidence of critical day length recognition for florigen gene expression and floral transition under long-day conditions in rice. Mol Plant 6:635–649

Kojima S, Takahashi Y, Kobayashi Y, Monna L, Sasaki T, Araki T, Yano M (2002) Hd3a, a rice ortholog of the Arabidopsis FT gene, promotes transition to flowering downstream of Hd1 under short-day conditions. Plant Cell Physiol 43:1096–1105

Lee S, Choi SC, An G (2008) Rice SVP-group MADS-box proteins, OsMADS22 and OsMADS55, are negative regulators of brassinosteroid responses. Plant J 54:93–105

Li C, Luo L, Fu Q, Niu L, Xu1 ZF (2014) Isolation and functional characterization of *JcFT*, a *FLOWERING LOCUS T* (*FT*) homologous gene from the biofuel plant *Jatropha curcas*. BMC Plant Biology 14:125

Li C, Luo L, Fu Q, Niu L, Xu ZF (2015) Identification and characterization of the FT/TFL1 gene family in the biofuel plant *Jatropha curcas*. Plant Mol Biol Rep 33:326–333

Liu L, Zhu Y, Shen L, Yu H (2013) Emerging insights into florigen transport. Curr Opin Plant Biol 16:607–613

Nakajima M, Shimada A, Takashi Y, Kim YC, Park SH, Ueguchi-Tanaka M, Suzuki H, Katoh E, Iuchi S, Kobayashi M, Maeda T, Matsuoka M, Yamaguchi I (2006) Identification and characterization of Arabidopsis gibberellin receptors. Plant J 46:880–889

Rijpkema AS, Gerats T, Vandenbussche M (2007) Evolutionary complexity of MADS complexes. Curr Opin Plant Biol 10:32–38

Saitou N, Nei M (1987) The neighbor-joining method: a new method for reconstructing phylogenetic trees. Mol Biol Evol 4:406–425

Sato S, Hirakawa H, Isobe S, Fukai E, Watanabe A, Kato M, Kawashima K, Minami C, Muraki A, Nakazaki N, Takahashi C, Nakayama S, Kishida Y, Kohara M, Yamada M, Tsuruoka H, Sasamoto S, Tabata S, Aizu T, Toyoda A, Shin-i T, Minakuchi Y, Kohara Y, Fujiyama A, Tsuchimoto S, Kajiyama S, Makigano E, Ohmido N, Shibagaki N, Cartagena JA, Wada N, Kohinata T, Atefeh A, Yuasa S, Matsunaga S, Fukui K (2011) Sequence analysis of the genome of an oil-bearing tree, *Jatropha curcas* L. DNA Res 18:65–76

Simpson GG, Dean C (2002) Arabidopsis, the Rosetta stone of flowering time? Science 296:285–289

Tamaki S, Matsuo S, Wong HL, Yokoi S, Shimamoto K (2007) Hd3a protein is a mobile flowering signal in rice. Science 316:1033–1036

Thomas B, Carré I, Jackson S (2006) Photoperiodism and flowering. In: Jordan BR (ed) The molecular biology and biotechnology of flowering. C a B Intl, Hardback, p 416

Thompson JD, Higgins DG, Gibson TJ (1994) CLUSTAL W: improving the sensitivity of progressive multiple sequence alignment through sequence weighting, position-specific gap penalties and weight matrix choice. Nucleic Acids Res 122:4673–4680

Turck F, Fornara F, Coupland G (2008) Regulation and identity of florigen: FLOWERING LOCOS T moves center stage. Annu Rev Plant Biol 59:573–594

Wu RM, Walton EF, Richardson AC, Wood M, Hellens RP, Varkonyi-Gasic E (2011) Conservation and divergence of four kiwifruit SVP-like MADS-box genes suggest distinct roles in kiwifruit bud dormancy and flowering. J Exp Bot 63:797–807

Ye J, Geng Y, Zhang B, Mao H, Qu J, Chua N-H (2014) The Jatropha FT ortholog is a systemic signal regulating growth and flowering time. Biotechnol Biofuels 7:91–101

Yu H, Ito T, Zhao Y, Peng J, Kumar P, Meyerowitz EM (2004) Floral homeotic genes are targets of gibberellin signaling in flower development. Proc Natl Acad Sci USA 101:7827–7832

The Genome-Wide Association Study 10

Haiyan Li, Suguru Tsuchimoto, Kyuya Harada
and Kiichi Fukui

Abstract

Jatropha (*Jatropha curcas* L.) is a promising biofuel producer for supporting the fight against global warming and has thus attracted the attention in recent years. However, as yet, we do not have a deep and comprehensive understanding of this species. Genome-wide association studies (GWASs) have been used to improve our understanding of various plant species. Therefore, it has been suggested that a GWAS on jatropha may help to identify several quantitative trait loci or candidate genes. Such a study would require the correct selection of material, specific simple sequence repeat markers, and a single nucleotide polymorphism array designed from the published whole jatropha genome and transcriptome sequences, as well as trait data collection and statistical model selection. It would provide a comprehensive reference and a scientific basis for jatropha breeding, particularly those with the high and stable oil yield and improved stress tolerance.

10.1 Introduction

Plant breeding involves the creation, selection, and fixation of superior plant phenotypes for the further development of improved varieties that meet the requirements of consumers. To answer the increasing demand for food and energy, there has been a continuous and urgent requirement to develop improved varieties of many kinds of crops; this situation has recently become even more severe due to the limited availability of land for cultivation and climate changes. Plant breeding generally uses one of two approaches: the construction of genetically modified plants

H. Li · K. Fukui (✉)
Department of Biotechnology, Graduate School of Engineering, Osaka University, Suita, Osaka 565-0871, Japan
e-mail: kfukui@bio.eng.osaka-u.ac.jp

S. Tsuchimoto (✉) · K. Harada
Plant Bioengineering for Bioenergy Laboratory, Graduate School of Engineering, Osaka University, Suita, Osaka 565-0871, Japan
e-mail: tsuchimoto@bio.eng.osaka-u.ac.jp

S. Tsuchimoto (ed.), *The Jatropha Genome*, Compendium of Plant Genomes,
DOI 10.1007/978-3-319-49653-5_10

(GM plants) or the use of traditional plant breeding strategies.

In GM plants, the adequate genetic information is introduced artificially to enhance resistance to certain pests, diseases, environmental stresses and chemical treatments, or to improve the biological parameters of the plants, such as the nutrient profile. The establishment of transformation methods for various plant species, including important crops such as rice (*Oryza sativa*), soybean (*Glycine max*), and cotton (*Gossypium* spp.), has allowed the successful creation of many kinds of transgenic crops, and the cultivation of GM crops has expanded rapidly in recent years (James 2011).

Traditional breeding methods involve the selection of plants that exhibit the desired traits and hybridization between different lines to combine these desired traits into one plant. The widespread application of molecular markers in plant breeding has revolutionized the pace and precision of selection in the progenies. One example is the linkage mapping of quantitative trait loci (QTLs) in plants, which is a well-established genetic method; for example, the key components of the pathway controlling flowering time in *Arabidopsis* (*Arabidopsis thaliana*) have been dissected using QTL mapping (Alonso-Blanco et al. 1998; Kowalski et al. 1994; Clarke et al. 1995). However, this method has two major limitations: only the limited allelic diversity that segregates in the particular F_2 population or a recombinant inbred line (RIL) population can be used (Borevitz and Nordborg 2003); the limited amount of recombination occurs during the establishment of an F_2 or RIL population, resulting in a low mapping resolution (Korte and Farlow 2013).

To overcome these limitations, genome-wide association studies (GWASs) are becoming increasingly popular, as they directly examine the statistical association between genetic markers and phenotypes in a broader germplasm context than linkage mapping, enabling us to analyze the genes or genomic regions that are responsible for complex traits and to identify superior alleles. Therefore, in this review, we outline the current status of and issues with GWAS in plants and discuss the need to use GWAS in the breeding of the important biofuel producer, jatropha.

10.2 Molecular Markers for GWAS

A comprehensive GWAS should rely on high-density genetic markers across several genomes that can guarantee the identification of significant associations between marker genotypes and the phenotypes of interest. The majority of GWASs have been conducted using single nucleotide polymorphism (SNP) markers, and some association studies have used simple sequence repeat (SSR) markers (Wang et al. 2015b; Zhang et al. 2015c).

10.2.1 Single Nucleotide Polymorphism Markers

SNP is a single base pair change in the DNA sequence that can be effectively used for germplasm fingerprinting, genetic linkage analysis, population studies, and GWAS. SNP analysis is widely recognized as a powerful marker technique, largely owing to the high frequency of SNPs within the genome and their high conservation throughout evolution or within a population. The use of SNPs is also more efficient, faster, and relatively less expensive than the other DNA markers. Furthermore, the decreasing cost of high-throughput sequencing has allowed GWASs that utilize SNPs (usually an SNP array) to be conducted on a very simplified genotyping platform.

To be successful, a GWAS that is based on SNPs usually requires an efficient and high-quality SNP array. In a recent GWAS of salinity tolerance in rice by Kumar et al. (2015), the genotyping array was first designed by selecting SNPs from differentially expressed transcripts in two *indica* rice genotypes, and these gene sequences were then co-aligned with the whole-genome sequencing data of six contrasting rice genotypes; they then selected SNPs that were evenly distributed across 12 rice chromosomes with a physical interval of <100 kb and finally used the selected SNPs with a high design score to make an SNP array. This specific SNP array had the higher success and lower missing genotype rates, suggesting its reliability and utility for high-throughput SNP genotyping in rice. In addition, owing to the expressed transcripts in rice, the SNPs in the array were annotated as either coding SNPs or regulatory SNPs, indicating that they should have the higher potential to affect gene function and, thus, be more informative.

10.2.2 Simple Sequence Repeat Markers

SSRs or microsatellite markers are short motifs (2–6 nucleotides long) that make up a significant portion of higher eukaryote genomes. SSR markers are codominant and have a high level of reproducibility. Furthermore, they are multi-allelic and so have higher levels of genetic variation or are more informative than the other molecular markers. SSR markers have been widely applied in genotyping, diversity assessments, and GWAS. For example, a GWAS was recently performed that aimed to improve the taste of tomato (*Solanum lycopersicum*) by identifying the association among 182 SSR markers and 28 volatiles of 174 tomato accessions, which detected 125 significant associations, several of which were co-localized with the previously identified QTLs (Zhang et al. 2015c).

SSR markers can be identified from both genome and transcript sequences. The latter tend to be rich in informative markers, because such markers are expressed in the organism and likely to be present in the genes, allowing us to expect a direct association between the molecular markers and the phenotypes. Considerable transferability of microsatellite markers has been observed in many plant species, with SSRs that are developed from expressed sequences generally exhibiting higher transferability than genomic SSRs. In general, SSRs in coding regions are less abundant than those in non-coding regions (Liu et al. 2013). It remains unclear how the repeat number variation of an SSR motif affects gene action, but there have been reports that the repeat number variation in the 5′ untranslated region (UTR) of the waxy gene affects the amylase content in rice (Bao et al. 2002) while that in the ribosomal protein genes affects fertilization in maize (*Zea mays*; Dresselhaus et al. 1999). It has been noted that the SSRs may reside in genomic regions that are involved in development, adaptation, or evolution. Thus, the expansion or contraction of SSRs in coding regions or UTRs would inactivate the gene via frameshift mutation or the destabilization of mRNA, and SSRs within introns would induce changes in their regulation and/or splicing.

10.2.3 Other Molecular Markers

Other kinds of DNA molecular markers are also widely used in plant genotyping and gene mapping. Restriction fragment length polymorphism (RFLP) markers were one of the first DNA-based genetic markers to be used to detect variation among individuals. This technique is based on the distance between restriction sites, and the specific banding patterns of the RFLPs are visualized by hybridization with a labeled probe. However, RFLP analysis is more complicated and expensive than the other types of analyses and requires more time and expertise, making it less useful nowadays.

The widespread applications of the polymerase chain reaction (PCR) led to a shift from RFLP markers to PCR-based DNA markers. The random amplified polymorphic DNA (RAPD) marker is one of the most popular PCR-based markers, the primers of which are usually short (10 base pairs long) and consist of random sequences. Similarly, amplified fragment length polymorphism (AFLP) markers can be assayed quickly using PCR; however, AFLP analysis can generate huge quantities of polymorphisms and so may require automated analysis with a computer.

Both RAPD and AFLP are dominant markers, however, which are less reliable and informative than codominant markers such as SSRs.

Relatively few GWAS studies have been conducted using RAPD or AFLP markers. For example, Hansen et al. (2001) assessed the linkage disequilibrium (LD) in sea beets (*Beta vulgaris* subsp. *maritima*) using 137 mapped AFLP markers with 106 plants and found that two AFLP markers showed significant LD with the B gene, which was involved in the annual growth habit.

10.3 GWAS in Plants

Unlike transgenic techniques, which require an understanding of the gene function and in some cases the biological pathway of the traits, GWAS can be used to find novel QTLs or candidate genes for various traits simultaneously over a large scale, which can lead to the identification of useful genes that make different contributions and can allow their functional characterization. Biological processes, such as stress tolerance, are controlled by a network that includes a considerable number of genes with various functions. GWAS can help us not only to identify the major genes that are involved in these processes, but also in some cases to rule out specific genetic associations, giving us a new insight into the functional mechanism of a particular process. This then serves as a basis to examine the genetic architecture of the trait and provides us with a new genetic approach by engineering sets of functional genes. Furthermore, GWAS can provide a genetic basis for selecting parents for QTL analysis and can suggest candidates for mutagenesis or genetic transformation, including genome editing.

Once a plant population with genetic diversity has been genotyped, it can be used for various GWASs by phenotyping any trait of interest. Such continuous GWASs can then be used to generate a comprehensive map of QTLs and candidate genes of the plant in the mapped population. In plants, GWASs have been employed to improve many crop species, and sets of important QTLs and candidate genes have been identified that will promote breeding programs through marker-assisted selection.

Arabidopsis is an ideal model plant for genetic and molecular studies due to its self-fertility, short generation time, small genome size, and wide geographical distribution. Recently, Morrison and Linder (2014) identified a total of 55 markers associated with germination traits in *Arabidopsis*, which were linked to 71 unique genes, four of which (*ZIGA4*, *PS1*, *TOR*, and *TT12*) appeared to be good candidates for further study of germination variation under different environmental conditions. Environmental conditions are known to affect plant growth, with the type of effect depending on the developmental stage. Bac-Molenaar et al. (2015) conducted a GWAS using 250 natural *Arabidopsis* accessions to explore the genetic loci that account for heat stress sensitivity in the flowering, vegetative, seeding, and fruiting stages, using the silique length to quantify the effect of heat stress. They found that two developmental stages (during meiosis and from fertilization to early embryo development) were sensitive to heat stress and also identified four QTLs that were related to the heat stress sensitivity in each developmental stage. Following this, T-DNA insertion knockout lines were constructed for some genes with associated SNPs to validate their actions, which confirmed that these genes were involved in the response to heat stress in *Arabidopsis*.

Rice, soybean, wheat, and maize are major crops that are cultivated worldwide, and so a large amount of information has been accumulated about their genomic sequences and biological mechanisms, including several GWASs. The GWASs differed between species, mainly due to differences in the power and resolution of mapping inbreeding and outcrossing species. *Arabidopsis*, rice, soybean, and wheat exhibit self-fertilization, which leads to a slower LD decay, whereas maize, which is a typical outcrossing species, shows a rapid LD decay and, thus, greater genetic diversity. This faster LD decay enables us to find the causal genes but requires high coverage sequencing, while the slower LD decay makes it difficult to find the target genes, although it usually only requires low coverage sequencing. Association mapping or GWAS has allowed a lot of phenotypic variations in these crops to be explained, and these studies provide models to perform association studies and further GWASs in the future, as shown in Table 10.1.

In rice, Yang et al. (2015) performed a GWAS for 29 leaf traits that were related to leaf size,

Table 10.1 Examples of association mapping studies in various plant species

Plant species	Population	Sample size	DNA marker	Trait	Reference
Arabidopsis	Diverse accessions	100	213,624 SNPs	Germination	Morrison and Linder (2014)
Arabidopsis	Diverse accessions	285	~215k SNPs	Fertility under heat stress	Bac-Molenaar et al. (2015)
Rice	Diverse accessions	533	4,358,600 SNPs (late tillering stage), 2,863,169 SNPs (late booting stage), 1,959,460 SNPs (milk grain stage)	Leaf number, area, shape, and color	Yang et al. (2015)
Rice	Diverse accessions	366	0.8 billion SNPs	Pia-blast disease resistance	Wang et al. (2014)
Rice	Diverse accessions	151	156 SSRs	Pia-blast disease resistance, heading date, plant height, paddy and brown seed weight under greenhouse and field conditions	Wang et al. (2015b)
Rice	Diverse accessions and landraces	237	1,019,883 SNPs	Ratio of deep rooting (RDR)	Lou et al. (2015)
Rice	Diverse accessions (traditional and improved japonica accessions)	167	9727 DArTs (Diversity Arrays Technology), 6717 SNPs	Deep root traits, plant biomass, shoot biomass, number of tillers, and root to shoot ratio, etc.	Courtois et al. (2013)
Soybean	Diverse accessions	309	31,045 SNPs	Days to flowering, maturity, duration of flowering-to-maturity, and plant height	Zhang et al. (2015b)
Soybean	Diverse accessions	309	31,045 SNPs	Seed weight	Zhang et al. (2015a)
Wheat	Hexaploid winter wheat	100	5525 SNPs and PAVs (presence/absence variation DArT markers)	Plant height, grain yield, and biomass potential for bioethanol production	Bellucci et al. (2015)
Wheat	Elite facultative winter wheat and varieties	120	3051 DArTs	Yield, quality, and other agronomic traits under rain-fed and irrigated conditions	Tadesse et al. (2015)
Maize	Recombinant inbred lines	4892	836 SNPs	Male inflorescence (tassel) and female inflorescence (ear)	Brown et al. (2011)

shape, and color at three growth stages using a panel of 533 rice accessions. From this, a total of 73, 123, and 177 new loci were detected to be associated with leaf size, color, and shape, respectively, nine of which contained known leaf-related genes, such as *Nal1* for controlling leaf width. In the other GWAS that genotyped 0.8 million SNP variants across 366 diverse *indica* accessions, Wang et al. (2014) found candidate genes that were associated with the rice *Pia*-blast disease and identified 21 markers that were associated with disease resistance and 16 markers that were associated with yield-related components, with a complex relationship between the two. In addition, Lou et al. (2015) found that nine SNPs were associated with the ratio of deep rooting (RDR) in rice, which is associated with drought avoidance, seven of which were verified as occurring on chromosomes 1 and 2 in two RDR extreme groups; Courtois et al. (2013) performed a GWAS for root traits using a panel of 167 *japonica* accessions that were genotyped at an average density of one marker per 22.5 kb and found 19 significant associations, as well as several co-localized QTLs, some of which were also co-localized with meta-QTLs for root traits.

In soybean, Zhang et al. (2015a, b) found that the LD decayed slowly and identified a total of 27 loci for days to flowering (DTF), 6 loci for days to maturity (DTM), 18 loci for the duration of flowering-to-maturity (DFTM), and 27 loci for plant height (PH). Among the genes identified, *Dt1* was identified at the locus that was strongly associated with both DTM and PH. In addition, they found 22 loci for soybean seed weight and candidate genes for conditioning seed development in a population of 309 soybean germplasm accessions that were genotyped with 31,045 SNPs.

In wheat, Bellucci et al. (2015) conducted a GWAS for plant height (PH), grain yield (GY), biomass, and the release of monomeric sugars related to bioethanol production using 100 varieties, and six QTLs for GY, PH, and glucose (GLU) content were subsequently found and mapped. One QTL for PH had previously been reported, while the remaining five QTLs were new genetic regions that explained the phenotypic variation. In addition, Tadesse et al. (2015) evaluated the yield, quality, and other agronomic traits in 120 elite facultative/winter wheat genotypes under rain-fed and irrigated conditions, which showed that five markers were associated with yield under rain-fed conditions, while three markers were associated with yield under irrigated conditions.

In maize, Brown et al. (2011) performed a GWAS based on four traits from the male inflorescence (tassel) and three traits from the female inflorescence (ear) in a large population of recombinant inbred lines, which detected some pleiotropic loci that control elongation of the ear and tassel.

10.4 *Jatropha curcas* L.

Jatropha is a tropical shrub belonging to family Euphorbiaceae that is currently being considered as an important producer of the next-generation biofuel. Jatropha oil is regarded as a candidate substitute for fossil fuel and so may contribute to preventing greenhouse gas emissions, which are now a serious global issue. However, although jatropha has recently attracted much attention as a biofuel, its oil yield is still too low for sustainable biodiesel production because it has not been domesticated as a biofuel plant, with some cultivation projects giving low productivity that is a long way from potential commercialization. Thus, the integration of knowledge about the breeding and performance of various jatropha lines in different environments is important for improving this species, and further scientific research is required to gain a deeper understanding of its biology.

10.4.1 Breeding of Jatropha

The main aim of jatropha breeding is the generation of higher yielding cultivars. There are two possible approaches to achieve this: to generate jatropha plants that produce more seeds or oil to be transformed into biofuel; or to enlarge the jatropha cultivation area without affecting food production.

The identification and isolation of target genes that are involved in seed number, seed size, and fatty acid (FA)/lipid biosynthesis in seeds would be important to genetically increase the oil yield or alter the seed oil composition to achieve the desired characteristics for biofuel production. However, these approaches have had limited success in jatropha to date. Several studies have successfully introduced gene constructs that express key enzymes involved in oil synthesis into model plants such as *Arabidopsis*, resulting in an altered oil composition or oil quality in the seeds of transgenic plants (Wu et al. 2009, 2013; Misra et al. 2013). However, there have been few transgenic or functional studies in jatropha plants to improve oil yield or composition (Kim et al. 2014), and we still have a limited genetic knowledge of this species to allow the generation of varieties with high oil productivity.

It is well known that jatropha can survive under drought conditions. Therefore, it could be cultivated in arid or semiarid areas, where it would have no competition with crop cultivation, and its cultivation could provide a reliable source of income for rural farmers in African countries (Chikara and Jaworsky 2007). However, its growth and yield can be greatly affected by various biotic and abiotic stresses, with seed production being lower under drought conditions (Niu et al. 2012). Therefore, it will be important to breed jatropha with high drought tolerance to produce a good supply of jatropha oil in arid areas, as well as an improved performance under other stress conditions, such as high salinity.

The biological mechanisms behind the tolerance of jatropha to drought or other stresses are poorly understood. Some studies have successfully developed transgenic *Arabidopsis* or tobacco (*Nicotiana* spp.) plants with improved tolerance against drought or salinity stresses using several jatropha genes, including those encoding transcription factors (Liang et al. 2013; Wang et al. 2015a), but few have attempted genetic transformation to improve the stress tolerance of jatropha itself (Tsuchimoto et al. 2012), despite several genetic transformation methods via *Agrobacterium tumefaciens* infection having been established and proposed (Li et al. 2008; Kumar et al. 2010; Pan et al. 2010; Khemkladngoen et al. 2011a, b; Subramanyam et al. 2011). Consequently, no transgenic stress-tolerant line has been developed for practical use in jatropha plantations to date.

10.4.2 Whole-Genome Sequencing Projects of Jatropha

Jatropha is a diploid species with a chromosome number ($2n$) of 22 (Nair et al. 2009) and a genome size of 418 Mb (Carvalho et al. 2008). Sato et al. (2011) performed the first whole-genome sequence analysis of jatropha, in which they obtained up to 285,858,490 bp of non-redundant sequences, consisting of 120,586 contigs and 29,831 singlets, and accounting for approximately 95% of the gene-containing regions. In addition, 4% of the putative protein-encoding genes were found to be specific to the Euphorbiaceae family, and a high degree of microsynteny with the genome of castor bean (*Ricinus communis*) was observed. Later, Hirakawa et al. (2012) upgraded the genome sequence information of jatropha to give a total length of 297,661,187 bp consisting of 39,277 contigs and deduced 30,203 complete and partial structures of protein-encoding genes. Still, three times of the gene number identified by Hirakawa et al. (2012) with complete structures was obtained (see Chap. 1).

Recently, Wu et al. (2015) reported the draft genome sequences, the assembled length of which has 320.5 Mb containing 27,172 putative protein-coding genes. They also established and anchored sequences to an interspecific linkage map containing 1208 markers using *Jatropha curcas* and *Jatropha integerrima*. Through gene family clustering, 15,268 families were identified, 13,887 of which existed in the castor bean genome.

The whole-genomic sequence and linkage map for jatropha will enable us to find as many representative and diverse molecular markers as possible. It will then provide a valuable resource for GWAS in the future.

10.4.3 Molecular Marker Design in Jatropha

The whole-genome sequence (Sato et al. 2011) and the draft anchored linkage map (Wu et al. 2015) make it possible to undertake a wide exploration of SSR and SNP markers in jatropha. The genomic sequence combined with the whole-transcriptome sequence would enable us to find association signals within the coding sequence. Recently, mRNA sequencing data have been used to conduct a GWAS in rapeseed (*Brassica napus*), which identified two QTLs responsible for the glucosinolate content of the seeds (Harper et al. 2012). Furthermore, it was also found that the genomic deletion of the orthologous gene of the transcription factor HAG1 (At5g61420), which controls aliphatic glucosinolate biosynthesis in *Arabidopsis*, reduced the seed glucosinolate phenotype in rapeseed.

The identification of a large number of SSR markers that are distributed evenly along the chromosomes and the construction of a specific genotyping array for jatropha that covers thousands of SNPs would allow GWASs to be undertaken, with the guarantee of a number of significant marker–trait associations at a large scale. To develop SNP markers at a large scale, it will be necessary to undertake sequencing and transcriptome analyses of inbred lines from Mexico or Central America, which have high levels of genetic diversity (see Sect. 10.4.5). In addition, genome-wide characterization of the LD decay is also required to determine the genome-wide linkage distance and to understand how many markers are needed. There is no information about the LD decay of jatropha to date.

10.4.4 Linkage and Association Mapping in Jatropha

The earliest published interspecific linkage map was generated using *J. curcas* and *J. integerrima* (Wang et al. 2011). This map was then used to conduct a whole-genome scan to map QTLs and expression QTLs (eQTLs) that affect the seed oil traits, which identified 18 QTLs underlying the oil traits and three eQTLs of the oleosin acid genes (Liu et al. 2011). QTL mapping of growth and yield-related traits was also performed, which detected two QTLs that control seed yield in two QTL clusters harboring several yield-related traits in different linkage groups, respectively (Sun et al. 2012; also see Chap. 2).

Later, an intraspecific linkage map of jatropha was developed by King et al. (2013), which was recently used to perform QTL mapping (King et al. 2015). This study detected a possible locus that confers a pleiotropic heterosis effect and also positioned a large number of genes involved in the biosynthesis of storage lipids onto the genetic map using a candidate gene approach and by integrating physical mapping data from the jatropha genome that was published by Wu et al. (2015). By comparing the positions of these genes with QTLs, a number of genes that potentially control seed traits were detected, including phosphatidate phosphatase genes.

10.4.5 Genetic Diversity of Jatropha

The collection of jatropha plants with a high level of variation would greatly help its improvement by plant breeding. However, the collection, characterization, and evaluation of global jatropha germplasms for growth, morphology, seed characteristics, and yield-related traits are still in the infancy.

Montes Osorio et al. (2014) calculated Jaccard's similarity and polymorphism information content (PIC) values for jatropha and performed a cluster analysis with the unweighted paired group method (UPGM) by using SSR, target region amplification polymorphism (TRAP), and AFLP markers in 182 accessions from Asia, Africa, South America, and Central America. Their findings suggested that jatropha in South America, Africa, and Asia had lower genetic variation than plants in Central America. They also compared the level of phenotypic variation between these localities, including seed morphological characteristics, seed oil content and fatty acid composition, and early growth traits, which showed that Central American accessions had the highest phenotypic variation and, thus, should be considered the most important source for plant breeding.

Some reports have indicated that jatropha accessions in Mexico have the highest genetic variability. For example, Basha et al. (2009) used three-dimensional principal coordinate analysis (PCoA) using data from RAPD and inter-simple sequence repeat (ISSR) marker systems to assess 72 accessions of jatropha derived from 13 countries, which showed close clustering of the accessions from all countries except Mexico, the accessions from Mexico were separated into different clusters. They also detected a high level of variation in the phorbol ester content in Mexican accessions, indicating rich genetic resources. Furthermore, non-toxic types only exist in Mexico, which strongly suggests that this should be the domestication center for this species, due to the high level of genetic diversity there (Dias et al. 2012).

Chiapas, which is the southernmost state in Mexico bordering Guatemala, is now considered an important place for jatropha improvement because a high level of genetic diversity has been detected there using AFLP markers and important agronomic traits (Pecina-Quintero et al. 2011). In addition, a high level of variation in oil content and other characters associated with productivity have been observed there, and an analysis of molecular variance (AMOVA) and the diversity index confirmed a broad genetic pool in the jatropha germplasm from Chiapas. In an analysis of a representative set of 175 accessions of jatropha from nine central and southeastern Mexican states using AFLP markers, the germplasm from Chiapas showed the highest genetic diversity and clearly differed from the accessions from the other states (Pecina-Quintero et al. 2014).

Jatropha in Africa and Asia were found to possess a narrow genetic background. For example, by applying SSRs and SNPs in a diverse, worldwide germplasm panel of 70 accessions from Central America, Mexico, Africa, Asia, and South America, Montes Osorio et al. (2014) found that African and Asian jatropha possessed very low levels of genetic polymorphism and showed similarly, using a wide jatropha collection from three Asian countries, two African countries, and different geographical regions in China. Zhang et al. (2011) found that jatropha in China had a narrow genetic diversity. Furthermore, in a study that used 29 microsatellites located on 11 linkage groups and 276 accessions of jatropha collected from nine locations in five countries in South America, Asia, and Africa, Yue et al. (2014) surprisingly found that these plants were homozygous at all loci and that all accessions shared the same genotype at each locus, suggesting no variation in the genome.

In summary, jatropha in Central America and Mexico has been shown to have the highest level of genetic diversity, while plants in Africa and Asia possess almost no variation. Narrow genetic variability will result in weak adaption to various environmental changes, and so it is critical that we collect all of the genotypes to preserve and make full use of the genetic resources of jatropha and to help in further breeding. Different types of core collections have been developed from the entire collection of soybean germplasm, using a combination of passport data and morphological traits (Qiu et al. 2003; Song et al. 2010). Therefore, the establishment of such a core collection is also desirable for jatropha.

10.4.6 Association Study in Jatropha

Since some genetic information is now available for jatropha, it is possible to perform a GWAS, as in other plant species. However, a GWAS of the seed, oil, or other agronomic traits of jatropha has not yet been reported, and there has also been no GWAS of jatropha under stress conditions. There has only been one marker–trait association study on jatropha, which used a limited number of genotypes and DNA molecular markers (Sharma and Chauhan 2012).

Jatropha and castor bean share a high level of microsynteny, have several taxonomic similarities, and produce seeds that contain a large amount of oil. Therefore, to assess the genetic diversity of 54 jatropha genotypes collected from India that are categorized by oil content, 48 primer sets were designed from the 32 candidate genes that are responsible for fatty acid biosynthesis also in the castor bean genome sequence, including 12 SSR markers (Sharma and Chauhan 2012). The transferability of markers was found to be approximately 80% from exons and 78% from exon–intron junctions, and there was a higher level of genetic polymorphism in markers from exon–intron junctions. It was also found that only 36 genotypes with high oil content had amplicons of 700 bp, designated as marker JJM1 from the exon–intron junction of the stearoyl desaturase gene, while there were no amplicons in 11 jatropha genotypes with low oil content. Thus, the two SNP sequence changes in the exonic region leads to the substitution of leucine with glutamine in the stearoyl desaturase polypeptide, causing the transition from low to high oil content genotypes.

10.4.7 GWAS Based on Epigenetic Markers

It was somewhat surprising that very limited genetic diversity was detected in African and Asian jatropha, as considerable phenotypic variations have been observed in these regions. Therefore, epigenetic factors are likely responsible for this variation (Montes Osorio et al. 2014). In addition to using DNA markers in GWAS, methylation markers can also be used to assess the relationship between epigenetic factors and phenotypic variation, which is a promising approach for revealing relationships that cannot be elucidated from genetic frame changes. For example, the application of this method has successfully shown that variation in genomic methylation is associated with leaf shape and photosynthetic traits in *Populus simonii* (Ci et al. 2015). Thus, it may be possible to use this approach to determine what has caused the phenotypic variation in African and Asian jatropha. Furthermore, a GWAS based on epigenetic markers may also help to broaden our knowledge of how to improve jatropha yield in the future by controlling the epigenetic variation.

10.5 Populations for GWAS

To ensure the success of GWAS, the most important concern for researchers is the appropriate selection of genetic materials. A worldwide collection of diverse accessions is usually preferable for capturing as much genetic diversity as possible, as shown in a recent GWAS on common bean (*Phaseolus vulgaris*), where the authors collected 259 Andean bean genotypes from Africa, South America, North America, Central America, the Caribbean, Asia, and Europe (Kamfwa et al. 2015).

For the collection and preservation of plant resources, it is important to characterize the plants at the center of origin, as this is where the most diverse genotypes tend to exist. The origin of jatropha has been extensively debated over the years, and the genetic diversity and phenotypic diversity of jatropha accessions have been detected and discussed in several reports. Montes Osorio et al. (2014) argued that Central America is the center of genetic variability. It has also been suggested that Mexico is the center of origin and should be considered an important place for the future breeding of jatropha (Basha et al. 2009; Dias et al. 2012), as jatropha accessions here have been found to have the highest genetic variability, and more recently, Pecina-Quintero et al. (2011, 2014) suggested that the state of Chiapas is the center of origin, as there is the highest genetic diversity within Mexico. Thus, it is believed that jatropha accessions from Mexico, especially Chiapas, will ensure high levels of genetic polymorphism and phenotypic variation, and thus a high number of marker–trait associations. By contrast, utilization of only African or Asian jatropha would lead to a low level of genetic variation, and only a low number of marker–trait associations have been detected from here, as shown in the study by Sharma and Chauhan (2012).

When collecting jatropha accessions for GWAS, it is important to consider the population structure and unequal relatedness among individuals. If accessions disproportionately form a genetic subgroup, any loci with different allele proportions between the subgroups and the general population will be associated with traits. Thus, the presence of population structure or kinship among individuals will lead to the detection of false-positive associations.

Over the years, association studies have remarkably improved. Many comprehensive statistical models have been established to distinguish true biological associations different from false positives that have arisen from population structure or pairwise individual relatedness. For example, it is considered that mixed linear models (MLMs) (Yu et al. 2006), rather than general linear models (GLMs), can be used to remove the confounding factors caused by population structure and unequal relatedness among pairs of individuals. Several MLM-based methods have also been developed to deal with large datasets, including the "compressed MLM" of Zhang et al. (2010), which can reduce the effective sample size of such datasets by clustering individuals into groups. Zhang et al. (2010) also presented a complementary approach called "population parameters previously determined" or P3D, which eliminates the need to re-compute various components. It has been shown that the implementation of these two methods significantly reduces the computing time while retaining or even improving the statistical power. Approaches have also been developed to quantify the population structure in a diverse population. For example, genomic control, structured population association, and principle components can be included as fixed effects to account for population structure, the latter two of which are most commonly used in GWAS. An appropriate MLM model usually requires a comparison of several different combinations of approaches, following which the best and most reliable model is selected. One approach that is widely used to assess the reliability of the model is the quantile–quantile plot (Q–Q plot), which plots the observed *p* values against the expected *p* values, whereby a strong deviation of the observed *p* values from the uniform distribution may suggest a high rate of false positives. Li et al. (2014) compared six combinations of MLM models (a naive model, PCA, Q matrix (Q), kinship matrix (K), PCA + K, and Q + K), from which they selected PCA and Q + K models to be better in identification of association signals.

10.6 Phenotyping and Data Collection for Association Mapping

The collection of trait data should be carried out over a long time period, because the phenotypes of plants are influenced by environmental factors, which is known as phenotypic plasticity (Chapin III et al. 1993; Valladares et al. 2000). Therefore, since the phenotypes of jatropha may also be influenced by environmental factors (Gairola et al. 2011), phenotypic data should be collected across multiple years and locations, and environmental conditions should also be recorded to allow any environmental effects to be excluded. Estimation of the correlation among traits and their heritability can then help to determine to what extent genetic improvement is possible through selection. To ensure high-quality data, robust and correct methods for phenotyping are also necessary. A bar coding system and scanner-based data collection would facilitate the data collection process (http://www.maizegenetics.net), and similar systems for data collection are currently being developed in many plants. For data storage and analysis, different models have been developed in maize (http://www.maizegenetics.net/gdpdm) and barley (http://www.barleycap.org).

10.7 Issues in Association Mapping

GWAS involves the use of more markers than traditional linkage mapping or QTL mapping. Thus, it is necessary to define an appropriate significance threshold that can take into account multiple comparisons (Gupta et al. 2014). It is also important to correct the *p* values for each marker when performing a statistical test on a set of markers. The most commonly used corrections are the false discovery rate (FDR) (Benjamini and Hochberg 1995) and the Bonferroni (1936) correction, among which the Bonferroni correction is the most stringent. Li et al. (2014) used this correction to set the threshold in their GWAS of rapeseed.

The occurrence of a rare allele also poses a problem in a GWAS. Because the frequency of the rare allele is too low to be included in the statistical analysis and cannot represent the normal case, it is always ignored in the study. However, in most cases in plants, the desired traits appear in the rare occasions, so this would result in the loss of the marker association with such rare desirable traits and missing heritability (Gupta et al. 2014). This problem could be addressed by using a combination of QTL analysis or linkage mapping with association mapping, a large population size in which the rare allele may be present in a sufficient number, and an improved statistical method, as well as several other new methods that have recently been proposed (Li and Leal 2008; Zhu et al. 2011; Gibson 2012).

Continuous efforts have been made to improve the power and resolution of association mapping, and several new approaches have been developed, including joint linkage association mapping (JLAM) and family-based association mapping (family-based AM) (Lu et al. 2010). The former method can involve either biparental populations or one or more multi-parental populations to combine both linkage analysis and association mapping; by using the same markers to genotype one population that consists of natural germplasm and another that contains biparental populations, the results of linkage mapping and association mapping could be compared. By contrast, the latter method can correct for spurious associations using the logistic regression ratio test (Stich et al. 2006), giving it an improved precision and power.

10.8 Conclusions

Our current understanding of jatropha prevents us from breeding better variety with a high yield even under stress conditions. GWAS is an ideal method to improve nowadays situation. However, to ensure the success of GWAS, it will be important to use jatropha accessions from Mexico, especially Chiapas, to choose a suitable MLM model for individual cases, to use a combination of SSR and a specific SNP array designed from the genome and transcriptome sequences of jatropha, and to collect phenotypic data with high accuracy. It would be also necessary to understand limitations of this approach at the same time. Useful genes and genomic region identified by a successful GWAS could be used and integrated to breed a high production jatropha.

Acknowledgements Plant Bioengineering for Bioenergy Laboratory was supported by Corporate Social Responsibility (CSR) Foundation of the Sumitomo Electric Industries (SEI), Ltd. This work was conducted and funded in part under the Joint Research Program of the Arid Land Research Center, Tottori University, Japan.

References

Alonso-Blanco C, El-Assal SE, Coupland G, Koornneef M (1998) Analysis of natural allelic variation at flowering time loci in the Landsberg erecta and Cape Verde Islands ecotypes of *Arabidopsis thaliana*. Genetics 149(2):749–764

Bac-Molenaar JA, Fradin EF, Becker FF, Rienstra JA, van der Schoot J, Vreugdenhil D, Keurentjes JJ (2015) Genome-wide association mapping of fertility reduction upon heat stress reveals developmental stage-specific QTLs in *Arabidopsis thaliana*. Plant Cell 27:1857–1874

Bao J, Corke H, Sun M (2002) Microsatellites in starch-synthesizing genes in relation to starch physicochemical properties in waxy rice (*Oryza sativa* L.). Theor Appl Genet 105:898–905

Basha SD, Franis G, Makkar HPS, Becker K, Sujatha M (2009) A comparative study of biochemical traits and molecular markers for assessment of genetic relationships between *Jatropha curcas* L. germplasm from different countries. Plant Sci 176:812–823

Bellucci A, Torp AM, Bruun S, Magid J, Andersen SB, Rasmussen SK (2015) Association mapping in scandinavian winter wheat for yield, plant height, and traits important for second-generation bioethanol production. Front Plant Sci 6:1046

Benjamini Y, Hochberg Y (1995) Controlling the false discovery rate: a practical and powerful approach to multiple testing. J R Stat Soc Ser B (Methodol) 57 (1):289–300

Bonferroni CE (1936) Teoria statistica delle classi e calcolo delle probabilità, Pubblicazioni del R. Istituto Superiore di Scienze Economiche e Commerciali di Firenze 8:3–62

Borevitz JO, Nordborg M (2003) The impact of genomics on the study of natural variation in *Arabidopsis*. Plant Physiol 132(2):718–725

Brown PJ, Upadyayula N, Mahone GS, Tian F, Bradbury PJ, Myles S, Holland JB, Flint-Garcia S, McMullen MD, Buckler ES, Rocheford TR (2011) Distinct genetic architectures for male and female inflorescence traits of maize. PLoS Genet 7:e1002383

Carvalho CR, Clarindo WR, Praca MM, Araujo FS, Carels N (2008) Genome size, base composition and karyotype of *Jatropha curcas* L., an important biofuel plant. Plant Sci 174:613–617

Chapin FS III, Autumn K, Pugnaire F (1993) Evolution of suites of traits in response to environmental stress. American Naturalist 142:S78–S92

Chikara J, Jaworsky G (2007) The little shrub that could—maybe. Nature 449:652–655

Ci D, Song Y, Du Q, Tian M, Han S, Zhang D (2015) Variation in genomic methylation in natural populations of *Populus simonii* is associated with leaf shape and photosynthetic traits. J Exp Bot 67(3):723–737

Clarke JH, Mithen R, Brown JK, Dean C (1995) QTL analysis of flowering time in *Arabidopsis thaliana*. Mol Gen Genet 248(3):278–286

Courtois B, Audebert A, Dardou A, Roques S, Ghneim-Herrera T, Droc G, Frouin J, Rouan L, Goze E, Kilian A, Ahmadi N, Dingkuhn M (2013) Genome-wide association mapping of root traits in a *Japonica* rice panel. PLoS ONE 8:e78037

Dias LAS, Missio RF, Dias DCFS (2012) Antiquity, botany, origin and domestication of *Jatropha curcas* (Euphorbiaceae), a plant species with potential for biodiesel production. Genet Mol Res 11:2719–2728

Dresselhaus T, Cordts S, Heuer S, Sauter M, Lorz H et al (1999) Novel ribosomal genes from maize are differentially expressed in the zygotic and somatic cell cycles. Mol Gen Genet 261:416–427

Gairola KC, Nautiyal AR, Sharma G, Dwivedi AK (2011) Variability in seed characteristics of *Jatropha curcas* Linn. from hill region of Uttarakhand. Bull Environ Pharmacol Life Sci 1:64–69

Gibson G (2012) Rare and common variants: twenty arguments. Nat Rev Genet 13(2):135–145

Gupta PK, Kulwal KL, Jaiswal V (2014) Association mapping in crop plants: opportunities and challenges. Adv Genet 85:109–147

Hansen M, Kraft T, Ganestam S, Säll T, Nilsson NO (2001) Linkage disequilibrium mapping of the bolting gene in sea beet using AFLP markers. Genet Res 77 (1):61–66

Harper AL, Trick M, Higgins J, Fraser F, Clissold L, Wells R, Hattori C, Werner P, Bancroft I (2012) Associative transcriptomics of traits in the polyploid crop species *Brassica napus*. Nat Biotechnol 30 (8):798–802

Hirakawa H, Tsuchimoto S, Sakai H, Nakayama S, Fujishiro T et al (2012) Upgraded genomic information of *Jatropha curcas* L. Plant Biotechnol 29:123–130

James C (2011) ISAAA Brief 43, Global status of commercialized biotech/GM crops: 2011. ISAAA Briefs. ISAAA, Ithaca

Kamfwa K, Cichy KA, Kelly JD (2015) Genome-wide association analysis of symbiotic nitrogen fixation in common bean. Theor Appl Genet 128:1999–2017

Khemkladngoen N, Cartagena J, Fukui K (2011a) Physical wounding-assisted Agrobacterium-mediated transformation for juvenile cotyledons of a biodiesel producing plant, *Jatropha curcas* L. Plant Biotechnol Rep 5:235–243

Khemkladngoen N, Cartagena J, Shibagaki N, Fukui K (2011b) Adventitious shoot regeneration from juvenile cotyledons of a biodiesel producing plant, *Jatropha curcas* L. J Biosci Bioeng 111(1):67–70

Kim MJ, Yang SW, Mao HZ, Veena SP, Yin JL, Chua NH (2014) Gene silencing of sugar-dependent 1 (*JcSDP1*), encoding a patatin-domain triacylglycerol lipase, enhances seed oil accumulation in *Jatropha curcas*. Biotechnol Biofuels 7(1):36

King AJ, Montes LR, Clarke JG, Affleck J, Li Y et al (2013) Linkage mapping in the oilseed crop *Jatropha curcas* L. reveals a locus controlling the biosynthesis of phorbol esters which cause seed toxicity. Plant Biotechnol J 11:986–996

King AJ, Montes LR, Clarke JG, Itzep J, Perez CA et al (2015) Identification of QTL markers contributing to plant growth, oil yield and fatty acid composition in the oilseed crop *Jatropha curcas* L. Biotechnol Biofuels 8:160

Korte A, Farlow A (2013) The advantages and limitations of trait analysis with GWAS: a review. Plant Methods 9:29

Kowalski SP, Lan TH, Feldmann KA, Paterson AH (1994) QTL mapping of naturally-occurring variation in flowering time of *Arabidopsis thaliana*. Mol Gen Genet 245(5):548–555

Kumar N, Anand KGV, Pamidimarri DVNS, Sarkar T, Reddy MP et al (2010) Stable genetic transformation of *Jatropha curcas* via *Agrobacterium tumefaciens*-mediated gene transfer using leaf explants. Ind Crop Prod 32:41–47

Kumar V, Singh A, Mithra SV, Krishnamurthy SL, Parida SK et al (2015) Genome-wide association mapping of salinity tolerance in rice (*Oryza sativa*). DNA Res 22(2):133–145

Li B, Leal SM (2008) Methods for detecting associations with rare variants for common diseases: application to analysis of sequence data. Am J Hum Genet 83 (3):311–321

Li F, Chen B, Xu K, Wu J, Song W et al (2014) Genome-wide association study dissects the genetic architecture of seed weight and seed quality in rapeseed (*Brassica napus* L.). DNA Res 21:355–367

Li M, Li H, Jiang H, Pan X, Wu G (2008) Establishment of an Agrobacterium-mediated cotyledon disc transformation method for *Jatropha curcas*. Plant Cell Tiss Organ Cult 92:173–181

Liang J, Zhou M, Zhou X, Jin Y, Xu M, Lin J (2013) JcLEA, a novel LEA-like protein from *Jatropha curcas*, confers a high level of tolerance to dehydration and salinity in *Arabidopsis thaliana*. PLoS ONE 8:e83056

Liu P, Wang CM, Li L, Sun F, Liu P, Yue GH (2011) Mapping QTLs for oil traits and eQTLs for oleosin genes in jatropha. BMC Plant Biol 11:132

Liu SR, Li WY, Long D, Hu CG, Zhang JZ (2013) Development and characterization of genomic and expressed SSRs in citrus by genome-wide analysis. PLoS ONE 8:e75149

Lou Q, Chen L, Mei H, Wei H, Feng F et al (2015) Quantitative trait locus mapping of deep rooting by linkage and association analysis in rice. J Exp Bot 66:4749–4757

Lu Y, Zhang S, Shah T, Xie C, Hao Z et al (2010) Joint linkage–linkage disequilibrium mapping is a powerful approach to detecting quantitative trait loci underlying drought tolerance in maize. Proc Natl Acad Sci USA 107(45):19585–19590

Misra A, Khan K, Niranjan A, Nath P, Sane VA (2013) Over-expression of *JcDGAT1* from *Jatropha curcas* increases seed oil levels and alters oil quality in transgenic *Arabidopsis thaliana*. Phytochemistry 96:37–45

Montes Osorio LR, Torres Salvador AF, Jongschaap RE, Azurdia Perez CA, Berduo Sandoval JE et al (2014) High level of molecular and phenotypic biodiversity in *Jatropha curcas* from Central America compared to Africa, Asia and South America. BMC Plant Biol 14:77

Morrison GD, Linder CR (2014) Association mapping of germination traits in *Arabidopsis thaliana* under light and nutrient treatments: searching for G × E effects. G3 (Bethesda) 4:1465–1478

Nair D, Maria TSW, Luiz ASD (2009) Chromosome numbers of *Jatropha curcas* L.: an important agrofuel plant. Crop Breed Appl Biotechnol 9:386–389

Niu GH, Rodriguez D, Mendoza M, Jifon J, Ganjegunte G (2012) Responses of *Jatropha curcas* to salt and drought stresses. Int J Agron. doi:10.1155/2012/632026

Pan J, Fu Q, Xu ZF (2010) *Agrobacterium tumefaciens*-mediated transformation of biofuel plant *Jatropha curcas* using kanamycin selection. Afr J Biotechnol 9:6477–6481

Pecina-Quintero V, Anaya-Lopeza JL, Zamarripa-Colmenero A, Montes-Garcia N, Nunez-Colina CA et al (2011) Molecular characterisation of *Jatropha curcas* L. genetic resources from Chiapas, Mexico through AFLP markers. Biomass Bioenergy 35:1897–1905

Pecina-Quintero V, Anaya-Lopeza JL, Zamarripa-Colmenero A, Nunez-Colina CA, Montes-Garcia N et al (2014) Genetic structure of *Jatropha curcas* L. in Mexico and probable center of origin. Biomass Bioenergy 60:147–155

Qiu LJ, Cao YS, Chang RZ, Zhou XA, Wang GX et al (2003) Establishment of Chinese soybean (*G. max*) core collection: sampling strategy. Sci Agric Sin 36:1442–1449

Sato S, Hirakawa H, Isobe S, Fukai E, Watanabe A et al (2011) Sequence analysis of the genome of an oil-bearing tree, *Jatropha curcas* L. DNA Res 18:65–76

Sharma A, Chauhan RS (2012) Identification and association analysis of castor bean orthologous candidate gene-based markers for high oil content in *Jatropha curcas*. Plant Mol Biol Rep 30:1025–1031

Song XE, Li YH, Chang RZ, Guo PY, Qiu LJ (2010) Population structure and genetic diversity of mini core collection of cultivated soybean (*Glycine max* (L.) Merr.) in China. Sci Agric Sin 43:2209–2219

Stich B, Melchinger AE, Piepho HP, Heckenberger M, Maurer HP, Reif JC (2006) A new test for family-based association mapping with inbred lines from plant breeding programs. Theor Appl Genet 113:1121–1130

Subramanyam K, Subramanyam K, Sailaja KV, Srinivasulu M, Lakshmidevi K (2011) Highly efficient Agrobacterium–mediated transformation of banana cv. Rasthali (AAB) via sonication and vacuum infiltration. Plant Cell Rep 30:425–436

Sun F, Liu P, Ye J, Lo LC, Cao S et al (2012) An approach for jatropha improvement using pleiotropic QTLs regulating plant growth and seed yield. Biotechnol Biofuels 5:42

Tadesse W, Ogbonnaya FC, Jighly A, Sanchez-Garcia M, Sohail Q et al (2015) Genome-wide association mapping of yield and grain quality traits in winter wheat genotypes. PLoS ONE 10(10):e0141339

Tsuchimoto S, Cartagena J, Khemkladngoen N, Singkaravanit S, Kohinata T et al (2012) Development of transgenic plants in jatropha with drought tolerance. Plant Biotechnol 29:137–143

Valladares F, Martinez-Ferri E, Balaguer L, Perez-Corona E, Manrique E (2000) Low leaf-level response to light and nutrients in Mediterranean evergreen oaks: a conservative resource-use strategy? New Phytol 148:79–91

Wang C, Yang Y, Yuan X, Xu Q, Feng Y et al (2014) Genome-wide association study of blast resistance in *indica* rice. BMC Plant Biol 14:311

Wang CM, Liu P, Yi C, Gu K, Sun F et al (2011) A first generation microsatellite- and SNP-based linkage map of Jatropha. PLoS ONE 6:e23632

Wang X, Han H, Yan J, Chen F, Wei W (2015a) A new AP2/ERF transcription factor from the oil plant *Jatropha curcas* confers salt and drought tolerance to transgenic tobacco. Appl Biochem Biotechnol 176(2):582–597

Wang X, Jia MH, Ghai P, Lee FN, Jia Y (2015b) Genome-wide association of rice blast disease resistance and yield-related components of rice. Mol Plant Microbe Interact 28(12):1383–1392

Wu P, Zhang S, Zhang L, Chen Y, Li M et al (2013) Functional characterization of two microsomal fatty acid desaturases from *Jatropha curcas* L. J Plant Physiol 170(15):1360–1366

Wu P, Zhou C, Cheng S, Wu Z, Lu W et al (2015) Integrated genome sequence and linkage map of physic nut (*Jatropha curcas* L.), a biodiesel plant. Plant J 81:810–821

Wu PZ, Li J, Wei Q, Zeng L, Chen YP et al (2009) Cloning and functional characterization of an acyl–acyl carrier protein thioesterase (JcFATB1) from *Jatropha curcas*. Tree Physiol 29(10):1299–1305

Yang W, Guo Z, Huang C, Wang K, Jiang N et al (2015) Genome-wide association study of rice (*Oryza sativa* L.) leaf traits with a high-throughput leaf scorer. J Exp Bot 66(18):5605–5615

Yu J, Pressoir G, Briggs WH, Bi VI, Yamasaki M et al (2006) A unified mixed-model method for association mapping that accounts for multiple levels of relatedness. Nat Genet 38:203–208

Yue GH, Lo LC, Sun F, Cao SY, Yi CX et al (2014) No variation at 29 microsatellites in the genome of *Jatropha curcas*. J Genom 2:59–63

Zhang J, Song Q, Cregan PB, Jiang GL (2015a) Genome-wide association study, genomic prediction and marker-assisted selection for seed weight in soybean (*Glycine max*). Theor Appl Genet 129 (1):117–130

Zhang J, Song Q, Cregan PB, Nelson RL, Wang X et al (2015b) Genome-wide association study for flowering time, maturity dates and plant height in early maturing soybean (*Glycine max*) germplasm. BMC Genom 16:217

Zhang J, Zhao J, Xu Y, Liang J, Chang P et al (2015c) Genome-wide association mapping for tomato volatiles positively contributing to tomato flavor. Front Plant Sci 6:1042

Zhang Z, Ersoz E, Lai CQ, Todhunter RJ, Tiwari HK et al (2010) Mixed linear model approach adapted for genome-wide association studies. Nat Genet 42:355–360

Zhang Z, Guo X, Liu B, Tang L, Chen F (2011) Genetic diversity and genetic relationship of *Jatropha curcas* between China and Southeast Asian revealed by amplified fragment length polymorphisms. Afr J Biotechnol 10:2825–2832

Zhu Q, Ge D, Maia JM, Zhu M, Petrovski S et al (2011) A genome-wide comparison of the functional properties of rare and common genetic variants in humans. Am J Hum Genet 88(4):458–468

Part IV
Breeding and Application

Towards Varietal Improvement of Jatropha by Genetic Transformation

11

Joyce Cartagena

Abstract

Since the 2008 oil crisis, the need for alternative sources of biofuels is becoming more and more apparent. Jatropha is considered as one of the important candidate sources of renewable energy because it is a non-edible plant and it contains high amounts of oil that is of good quality. Furthermore, Jatropha is said to be drought tolerant and can be cultivated in marginal lands, thus avoiding competition with food crops for agricultural lands. However, disappointing result in Jatropha plantations caused a decline in the general interest toward this biofuel crop. Since Jatropha has not been domesticated, there will be more possibilities for varietal improvement either by conventional breeding or genetic modification. Several traits that need improvement in Jatropha include toxicity, low yield, and low tolerance to abiotic stress. The development of transgenic Jatropha will be a more practical way to address the growing need for a cheap and renewable source of energy. Moreover, Jatropha is a suitable model plant to study drought tolerance and its applicability for cultivation in arid land. This chapter presents the current progress in the development of transgenic Jatropha with special emphasis on gene discovery and transformation methods.

11.1 Introduction

With the diminishing source of fossil fuel, there is now an urgent need for a renewable source of alternative energy. Renewable energy can be any of the following forms: solar, wind power, hydrogen and fuel cells, hydroelectric energy, geothermal power, and bioenergy. Bioenergy crops in general are better alternative sources of energy because they can balance out the CO_2

J. Cartagena (✉)
Graduate School of Bioagricultural Sciences, Nagoya University, Furo-cho, Chikusa-ku, Nagoya 464-8601, Japan
e-mail: joyce@agr.nagoya-u.ac.jp

S. Tsuchimoto (ed.), *The Jatropha Genome*, Compendium of Plant Genomes,
DOI 10.1007/978-3-319-49653-5_11

produced during the burning of fuels. Jatropha, *Jatropha curcas* L., is a non-edible biofuel plant that can be grown even in marginal lands. Thus, it will not compete with food crops in terms of area for plantation and will not bring about problems related to food prices. The cultivation of Jatropha can also provide livelihood for farmers in the rural areas, such as in African countries (Eckart and Henshaw 2012; Sinkala and Johnson 2012). Among other candidate biofuel crops, the oil content in Jatropha seeds is relatively higher (Ong et al. 2011). Furthermore, Jatropha oil is considered of good quality (Adarsh 2011; Raja et al. 2011), and it has been used to conduct test flights done by major airline companies such as Air New Zealand (AIZ.AX), Japan Airlines (9205.T), and Continental Airlines (CAL) (Spielberg 2009).

The emergence of Jatropha as a promising source of biofuel is quite recent, but its use as traditional medicine has been known in different parts of the world for many decades now. Thomas et al. (2008), Heller (1996), and Debnath and Bisen (2008) presented extensive reviews on the traditional medicinal uses of Jatropha (see Chap. 15 for details). Aside from being a source of oil for bioenergy, Jatropha also contains other metabolites and substances that have pharmacological importance making Jatropha a very important crop both for energy and medical industries (see Chap. 5 for details).

Jatropha has an inherent drought tolerance and can be used to rehabilitate waste lands that will otherwise be useless. In the Philippines, biofuel crops are being used for reforestation efforts in order to derive economic productivity out of idle lands (Diaz 2009 personal communication). Using degraded lands to grow biofuel crops will result in lower greenhouse gas (GHG) emissions and lower carbon debt (Fargione et al. 2008). According to Fargione et al. (2008), carbon debt is the amount of CO_2 released during land clearing and conversion of undisturbed areas or agricultural lands to a biofuel plantation. This debt can be repaid if the net GHG emissions that result from the production and combustion of the biofuel products are less than that of fossil fuels. Based on the calculations by Fargione et al. (2008), it was concluded that using degraded and abandoned lands provides much benefit in terms of amount of GHG emissions and carbon debt. Therefore, utilizing Jatropha for cultivation in marginal lands will be beneficial as long as Jatropha can stay productive despite the unsatisfactory growing conditions.

Jatropha has not been domesticated and there has been slow progress in Jatropha breeding in the last several years. This is the main reason most of the investments on Jatropha have been called off, which gave Jatropha a bad reputation as a biofuel crop (Sanderson 2009). In the early 2000s, Jatropha was recognized as a promising alternative source of biofuel in several areas in Asia and Africa. The global awareness to the diminishing fossil fuel supply was a significant motivation for many countries to implement government policies on the use of alternative fuels, such as Jatropha biofuel (Renner et al. 2008). This was the driving force for the establishment of Jatropha plantations in Asian and African regions. Until 2008, Jatropha was hailed as the miracle crop that can be grown even in waste lands and produce high-quality oil in significant amounts. However, in a matter of just a few years of growing Jatropha, farmers realized that the miracle crop is not different from other

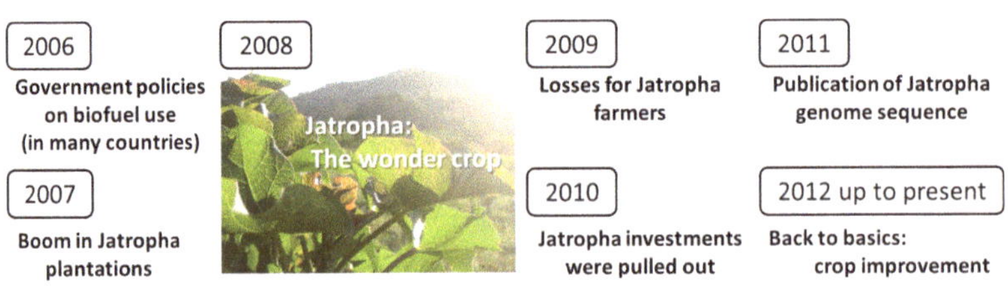

Fig. 11.1 The case of Jatropha: crop improvement is essential for sustainable biofuels production

plants in that Jatropha also needs sufficient care in order to be productive. Figure 11.1 shows that the failure in Jatropha plantations is due to the lack of a standard variety which could have been obtained by crop improvement through genetic modifications. After the publication of the whole-genome sequence of Jatropha, more researchers were able to carry out basic research especially toward the development of improved Jatropha lines.

11.2 The Need for Genetic Modification in Jatropha

In conventional plant breeding, it is very important to have a significant genetic variation among cultivars. However, it has been established by several studies that Jatropha lines collected from Asia and Africa are genetically similar, but shows variation from those found in Latin America (Basha and Sujatha 2007). Since Jatropha plants collected from many parts of the world showed very low genetic variation, genetic modification, instead of conventional plant breeding, will be an appropriate method to efficiently develop new lines with improved traits.

Jatropha has not been domesticated yet as a crop, thus extensive studies on its basic agronomic traits such as vegetative and reproductive growth and development are necessary to take full advantage of the benefits that can be obtained from this cash crop. In order to be applied for commercial plantation, an elite cultivar(s) has to be first identified, which can provide a starting point for varietal improvement. Among the limitations of using Jatropha for commercial-scale plantations include toxicity, low yield, and low tolerance to abiotic stress.

11.2.1 Jatropha Contains Toxic Compounds

Of the bioactive compounds produced by Jatropha, the ribosome-inactivating protein curcin and the toxic phorbol esters are the most popular. As early as 1974, studies on the toxicity of Jatropha seeds in mice have been reported (Adam 1974). When Jatropha became popular as a biofuel crop, more and more research groups became interested in analyzing the toxic compounds in different Jatropha plant parts especially seeds (Hallare et al. 2014) and leaves (Igbinosa et al. 2013). Health risks for farmers growing Jatropha and workers producing Jatropha oil, and environmental impact of using Jatropha biodiesel have become matters of public concern. The Jatropha oil was shown to have some adverse effects on mice when given orally at a dose of 500 mg/kg per day (Poon et al. 2013). Hallare et al. (2014) demonstrated the toxicity effect of Jatropha seed cake using zebrafish embryos. Similar studies are important to assess the effects of Jatropha biodiesel by-products on the environment and establish guidelines for disposal or reuse of such by-products. The toxicity of Jatropha seeds and leaves are attributed to phorbol esters (Li et al. 2010; Devappa et al. 2013; Hallare et al. 2014) and curcin (Lin et al. 2010), and this has been shown in mice (Li et al. 2010; Lin et al. 2010) and other animals (Becker and Makkar 1998), as well as in human tissues (Devappa et al. 2013). Accidental ingestion of Jatropha seeds by children and adults presented symptoms such as nausea, vomiting, and diarrhea (Shah and Sanmukhani 2010; Chomchai et al. 2011). Although molecular markers were already developed to distinguish between toxic and non-toxic varieties (Sudheer Pamidimarri et al. 2009), further work is necessary to establish a collection of non-toxic Jatropha that will be more suitable for use in actual plantations (see Chap. 6 for detailed description of Jatropha toxic compounds).

11.2.2 Female-to-Male Flower Ratio Is Low

Jatropha produces flowers that are either male or female (Heller 1996), with very small number of female flowers relative to male flowers. The low female-to-male flower ratio of 1:29 (Raju and Ezradanam 2002) result in low seed yields in Jatropha. Since the oil has to be extracted from the seeds, suitable Jatropha varieties should be producing more female flowers. In order to address this issue, Pan and Xu (2011) investigated the effect of benzyladenine (BA), a synthetic plant hormone with cytokinin activity, on the female-to-male flower ratio and seed yield in Jatropha. When BA was sprayed on Jatropha inflorescences, bisexual flowers were induced while control inflorescences had the usual unisexual flowers. Furthermore, a significant increase in the female-to-male flower ratio was seen in the treated inflorescences (Pan and Xu 2011). From this work, the same research group identified the key genes involved in the response to cytokinin responsible for changes in floral organ identity and increased flower number in cytokinin-treated Jatropha (Pan et al. 2014). The results of this recent work will be very useful for the future genetic modification efforts to develop high-yielding Jatropha varieties.

11.2.3 Tolerance to Abiotic Stress Such as Drought

Many reports showed that Jatropha is drought tolerant to a certain extent (Kheira and Atta 2009; Achten et al. 2010; Niu et al. 2012; Sapeta et al. 2013). However, the level of drought tolerance is not very different from other crops such as maize and soybean (Fujimaki and Kikuchi 2010). Therefore, in order to have a drought-tolerant Jatropha, further research has to be done, particularly a basic understanding of drought response mechanism is important. Several works on the evaluation of morphological and physiological traits of Jatropha during exposure to drought stress were carried out (Kheira and Atta 2009; Maes et al. 2009; Fujimaki and Kikuchi 2010; Krishnamurthy et al. 2012; Niu et al. 2012; Sapeta et al. 2013). From the works of Krishnamurthy et al. (2012), it was established that Jatropha's mechanism for drought response is categorized as drought avoidance. This is characterized by drought-induced leaf drop and limited root growth as the plants enter the dormant state. Furthermore, Niu et al. (2012) showed that drought significantly reduced the growth and leaf development in Jatropha plants grown in the greenhouse. In another work, similar symptoms were observed in drought-stressed Jatropha plants that were not watered for three weeks, such as wilting and reduced shoot growth; while prolonged drought treatment caused all the mature leaves to drop (Cartagena et al. 2014). However, the contribution of root branching to drought tolerance in Jatropha is not clear (Krishnamurthy et al. 2012). Gene discovery studies on Jatropha's response to drought stress have been carried out (Tang et al. 2007; Zhang et al. 2007; Cartagena et al. 2014; Zhang et al. 2015), and the results are described in the succeeding section.

11.3 Identification of Target Genes

Gene discovery is very important to identify key players in Jatropha growth and development, as well as in oil biosynthesis and survival under stressed conditions. Moreover, the identification and characterization of key genes is a prerequisite to carry out genetic modification. Genome and transcriptome analyses such as gene cloning, genome sequencing, and microarray analysis have been applied to discover genes in Jatropha.

11.3.1 Gene Cloning

Gene cloning is the most common and basic approach in gene discovery. One of the first Jatropha genes that was cloned is the curcin gene (Lin et al. 2003), indicating that toxicity is of major concern for Jatropha. Table 11.1 lists some of the Jatropha genes that were recently cloned and characterized, including especially those that play a role in oil biosynthesis and stress tolerance. Some of these genes have been used for the creation of transgenic Jatropha plants and are discussed in Sect. 11.5.

11.3.2 Sequencing

In 2010, three different research groups reported their results of expressed sequence tag

Table 11.1 Representative genes cloned and characterized in Jatropha

Gene name	Function	Reference
JcAPX (ascorbate peroxidase)	Salt tolerance shown in tobacco	Liu et al. (2014)
JcR1MYB1 (R1-MYB transcription factor)	Expression in tobacco enhanced by PEG, NaCl, cold treatment, ABA, JA, and ethylene treatment	H-L. Li et al. (2014)
JcWRKY (WRKY transcription factor)	Enhance systemic acquired resistance for disease tolerance and stress response	Agarwal et al. (2014)
JcARF19 and *JcIAA9* (auxin response factor and indole-3-acetic acid)	Involved in auxin signal transduction; contribute to seed length	Ye et al. (2014a)
JcMFT1 (mother of FT and TFL1)	Functions in seed development in Arabidopsis	Tao et al. (2014)
JcSDP1 (sugar dependent 1)	Oil biosynthesis; involved in breakdown of TAG to glycerol and FFAs	Kim et al. (2014)
JcCBF2 (C-repeat binding factor)	Transcription factor targeting cold-responsive (COR) genes such as *RD29A*, *COR105A*, and *COR6.6* in Arabidopsis	Wang et al. (2014)
JcFAD2-1 (delta 12 fatty acid desaturase)	Oil biosynthesis; conversion of oleic acid to linoleic acid	Qu et al. (2012)
JcNF-YB (B subunit of nuclear factor Y)	Transcription factor; enhanced drought tolerance in corn	Tsuchimoto et al. (2012)
JcBD1 (betaine aldehyde dehydrogenase 1)	Isolated from stress-treated Jatropha; conferred salt tolerance to *E. coli*	Zhang et al. (2008)
JcPIP2 (plasma membrane intrinsic protein 2)	Accumulate in Jatropha seedlings subjected to drought stress	Zhang et al. (2007)
JcERF (ethylene-responsive element-binding factor)	Isolated in Jatropha plants that were exposed to abiotic stresses such as cold, drought, high salt, ethylene, and wounding	Tang et al. (2007)

(EST) sequencing from cDNA isolated from seeds (Costa et al. 2010; Gomes et al. 2010; Natarajan et al. 2010). From developing seeds, Natarajan et al. (2010) reported 12,084 ESTs and 6,233 unigenes, while Gomes et al. (2010) identified 13,193 ESTs and 5,841 unigenes. On the other hand, Costa et al. (2010) reported 13,249 ESTs and 4,622 unigenes from developing and germinating endosperm. In a further study, roots from Jatropha plants subjected to salinity stress were used by Eswaran et al. (2012) to generate EST sequences. The *J. curcas S-adenosylmethionine-dependent methyltransferase* (*JcSAM*) gene was identified as one of the up-regulated genes in salt-stressed Jatropha plants (Eswaran et al. 2012). SAM-methyltransferases are involved in the synthesis of plant secondary metabolites, including those that are important in osmotic stress response (Moffatt and Weretilnyk 2001). Another gene shown by Eswaran et al. (2012) to be induced by high salinity in Jatropha leaf tissues was *J. curcas late embryogenesis abundant protein-5* (*JcLEA-5*). LEA genes contribute to stress tolerance in plants by protecting the membranes and the cellular proteins from damages that may be caused by stress from exposure to dry conditions (Wang et al. 2003, 2007).

The genome sequence of Jatropha was reported by our group in 2011 (Sato et al. 2011). A total of 40,929 genes were identified while 21,225 unigenes from leaf and callus transcriptome were found. Genes related to oil and toxin biosynthesis, disease resistance, and flowering are considered highly important for basic research in Jatropha and thus were further analyzed (Sato et al. 2011). The data initially reported by Sato et al. (2011) from the draft sequence was further evaluated and an upgraded genome information result in a total number of 30,203 genes with transcriptome data containing 19,454 sequences (Hirakawa et al. 2012). Following the initial sequencing works, Natarajan and Parani (2011) reported their pyrosequencing data of cDNAs from Jatropha roots, mature leaves, flowers, developing seeds, and mature embryo, which result in a transcriptome data set of 17,457 sequences, of which 27 sequences are directly involved in oil biosynthesis. Furthermore, King et al. (2011) used 454 sequencing to analyze the transcriptome of developing Jatropha seeds, which result in 12,419 contigs and 17,333 singletons. The enormous amount of genomic data will be useful for the development of Jatropha lines with improved agronomic traits, and more importantly increased oil yield (see Chap. 1 for detailed description of genomic sequencing).

RNA sequencing is the most current method for genome-wide transcriptome analysis. In Jatropha, RNA sequencing was carried out by Zhang et al. (2014) to analyze the transcription profiles of roots and leaves from plants grown in salt stress condition. It was reported that 1504 and 1115 genes were differentially expressed in roots and leaves, respectively, under salt stress. The major regulated genes detected in this transcriptomic data were related to trehalose synthesis and cell wall structure modification in roots, while related to raffinose synthesis and reactive oxygen scavenger in leaves (Zhang et al. 2014). A subsequent report by the same group presented their work on RNA sequencing for transcriptome analysis of drought-stressed Jatropha plants, and reported the identification of a total of 4103 differentially expressed genes (DEGs), majority of which were observed in leaves (Zhang et al. 2015). The number of DEGs increased as the plants were exposed to longer drought treatment, from one day to four and seven days. Genes that are expected to play significant roles during exposure to drought stress were identified including those involved in abscisic acid (ABA) synthesis and signaling in both Jatropha roots and leaves. Furthermore, several transcription factors and genes related to endoplasmic reticulum (ER) stress response, ethylene wax biosynthesis, and trehalose and raffinose biosynthesis were up-regulated in the leaves (Zhang et al. 2015). The identification of key genes conferring important traits in Jatropha will be useful in creating new transgenic Jatropha lines.

11.3.3 Microarray

A custom microarray for Jatropha was developed by our team and used for transcriptome profiling in drought-stressed plants (Cartagena et al. 2014). In addition to drought-responsive genes, the microarray analysis also evaluated gene expression in Jatropha plants recovering from drought stress. In this study, the duration of drought treatment was relatively longer (i.e., three weeks without irrigation) than that of the study presented by Zhang et al. (2015), and recovery was for one day after re-watering. A total of 332 genes were found to be associated with drought response in Jatropha, including some of the genes that were previously characterized such as *JcPIP2, JcARF19* and *JcIAA9* (Table 11.1), as well as *JcLEA* and *JcSAM* (Sect. 11.3.2). Moreover, there were 585 genes that are involved in recovery from drought stress, while 374 genes were found to be important during response to both drought and recovery.

Table 11.2 lists the most highly up-regulated genes in Jatropha after drought treatment. The list includes a variety of genes, ranging from those involved in ABA biosynthesis to sugar transporters. On the other hand, down-regulated genes in drought-stressed Jatropha include genes that code for a gibberellin-regulated family protein, transporter proteins, and enzymes involved in cell growth and development (Table 11.3). Not shown in the tables are the genes with no functional annotations, which were found to be differentially expressed in Jatropha plants during stress treatment. These genes represent the unique set of regulators to stress tolerance that are not conserved in other plant species. By utilizing the data generated from transcriptome profiling of Jatropha subjected to drought stress,

Table 11.2 Up-regulated genes in drought-treated Jatropha

Jatropha gene ID	Average expression ratio	AGI gene code	Description
Jcr4S00724.90	53.80	AT3G22840.1	Chlorophyll A-B binding family protein (ELIP1, ELIP)
Jcr4S00476.30	49.56	AT1G11530.1	C-terminal cysteine residue is changed to a serine 1 (ATCXXS1, CXXS1)
Jcr4S00868.110	24.45	AT3G29575.4	ABI five binding protein 3 (AFP3)
Jcr4S02308.90	21.78	AT3G22490.1	Seed maturation protein
Jcr4S00434.50	21.19	AT1G16770.1	Unknown protein
Jcr4S00096.220	19.73	AT1G07430.1	Highly ABA-induced PP2C gene 2 (HAI2)
Jcr4U31006.10	17.98	AT1G76010.1	Alba DNA/RNA-binding protein
Jcr4S00610.130	17.02	AT1G77210.2	Sugar transporter 14 (STP14)
Jcr4S00096.220	15.34	AT5G59220.1	Highly ABA-induced PP2C gene 1 (HAI1)
–No_Hits–	13.50	AT2G21650.1	Homeodomain-like superfamily protein (MEE3, ATRL2, RSM1)
Jcr4S13027.20	13.39	AT1G24020.2	MLP-like protein 423 (MLP423)
Jcr4S00108.170	13.22	AT1G06620.1	2-Oxoglutarate (2OG) and Fe(II)-dependent oxygenase superfamily protein
Jcr4S00022.120	11.88	AT2G46240.1	BCL-2-associated athanogene 6 (BAG6, ATBAG6)
Jcr4S06853.20	10.34	AT5G50700.1	Hydroxysteroid dehydrogenase 1 (HSD1)
Jcr4S00434.50	10.32	AT1G16760.1	Protein kinase protein with adenine nucleotide alpha hydrolases-like domain
Jcr4S07211.40	9.63	AT2G05580.1	Glycine-rich protein family
Jcr4S05041.50	9.58	AT1G05460.1	P-loop containing nucleoside triphosphate hydrolases superfamily protein (SDE3)
Jcr4S20471.10	9.53	AT3G26210.1	Cytochrome P450, family 71, subfamily B, polypeptide 23 (CYP71B23)
Jcr4S00772.40	9.21	AT1G72100.1	Late embryogenesis abundant domain-containing protein/LEA domain-containing protein
Jcr4S02582.50	9.09	AT3G57520.2	Seed imbibition 2 (AtSIP2, SIP2)
Jcr4S02308.90	8.83	AT3G22490.1	Seed maturation protein
–No_Hits–	8.46	AT3G48770.1	DNA binding; ATP binding

it will be possible to elucidate the molecular mechanism of drought response and will be useful in designing a more drought-tolerant Jatropha variety.

11.4 Establishment of Transformation Method

Agrobacterium-mediated transformation is still the most convenient way of introducing genes into plant cells. This method has been applied to Jatropha and was first reported by Li et al. (2008). Succeeding reports were aimed at improving transformation efficiency either by using different types of plant tissues (Kumar et al. 2010) or modifying the infection process (Khemkladngoen et al. 2011a; Jaganath et al. 2014). Aside from the type of tissue and infection process, the age of plant tissues and the tissue culture method are also critical factors that influence transformation efficiency. While Li et al. (2008) used cotyledons from 2-week-old seedlings, the use of younger cotyledons that

Table 11.3 Down-regulated genes in drought-treated Jatropha

Jatropha gene ID	Average expression ratio	AGI gene code	Description
Jcr4S04740.10	0.004	AT1G22690.3	Gibberellin-regulated family protein
Jcr4S00051.20	0.01	AT5G20630.1	GLP3, GLP3A, GLP3B, ATGER3, GER3 | germin 3
Jcr4S01859.30	0.01	AT5G39130.1	RmlC-like cupins superfamily protein
Jcr4S08395.10	0.01	AT1G72610.1	GLP1, ATGER1, GER1 | germin-like protein 1
–No_Hits–	0.02	AT5G57560.1	TCH4, XTH22 | Xyloglucan endotransglucosylase/hydrolase family protein
Jcr4S01681.30	0.02	AT2G28950.1	ATEXPA6, ATEXP6, ATHEXP ALPHA 1.8, EXPA6 | expansin A6
–No_Hits–	0.02	AT5G59970.1	Histone superfamily protein
Jcr4S03495.20	0.02	AT2G42840.1	PDF1 | protodermal factor 1
–No_Hits–	0.02	AT5G57560.1	TCH4, XTH22 | Xyloglucan endotransglucosylase/hydrolase family protein
Jcr4S02325.90	0.02	AT1G70710.1	ATGH9B1, CEL1, GH9B1 | glycosyl hydrolase 9B1
Jcr4U39085.20	0.02	AT2G38080.1	IRX12, LAC4, ATLMCO4, LMCO4 | Laccase/Diphenol oxidase family protein
Jcr4S00340.130	0.02	AT1G04680.1	Pectin lyase-like superfamily protein
Jcr4S04329.20	0.02	AT5G22740.1	ATCSLA02, CSLA02, ATCSLA2, CSLA2 | cellulose synthase-like A02
–No_Hits–	0.02	AT2G41790.1	Insulinase (Peptidase family M16) family protein
Jcr4S08179.20	0.03	AT5G40020.1	Pathogenesis-related thaumatin superfamily protein
–No_Hits–	0.03	AT4G14960.2	TUA6 | Tubulin/FtsZ family protein
Jcr4S02434.20	0.03	AT5G18430.1	GDSL-like Lipase/Acylhydrolase superfamily protein
–No_Hits–	0.03	AT5G59320.1	LTP3 | lipid transfer protein 3
Jcr4S00412.40	0.03	AT1G02180.1	Ferredoxin-related
Jcr4S01605.60	0.03	AT3G45010.1	scpl48 | serine carboxypeptidase-like 48
Jcr4S01270.10	0.03	AT1G71691.1	GDSL-like Lipase/Acylhydrolase superfamily protein
–No_Hits–	0.03	AT3G16920.1	CTL2, ATCTL2 | chitinase-like protein 2
–No_Hits–	0.03	AT5G52860.1	ABC-2 type transporter family protein

were precultured for one week showed a significant increase in transformation efficiency (Khemkladngoen et al. 2011a). Aside from using younger tissues, a major factor that contributed to higher transformation efficiency in the works of Khemkladngoen et al. (2011a) is the physical wounding step that involved sonicating the Agrobacterium solution containing the cotyledon tissues. Sonication result in localized, micro-wounds on plant cells which facilitated Agrobacterium infection. Furthermore, since the wounds created were just localized on the surface of tissues, the plant cells were not completely damaged and made it possible to regenerate when cultured in medium. The optimized medium composition for tissue culture (Khemkladngoen et al. 2011b) was applied to the transformed explants; and finally, stable transformants were obtained at a rate of 53% (Khemkladngoen et al. 2011a).

While majority of the reports on creation of transgenic Jatropha uses Agrobacterium, two groups presented their study on the establishment of a microprojectile bombardment technique for Jatropha (Purkayastha et al. 2010; Joshi et al. 2011). The method by Joshi et al. (2011) was later applied to create salt-tolerant transgenic Jatropha plants (Jha et al. 2013). Just like the Agrobacterium-mediated transformation method, microprojectile bombardment also relies on tissue regeneration of transformed explants.

The tissue culture of Jatropha has become one of the biggest limitations in creating transgenic Jatropha plants. Recently, in planta transformation was carried out in germinating Jatropha seeds (Patade et al. 2014) and plantlets (Jaganath et al. 2014), which involved pricking with a needle to create some wounds on the tissues prior to *Agrobacterium* infection. Both methods result in very high transformation efficiency and stably transformed transgenic lines. Furthermore, Jaganath et al. (2014) developed a grafting method to propagate transgenic Jatropha that result in 100% genetic stability between mother and grafted plants. Further improvements of the transformation technique in Jatropha include the creation of marker-free transgenic plants using the chemical inducible Crelox-mediated site-specific recombination system (Qu et al. 2012; Gu et al. 2014). The production of genetically modified plants free of an antibiotic selection marker is expected to significantly contribute to the acceptance of transgenic Jatropha plants by the public.

11.5 Progress in the Creation of Transgenic Jatropha Lines

11.5.1 Drought Tolerance

Using *Agrobacterium*-mediated transformation, three target genes were introduced into Jatropha (Tsuchimoto et al. 2012). The first one codes for the enzyme phosphopantetheine adenylyltransferase (PPAT), which helps the plant to increase overall metabolic activity during drought stress response. In Arabidopsis, Rubio et al. (2008) showed that *PPAT* overexpressing lines had about 1.6 higher levels of CoA + acetyl-CoA. Furthermore, vegetative and reproductive growth was also improved as well as tolerance to salt and osmotic stress (Rubio et al. 2008). For a biofuel crop such as Jatropha, PPAT will be very useful since *PPAT* overexpression in Arabidopsis produced seeds which contained between 35 and 50% more fatty acids than wild type (Rubio et al. 2008). In Jatropha, it is expected that metabolic activity and drought stress tolerance will be enhanced by overexpression of the *PPAT* gene.

The second gene is called *NF-YB* (B subunit of the transcriptional regulator NF-Y) that can control the expression of genes that improves drought stress response in crops. Plants that encounter drought stress tend to shrink result in smaller biomass. However, maize that constitutively overexpress the NF-YB protein was shown to have better drought tolerance without sacrificing plant growth (Nelson et al. 2007). Furthermore, the yield of the transgenic maize was reported to be higher than that of the wild type under dry condition. These results indicate that plants expressing NF-YB are capable of immediately responding to drought stress. The homolog of Arabidopsis and maize NF-YB in Jatropha was already identified by comparison with amino acid sequence of NF-Y family protein encoded in the Jatropha genome. This *JcNF-YB* gene was cloned and overexpressed in Jatropha (Tsuchimoto et al. 2012).

The third set of transgenic plants developed was allowed to overexpress the genes coding for GSMT (glycine sarcosine methyltransferase) and DMT (dimethylglycine methyltransferase) isolated from the cyanobacterium, *Synechococcus* sp. (Waditee et al. 2005). GSMT and DMT are key enzymes in the glycine betaine biosynthesis (GB) pathway in halophytes. Halophytes with extremely high salt resistance produce a low molecular weight compound called glycine betaine. This compound helps to maintain a high osmotic pressure in the cell and thus enabling efficient water absorption from salt water. In addition, crops such as wheat, maize, and sugar beet had been genetically modified to produce glycine betaine when exposed to a dry condition, which indicates that these plants can do so as a strategy which contributes to drought stress tolerance. It was also reported that plants overproducing glycine betaine not only showed salt and drought stress tolerance, but also improved their tolerance to low temperature. Thus, it is expected that the transformed Jatropha that overproduces glycine betaine can be grown in a wider area, including dry lands and temperate regions. The GB produced by Jatropha transformants were about three times more than the non-transformed plants (Tsuchimoto et al. 2012).

The ability to accumulate GB in plants allows them to have increased tolerance to high osmotic pressure. Under dry conditions, the underground salt concentration rises and consequently affects plant growth. A high salt concentration in the soil result in high osmotic pressure causing an imbalance of osmotic pressure in the plant cell, which eventually result in inefficient water absorption.

11.5.2 Salt Tolerance

Transgenic Jatropha were created by introducing the *SbNHX1* gene isolated from the extreme halophyte *Salicornia brachiata* (Jha et al. 2013). Using microprojectile bombardment of precultured embryo axes, the salt tolerance of transgenic Jatropha was shown to be enhanced. The identification of salt-induced genes such as *JcAPX*, *JcERF*, and *JcBD1* (Table 11.1) presents a source of potential genes that can be used to create salt-tolerant Jatropha plants.

11.5.3 Enhanced Oil Biosynthesis

Oil content in Jatropha was improved by targeting the *FAD2* gene which codes for the key enzyme responsible for conversion of oleic acid to linoleic acid during linoleic acid biosynthesis in plants (Qu et al. 2012). The cetane number of biofuels is an indicator of oil quality and is affected by the amount of oleic acid. The approach was to silence the expression of *JcFAD2* using RNAi to inhibit the conversion of oleic acid to linoleic acid, which will result in accumulation of oleic acid and an increase in cetane number. As a result, the transgenic Jatropha produced oil with 78% oleic acid, compared to that produced by the wild type having only 37% oleic acid (Qu et al. 2012). Moreover, the level of polyunsaturated fatty acids was significantly reduced to 3% from 41% (Qu et al. 2012). The oil quality produced by the transgenic Jatropha was tremendously improved by silencing only one gene, *JcFAD2*.

Another gene that was targeted for RNAi silencing was *JcSDP1* (*sugar dependent 1*) (Kim et al. 2014). In Arabidopsis, SDP1 is responsible for the first step of triacylglycerol (TAG) degradation during seed germination and was found to associate with other patatin-like TAG lipases (PTLs). TAG degradation acted upon by SDP1 produces glycerol and free fatty acids. Thus, *JcSDP1* silencing result in accumulation of seed oils in Jatropha from 13 to 30% seed storage lipid more than non-transgenic seeds (Kim et al. 2014). The free fatty acid content was threefold lower at 8.5% from the 27% contained in non-transgenic seeds (Kim et al. 2014) (see Chap. 7 for detailed description of oil metabolism).

11.5.4 Non-Toxicity

In order to develop non-toxic transgenic Jatropha plants, Patade et al. (2014) designed an RNAi vector to silence the *curcin precursor gene*. Using Agrobacterium-mediated transformation of seedlings, they were able to produce transgenic Jatropha with considerably reduced transcript levels for the *curcin precursor gene* (Patade et al. 2014). It was also shown that the transcript level of the *curcin precursor gene* increased when wild-type Jatropha plants were subjected to mechanical stress. This indicates that curcin is essential in response to stress and should be carefully considered not to be completely inactivated when designing transgenic Jatropha plants. Patade et al. (2014) noted that a seed-specific silencing construct is being considered, which will allow Jatropha plants to express curcin until before seed development. It is important for Jatropha seeds to be curcin-free so that the seed cake can be utilized as animal feeds (see Chap. 6).

11.5.5 Pest Resistance

Like any other plant, Jatropha is also susceptible to insect pests such as the tortrix moth, *Archips micaceanus*. Since the use of chemical pesticides will also kill insect pollinators for Jatropha, Gu et al. (2014) attempted to engineer transgenic Jatropha expressing the *cry* gene from *Bacillus thuringiensis* (*Bt*). Particularly, the overexpression of the hybrid *cry* gene *Cry1Ab/1Ac* in Jatropha was shown to be effective against *A. micaceanus* (Gu et al. 2014).

Using an RNAi approach, transgenic Jatropha plants resistant to the Indian cassava mosaic virus (ICMV) were successfully created by Ye et al. (2014b). Five ICMV genes involved in genome replication and gene expression as well as transcription factors negatively regulating genes involved in host defense system were used to construct a hairpin double-stranded (ds) RNA (Ye et al. 2014b). The resulting transgenic Jatropha plants were shown to be resistant to ICMV as indicated by the absence of virus particles. Furthermore, an ICMV-resistant T_1 generation line was produced, demonstrating that the resistance trait is heritable (Ye et al. 2014b).

11.5.6 Early Flowering

A homolog of the Arabidopsis *FT* gene was isolated in Jatropha and used to create an early flowering transgenic line (C. Li et al. 2014). Transgenic Jatropha that were cultured in vitro for seven weeks already produced flower buds. Morphological analysis of the flowers from transgenic Jatropha showed abnormal floral organs indicating that *JcFT* directly regulates genes involved in flowering. This was also demonstrated by the increased gene expression levels of flowering genes such as *JcSOC1*, *JcLFY*, and *JcAP1* in the *JcFT* transgenic Jatropha (C. Li et al. 2014).

11.6 Conclusion

Genetic modification provides an efficient means of obtaining improved Jatropha varieties since the need for the development of a reliable source of biofuels is urgent. With the progress in basic research, several transgenic lines have already been developed in Jatropha. The discovery of key genes has contributed to the increase in gene modification experiments. However, more work has to be done in order to establish a complete transformation method including tissue culture. Alternatively, the development of an *in planta* method will facilitate faster gene modification in Jatropha. Much progress was seen in the last 8 or 10 years, but more work is needed to obtain a more acceptable final product i.e., trees and oil that are safe for humans and the environment. It is imperative that scientists should develop Jatropha plants that will not only be productive, but also environmentally safe in order to attain sustainability in biofuel production.

References

Achten WMJ, Maes WH, Reubens B, Mathijs E, Singh VP, Verchotd L, Muys B (2010) Biomass production and allocation in *Jatropha curcas* L. seedlings under different levels of drought stress. Biomass Bioenergy 34:667–676

Adam SEI (1974) Toxic effects of Jatropha curcas in mice. Toxicology 2:67–76

Adarsh BV (2011) Synthetic Paraffinic Kerosene (SPK) from *Jatropha curcas*: overall impact on environment. Int J Res Chem Environ 1(2):123–126

Agarwal P, Dabi M, Agarwal PK (2014) Molecular cloning and characterization of a group II WRKY transcription factor from *Jatropha curcas*, an important biofuel crop. DNA Cell Biol 33(8):503–513

Basha SD, Sujatha M (2007) Inter and intra-population variability of *Jatropha curcas* (L.) characterized by RAPD and ISSR markers and development of population-specific SCAR markers. Euphytica 156:375–386

Becker K, Makkar HP (1998) Effects of phorbol esters in carp (*Cyprinus carpio* L.). Vet Hum Toxicol 40 (2):82–86

Cartagena JA, Seki M, Tanaka M, Yamauchi T, Sato S, Hirakawa H, Tsuge T (2014) Gene expression profiles in Jatropha under drought stress and during recovery. Plant Mol Biol Rep 33(4):1075–1087

Chomchai C, Kriengsunthornkij W, Sirisamut T, Nimsomboon T, Rungrueng W, Silpasupagornwong U (2011) Toxicity from ingestion of *Jatropha curcas* ('saboo dum') seeds in Thai children. Southeast Asian J Trop Med Public Health 42(4):946–950

Costa GG, Cardoso KC, Del Bem LE, Lima AC, Cunha MA, de Campos-Leite L, Vicentini R, Papes F, Moreira RC, Yunes JA, Campos FA, Da Silva MJ (2010) Transcriptome analysis of the oil-rich seed of the bioenergy crop *Jatropha curcas* L. BMC Genom 11:462. doi:10.1186/1471-2164-11-462

Debnath M, Bisen PS (2008) *Jatropha curcas* L., a multipurpose stress resistant plant with a potential for ethnomedicine and renewable energy. Curr Pharm Biotechnol 9(4):288–306

Devappa RK, Roach JS, Makkar HP, Becker K (2013) Occular and dermal toxicity of *Jatropha curcas* phorbol esters. Ecotoxicol Environ Saf 94:172–178

Eckart K, Henshaw P (2012) *Jatropha curcas* L. and multifunctional platforms for the development of rural sub-Saharan Africa. Energy Sustain Dev 16(3): 303–311

Eswaran N, Parameswaran S, Anantharaman B, Kumar GRK, Sathram B, Johnson TS (2012) Generation of an expressed sequence tag (EST) library from salt-stressed roots of *Jatropha curcas* for identification of abiotic stress-responsive genes. Plant Biol 14:428–437

Fargione J, Hill J, Tilman D, Polasky S, Hawthorne P (2008) Land clearing and the biofuel carbon debt. Science 319(5867):1235–1238

Fujimaki H, Kikuchi N (2010) Drought and salinity tolerances of young Jatropha. Int Agrophys 24:121–127

Gomes KA, Almeida TC, Gesteira AS, Lôbo IP, Guimarães ACR, de Miranda AB, Van Sluys MA, da Cruz RS, Cascardo JCM, Carels N (2010) ESTs from seeds to assist the selective breeding of *Jatropha curcas* L. for oil and active compounds. Genom Insights 3:29–56

Gu K, Mao H, Yin Z (2014) Production of marker-free transgenic *Jatropha curcas* expressing hybrid *Bacillus thuringiensis* δ-endotoxin Cry1Ab/1Ac for resistance to larvae of tortrix moth (*Archips micaceanus*). Biotechnol Biofuels 7:68. doi:10.1186/1754-6834-7-68

Hallare AV, Ruiz PLS, Cariño JCED (2014) Assessment of *Jatropha curcas* L. biodiesel seed cake toxicity using the zebrafish (*Danio rerio*) embryo toxicity (ZFET) test. Environ Sci Pollut Res 21:6044–6056

Heller J (1996) Physic Nut. *Jatropha curcas* L. Promoting the conservation and use of underutilized and neglected crops. International Plant Genetic Resources Institute, Rome, Italy

Hirakawa H, Tsuchimoto S, Sakai H, Nakayama S, Fujishiro T, Kishida Y, Kohara M, Watanabe A, Yamada M, Aizu T, Toyoda A, Fujiyama A, Tabata S, Fukui K, Sato S (2012) Upgraded genomic information of *Jatropha curcas* L. Plant Biotechnol 29:123–130

Igbinosa OO, Oviasogie EF, Igbinosa EO, Igene O, Igbinosa IH, Idemudia OG (2013) Effects of biochemical alteration in animal model after short-term exposure of *Jatropha curcas* (Linn) leaf extract. Sci World J. doi:10.1155/2013/798096

Jaganath B, Subramanyam K, Mayavan S, Karthik S, Elayaraja D, Udayakumar R, Manickavasagam M, Ganapathi A (2014) An efficient in planta transformation of *Jatropha curcas* (L.) and multiplication of transformed plants through in vivo grafting. Protoplasma 251(3):591–601

Jha B, Mishra A, Jha A, Joshi M (2013) Developing transgenic Jatropha using the *SbNHX1* gene from an extreme halophyte for cultivation in saline wasteland. PLoS ONE 8(8):e71136

Joshi M, Mishra A, Jha B (2011) Efficient genetic transformation of *Jatropha curcas* L. by microprojectile bombardment using embryo axes. Ind Crop Prod 33:67–77

Kheira AAA, Atta NMM (2009) Response of *Jatropha curcas* L. to water deficits: yield, water use efficiency and oil seed characteristics. Biomass Bioenergy 33:1343–1350

Khemkladngoen N, Cartagena J, Fukui K (2011a) Physical wounding-assisted Agrobacterium-mediated transformation for juvenile cotyledons of a biodiesel producing plant, *Jatropha curcas* L. Plant Biotechnol Rep 5:235–243

Khemkladngoen N, Cartagena J, Shibagaki N, Fukui K (2011b) Adventitious shoot regeneration from juvenile cotyledons of a biodiesel producing plant, *Jatropha curcas* L. J Biosci Bioeng 111:67–70

Kim MJ, Yang SW, Mao HZ, Veena SP, Yin JL, Chua NH (2014) Gene silencing of *Sugar-dependent 1* (*JcSDP1*), encoding a patatin-domain triacylglycerol lipase, enhances seed oil accumulation in *Jatropha curcas*. Biotechnol Biofuels 7(1):36. doi:10.1186/1754-6834-7-36

King AJ, Li Y, Graham IA (2011) Profiling the developing *Jatropha curcas* L. seed transcriptome by pyrosequencing. Bioenerg Res 4:211–221

Krishnamurthy L, Zaman-Allah M, Marimuthu S, Wani SP, Rao AVRK (2012) Root growth in Jatropha and its implications for drought adaptation. Biomass Bioenergy 39:247–252

Kumar N, Anand KGV, Pamidimarri DVNS, Sarkar T, Reddy MP, Radhakrishnan T, Kaul T, Reddy MK, Sopori SK (2010) Stable genetic transformation of *Jatropha curcas* via Agrobacterium tumefaciens-mediated gene transfer using leaf explants. Ind Crops Prod 32:41–47

Li M, Li H, Jiang H, Pan X, Wu G (2008) Establishment of an Agrobacterium-mediated cotyledon disc transformation method for *Jatropha curcas*. Plant Cell Tiss Org Cult 92:173–181

Li CY, Devappa RK, Liu JX, Lv JM, Makkar HP, Becker K (2010) Toxicity of *Jatropha curcas* phorbol esters in mice. Food Chem Toxicol 48(2):620–625

Li H-L, Guo D, Peng SQ (2014) Molecular characterization of the *Jatropha curcas JcR1MYB1* gene encoding a putative R1-MYB transcription factor. Genet Mol Biol 37(3):549–555

Li C, Luo L, Fu Q, Niu L, Xu ZF (2014) Isolation and functional characterization of *JcFT*, a *FLOWERING LOCUS T* (*FT*) homologous gene from the biofuel plant *Jatropha curcas*. BMC Plant Biol 14:125. doi:10.1186/1471-2229-14-125

Lin J, Li YX, Zhou XW, Tang KX, Chen F (2003) Cloning and characterization of a curcin gene encoding a ribosome inactivating protein from *Jatropha curcas*. DNA Seq 14(4):311–317

Lin J, Zhou X, Wang J, Jiang P, Tang K (2010) Purification and characterization of curcin, a toxic lectin from the seed of *Jatropha curcas*. Prep Biochem Biotechnol 40(2):107–118

Liu Z, Bao H, Cai J, Han J, Zhou L (2014) A novel thylakoid ascorbate peroxidase from *Jatropha curcas* enhances salt tolerance in transgenic tobacco. Int J Mol Sci 15:171–185

Maes WH, Achten WMJ, Reubens B, Raes D, Samson R, Muys B (2009) Plant–water relationships and growth strategies of Jatropha curcas L. saplings under different levels of drought stress. J Arid Environ 73:877–884

Moffatt BA, Weretilnyk EA (2001) Sustaining S-adenosyl-L-methionine-dependent methyltransferase activity in plant cells. Physiol Plant 113:435–442

Natarajan P, Kanagasabapathy D, Gunadayalan G, Panchalingam J, Shree N, Sugantham PA, Singh KK, Madasamy P (2010) Gene discovery from *Jatropha curcas* by sequencing of ESTs from normalized and full-length enriched cDNA library from developing seeds. BMC Genom 11:606. doi:10.1186/1471-2164-11-606

Natarajan P, Parani M (2011) De novo assembly and transcriptome analysis of five major tissues of *Jatropha curcas* L. using GS FLX titanium platform of 454 pyrosequencing. BMC Genom 12:191. doi:10.1186/1471-2164-12-191

Nelson DE, Repetti PP, Adams TR, Creelman RA, Wu J, Warner DC, Anstrom DC, Bensen RJ, Castiglioni PP, Donnarummo MG, Hinchey BS, Kumimoto RW, Maszle DR, Canales RD, Krolikowski KA, Dotson SB, Gutterson N, Ratcliffe OJ, Heard JE (2007) Plant nuclear factor Y (NF-Y) B subunits confer drought tolerance and lead to improved corn yields on water-limited acres. Proc Natl Acad Sci USA 104:16450–16455

Niu G, Rodriguez D, Mendoza M, Jifon J, Ganjegunte G (2012) Responses of *Jatropha curcas* to salt and drought stresses. Int J Agron. doi:10.1155/2012/632026

Ong HC, Mahlia TMI, Masjuki HH, Norhasyima RS (2011) Comparison of palm oil, *Jatropha curcas* and *Calophyllum inophyllum* forbiodiesel: a review. Renew Sustain Energy Rei 15:3501–3515

Pan BZ, Chen MS, Ni J, Xu ZF (2014) Transcriptome of the inflorescence meristems of the biofuel plant *Jatropha curcas* treated with cytokinin. BMC Genom 15:974. doi:10.1186/1471-2164-15-974

Pan BZ, Xu ZF (2011) Benzyladenine treatment significantly increases the seed yield of the biofuel plant *Jatropha curcas*. J Plant Growth Regul 30:166–174

Patade VY, Khatri D, Kumar K, Grover A, Kumari M, Gupta SM, Kumar D, Nasim M (2014) RNAi mediated curcin precursor gene silencing in Jatropha (*Jatropha curcas* L.). Mol Biol Rep 41:4305–4312

Poon R, Valli VE, Ratnayake WM, Rigden M, Pelletier G (2013) Effects of Jatropha oil on rats following 28-day oral treatment. J Appl Toxicol 33(7):618–625

Purkayastha J, Sugla T, Paul A, Solleti SK, Mazumdar P, Basu A, Mohommad A, Ahmed Z, Sahoo L (2010) Efficient in vitro plant regeneration from shoot apices and gene transfer by particle bombardment in *Jatropha curcas*. Biol Plant 54(1):13–20

Qu J, Mao HZ, Chen W, Gao SQ, Bai YN, Sun YW, Geng YF, Ye J (2012) Development of marker-free transgenic Jatropha plants with increased levels of seed oleic acid. Biotechnol Biofuels 5(1):10. doi:10.1186/1754-6834-5-10

Raja SA, Smart DSR, Lee CLR (2011) Biodiesel production from Jatropha oil and its characterization. Res J Chem Sci 1:81–87

Raju AJS, Ezradanam V (2002) Pollination ecology and fruiting behaviour in a monoecious species, *Jatropha curcas* L. (Euphorbiaceae). Curr Sci 83:1395–1398

Renner A, Zelt T, Gerteiser S (2008) Global market study on Jatropha. GEXSI, London. http://www.jatropha-alliance.org/fileadmin/documents/GEXSI_Global-Jatropha-Study_FULL-REPORT.pdf Accessed 12 Apr 2015

Rubio S, Whitehead L, Larson TR, Graham IA, Rodriguez PL (2008) The coenzyme a biosynthetic enzyme phosphopantetheine adenylyltransferase plays a crucial role in plant growth, salt/osmotic stress resistance, and seed lipid storage. Plant Physiol 148:546–556

Sanderson K (2009) Wonder weed plans fail to flourish. Nature 461:328–329

Sapeta H, Costa JM, Lourenço T, Maroco J, van der Linde P, Oliveira MM (2013) Drought stress response in *Jatropha curcas*: growth and physiology. Environ Exp Bot 85:76–84

Sato S, Hirakawa H, Isobe S, Fukai E, Watanabe A, Kato M, Kawashima K, Minami C, Muraki A, Nakazaki N, Takahashi C, Nakayama S, Kishida Y, Kohara M, Yamada M, Tsuruoka H, Sasamoto S, Tabata S, Aizu T, Toyoda A, Shin-I T, Minakuchi Y, Kohara Y, Fujiyama A, Tsuchimoto S, Kajiyama S, Makigano E, Ohmido N, Shibagaki N, Cartagena JA, Wada N, Kohinata T, Atefeh A, Yuasa S, Matsunaga S, Fukui K (2011) Sequence analysis of the genome of an oil-bearing tree *Jatropha curcas* L. DNA Res 18(1):65–76

Shah V, Sanmukhani J (2010) Five cases of *Jatropha curcas* poisoning. J Assoc Physicians India 58:245–246

Sinkala T, Johnson FX (2012) Small-scale production of Jatropha in Zambia and its implications for rural development and national biofuel policies. In: Janssen Rainer, Rutz Dominik (eds) Bioenergy for sustainable development in Africa, Part 1. Springer, Netherlands, pp 41–51

Spielberg GT (2009) Alternative jet fuel: the Jatropha plant? http://www.businessweek.com/bwdaily/dnflash/content/feb2009/db2009026_918710.htm. Accessed 17 Dec 2014

Sudheer Pamidimarri DV, Singh S, Mastan SG, Patel J, Reddy MP (2009) Molecular characterization and identification of markers for toxic and non-toxic varieties of *Jatropha curcas* L. using RAPD, AFLP and SSR markers. Mol Biol Rep 36:1357–1364

Tang M, Sun J, Liu Y, Chen F, Shen S (2007) Isolation and functional characterization of the *JcERF* gene, a putative AP2/EREBP domain containing transcription factor, in the woody oil plant *Jatropha curcas*. Plant Mol Biol 63:419–428

Tao YB, Luo L, He LL, Ni J, Xu ZF (2014) A promoter analysis of *MOTHER OF FT AND TFL1 1* (*JcMFT1*), a seed-preferential gene from the biofuel plant *Jatropha curcas*. J Plant Res 127(4):513–524

Thomas R, Sah NK, Sharma PB (2008) Therapeutic biology of *Jatropha curcas*: a mini review. Curr Pharm Biotechnol 9(4):315–324

Tsuchimoto S, Cartagena J, Khemkladngoen N, Singkaravanit S, Kohinata T, Wada N, Sakai H, Morishita Y, Suzuki H, Shibata D, Fukui K (2012) Development of transgenic plants in jatropha with drought tolerance. Plant Biotechnol 29:137–143

Waditee R, Bhuiyan MN, Rai V, Aoki K, Tanaka Y, Hibino T, Suzuki S, Takano J, Jagendorf AT, Takabe T, Takabe T (2005) Genes for direct methylation of glycine provide high levels of glycinebetaine and abiotic-stress tolerance in *Synechococcus* and *Arabidopsis*. Proc Natl Acad Sci USA 102:1318–1323

Wang L, Gao J, Qin X, Shi X, Luo L, Zhang G, Yu H, Li C, Hu M, Liu Q, Xu Y, Chen F (2014) *JcCBF2* gene from *Jatropha curcas* improves freezing tolerance of *Arabidopsis thaliana* during the early stage of stress. Mol Biol Rep. doi:10.1007/s11033-014-3831-0

Wang W, Vinocur B, Altman A (2003) Plant responses to drought, salinity and extreme temperatures: towards genetic engineering for stress tolerance. Planta 218:1–14

Wang XS, Zhu HB, Jin GL, Liu HL, Wu WR, Zhu J (2007) Genome scale identification and analysis of *LEA* genes in rice (*Oryza sativa* L.). Plant Sci 172:414–420

Ye J, Liu P, Zhu C, Qu J, Wang X, Sun Y, Sun F, Jiang Y, Yue G, Wang C (2014a) Identification of candidate genes *JcARF19* and *JcIAA9* associated with seed size traits in Jatropha. Funct Integr Genom 14(4):757–766

Ye J, Qu J, Mao HZ, Ma ZG, Rahman NE, Bai C, Chen W, Jiang SY, Ramachandran S, Chua NH (2014b) Engineering geminivirus resistance in *Jatropha curcus*. Biotechnol Biofuels 7(1):149. doi:10.1186/s13068-014-0149-z

Zhang C, Zhang L, Zhang S, Zhu S, Wu P, Chen Y, Li M, Jiang H, Wu G (2015) Global analysis of gene expression profiles in physic nut (*Jatropha curcas* L.) seedlings exposed to drought stress. BMC Plant Biol 15(1):17

Zhang FL, Niu B, Wang YC, Chen F, Wang SH, Xu Y, Jiang LD, Gao S, Wu J, Tang L, Jia YJA (2008) A novel betaine aldehyde dehydrogenase gene from *Jatropha curcas*, encoding an enzyme implicated in adaptation to environmental stress. Plant Sci 174:510–518

Zhang L, Zhang C, Wu P, Chen Y, Li M, Jiang H, Wu G (2014) Global analysis of gene expression profiles in physic nut (*Jatropha curcas* L.) seedlings exposed to salt stress. PLoS One 9(5):e97878. doi:10.1371/journal.pone.0097878

Zhang Y, Wang Y, Jiang L, Xu Y, Wang Y, Lu D, Chen F (2007) Aquaporin JcPIP2 is involved in drought responses in *Jatropha curcas*. Acta Biochim Biophys Sin (Shanghai) 39:787–794

Agrobacterium-Mediated Genetic Transformation for Larger Seed Size in Jatropha

12

Harumi Enoki, Akimitsu Funato, Yusei Nabetani, Shinya Takahashi, Takanari Ichikawa, Minami Matsui and Reiko Motohashi

Abstract

Jatropha, *Jatropha curcas*, is known as the oilseed plant that yields biofuel. An average of 30–40% of a single Jatropha seed is made of oils and fats and can be processed for use as diesel engine fuel. This chapter is concerned with using transgenic approaches to increase oil production in Jatropha. At present, we are attempting to make larger Jatropha seeds by transferring rice genes coupled with the *CaMV35S* promoter into Jatropha plants. We found candidate rice genes that produce increased seed size in other plants (*Arabidopsis thaliana*) by using rice Full-Length cDNA OvereXpressing gene 14 hunting system. We are transferring four genes (*LOC_Os08g41910* encoding Sua5/YciO/YrdC/YwlC family protein, *LOC_Os04g43210* encoding probable inositol transporter 2-like, *LOC_Os03g49180* encoding alkaline ceramidase, and *LOC_Os10g40934* encoding putative flavonol synthase/flavanone 3-hydroxylase or 2OG-Fe (II) oxygenase containing protein) to attempt to make larger Jatropha seeds. We already made some transgenic Jatropha. Here we also discuss

H. Enoki · A. Funato · Y. Nabetani · R. Motohashi (✉)
Department of Agriculture, Shizuoka University, 836 Ohoya, Suruga-Ku, Shizuoka, Shizuoka 422-8529, Japan
e-mail: armotoh@ipc.shizuoka.ac.jp

S. Takahashi · T. Ichikawa · M. Matsui
RIKEN Plant Science Center (now Synthetic Genomic Research Group, Center for Sustainable Resource Science), Yokohama, Japan

S. Takahashi
Alliance for Research on North Africa, University of Tsukuba, 1-1-1 Tennodai, Tsukuba, Ibaraki 305-8572, Japan

T. Ichikawa
Okinawa Institute of Science and Technology Graduate University, Okinawa, Japan

S. Tsuchimoto (ed.), *The Jatropha Genome*, Compendium of Plant Genomes,
DOI 10.1007/978-3-319-49653-5_12

an improved method of *Agrobacterium*-mediated transformation to increase seed growth efficiency. We used Gamborg'B5 medium containing 2% sucrose and 250 mg/l active charcoal to improve the condition for root induction.

12.1 Introduction

Jatropha (*Jatropha curcas* L.) seed oil can be easily processed to partially or fully replace petroleum-based diesel fuel (Forson et al. 2004). Manipulation of seed size is a direct strategy to improve the yield of seed crops (Martínez-Andújar et al. 2012; Ruan et al. 2012). In order to increase oil yields from Jatropha seeds, we need to find a way to make the seeds larger. Expansion of vegetative tissues (including the seed) and increased yield have already been reported (Gonzalez et al. 2007; Mizukami et al. 2000; Sun et al. 2012; Zhang et al. 2012; Ral et al. 2012).

An endocycle is a type of cell cycle in which DNA replication continues without cell division, so the ploidy of cells in the endocycle increases to 4C, 8C, 16C, 32C, etc. Endocycles are universally observed in both plants and animals. Endoreduplication is a type of DNA replication in endocycles, and while it increases ploidy levels, it also frequently makes cells larger than normal.

It is sometimes observed in plants that organs are composed of somatic cells of different ploidy levels. Cells from *Arabidopsis thaliana* hypocotyls contain as much as 8C nuclear DNA in light-grown seedlings and as much as 16C in dark-grown seedlings (Gendreau et al. 1997). The normally triploid endosperm cell can reach ploidy levels greater than 200× in some lines of maize, because after fertilization, the endosperm proceeds through several rounds of nuclear division without cytokinesis to form a multinucleated syncytium (Bauer and Birchler 2006). In an extreme example, embryo suspensors of *Trapaolum majusno* have a DNA content of 2048C (Nagel 1978).

Vegetative cell size tends to grow relative to the amount of DNA in the nucleus. As such it is anticipated that promoting endoreduplication will expand organ size. Cyclin-dependent kinases (CDKs) are the important regulators of the cell cycle phase transitions. WEE1 kinase negatively regulates CDKs through phosphorylation of threonine and tyrosine residues (Francis 2007). The *Solly;WEE1* gene participates in the control of cell size and/or the onset of the endoreduplication process putatively driving cell expansion. Impairing the expression of *Solly;WEE1* in transgenic tomato plants resulted in a reduction in plant and fruit size. The reduction of plant fruit and seed size originated from a reduction in cell size, which could be correlated with a decrease of DNA ploidy levels (Gonzalez et al. 2007). *AINTEGUMENTA* (*ANT*) encodes a transcription factor of the AP2-domain family that is unique to plants (Elliott et al. 1996; Klucher et al. 1996). *ANT* controls plant organ cell number and organ size throughout shoot development. Gain of *ANT* function, via ectopic expression of a 35S::*ANT* transgene, enlarges embryonic and all shoot organs without altering superficial morphology by increasing cell number in both Arabidopsis and tobacco plants (Mizukami et al. 2000). As mentioned above, several different genes have been reported to expand vegetative tissues. However, we focused only on genes that increase seed size. Constitutive expression of *Arabidopsis purple acid phosphatase 2* (*AtPAP2*) by the *Cauliflower mosaic virus 35S* (*CaMV35S*) promoter in Arabidopsis enhances plant growth and seed yield by modulating carbon metabolism (Sun et al. 2012). Overexpression of *AtPAP2* in leaves of *Camelina sativa* activated sucrose phosphate synthase and resulted in a higher photosynthetic rate. Transgenic *Camelia* resulted in longer hypocotyls, earlier flowering, faster growth rate, and increased seed yield and seed size in comparison with wild-type plants (Zhang et al. 2012). In wheat (*Triticum aestivum*), RNAi-mediated downregulation of glucan, water-dikinase (GWD), the primary enzyme required for starch phosphorylation, under the control of an endosperm-specific promoter resulted in a decrease in starch phosphate content and an increase in grain size (Ral et al. 2012).

Phytohormone auxins control seed and fruit development. Key genes in the auxin pathway including *ARF* and *IAA* families and downstream effectors were screened to identify candidate genes controlling seed size in Jatropha (Ye et al. 2014). By using expression quantitative trait loci (eQTL) analysis, it was observed that the seed traits were affected by the interaction of *JcARF19* and *JcIAA9*—both of which are involved in auxin signal transduction.

By introducing candidate genes related to seed size growth into Jatropha, we anticipate an increase in yield and size. Here, we would like to introduce the genes by which we are transforming Jatropha. About 32,000 genes were annotated in the complete genome of rice, *Oryza sativa* L., ssp. *Japonica* cultivar Nipponbare (Rice Annotation Project 2007, 2008). A total of 28,469 full-length complementary DNA clones were collected and completely sequenced from it (Rice Full-Length cDNA Consortium 2003). To understand the functions of unknown genes, a number of tagged lines or overexpressed lines were made (An et al. 2003; Hirochika et al. 2004; Miyao et al. 2007). For activation-tagging technology, a T-DNA vector possessing *CaMV35S* gene enhancers was developed (Hayashi et al. 1992). Nakazawa et al. made approximately 50,000 activation-tagging lines and isolated four dominant mutants (Nakazawa et al. 2003). Gain-of-function mutants produced by activation-tagging T-DNAs may have different spectra of mutants that have never been isolated as conventional loss-of-function mutants. A novel gain-of-function system that we have named the FOX hunting system has been developed (Ichikawa et al. 2006). Each normalized full-length cDNA was introduced into Arabidopsis via *Agrobacterium tumefaciens*-mediated transformation. For rapid and efficient elucidation of useful traits, rice full-length cDNAs were introduced into Arabidopsis plants, and ~30,000 independent Arabidopsis transgenic lines expressing rice full-length cDNAs (rice FOX Arabidopsis mutant lines) were generated (Sakurai et al. 2011). These rice FOX Arabidopsis lines were screened systematically for various criteria such as morphology, photosynthesis, UV resistance, plant hormone profile, and heat and salt tolerance. We focused on size of seeds in these rice FOX Arabidopsis lines.

12.2 An Attempt of Making Lager Seeds Using the FOX Hunting System

12.2.1 Rice cDNAs of FOX Arabidopsis Lines that Showed Larger Seeds

Anticipating larger seed production, we are attempting to transfer rice cDNAs of FOX Arabidopsis lines that showed larger seeds into Jatropha. The table below shows data of rice full-length cDNAs and genes used for making transgenic plants (Table 12.1).

Table 12.1 Rice full-length cDNAs and genes used for making transgenic Jatropha

Rice FOX Arabidopsis line	Full-length cDNA	Gene name
k42340	J013094J23	*LOC_Os08g41910* (MOsDB, Indica)/*Os08g0531300* (Japonica Group)
k06835	J023054A07	*LOC_Os04g43210*/*Os04g0511400*
k32722	J033024P22	*LOC_Os03g49180*/*Os03g0698900*
k15507	J013094J23	*LOC_Os10g40934*/*Os10g0558750*

LOC_Os08g41910 encoding Sua5/YciO/YrdC/YwlC family protein is a putative translation factor involved in translation, ribosomal structure, and biogenesis in *S. oneidensis* and the psychrophile *Colwellia psychrerythraea* 34H (Heidelberg et al. 2002; Methe et al. 2005). Moreover, Sua5/YciO/YrdC/YwlC family protein in *Shewanella piezotolerans* may assist gene translation under cold shock and/or function at the assembly of ribosomes under the low temperature (Li et al. 2008b).

LOC_Os04g43210 encodes probable inositol transporter 2-like. Arabidopsis inositol transporter 2 (AtINT2) is localized in the plasma membrane and mediates the symport of H^+ and several inositol epimers, such as myoinositol, scyllo-inositol, D-chiro-inositol, and muco-inositol (Schneider et al. 2007). Myoinositol is a precursor in the biosynthesis of UDP-GlcUA, GalUA, Xylapiose, and Ara (Loewus and Murthy 2000; Kanter et al. 2005), and it is used in the formation of galactinol, a myoinositol-linked, activated form of Gal that is used for the biosynthesis of raffinose and its derivatives (Kandler and Hopf 1982). It may be conjugated to auxins to prevent their biological activity and to allow long-distance transport (Cohen and Bandurski 1982).

LOC_Os03g49180 encodes alkaline ceramidase. Ceramidases hydrolyze ceramide into sphingosine and fatty acids. Sphingosine forms a primary part of sphingolipids. Sphingolipids are a group of lipids that are derived from long-chain 1,3-dihydroxy-2-amino bases and that are involved in important processes in plants. The developing sunflower seed kernel is a tissue rich in sphingolipids (Salas et al. 2011). The Arabidopsis ceramidase (AtACER), a homolog of human alkaline ceramidases, is related to biotic and abiotic stresses in plants (Wu et al. 2015). Arabidopsis TurgOr regulation Defect 1 (TOD1) is a Golgi-localized alkaline ceramidase and preferentially expressed in pollen tubes and silique guard cells (Chen et al. 2015). TOD1 acts in turgor pressure regulation in both guard cells and pollen tubes.

LOC_OS10g40934 encodes putative flavonol synthase/flavanone 3-hydroxylase or 2OG-Fe(II) oxygenase containing protein. This gene has various sizes of transcripts. Transcripts of *LOC_Os10g40934.3* and *LOC_Os10g40934.11* are the longest. We made two types of overexpressing constructs that transfer into Jatropha. One is from first to 1074 bp coding sequence (CDS) of *LOC_Os10g40934.3* (*LOC_Os10g 40934.11*) and the other is from 429 to 1074 bp of CDS. We made five overexpressing vectors for four genes and attempted to transfer them into Jatropha (Fig. 12.1).

12.2.2 Phenotype of Seeds in Four Rice FOX Arabidopsis Lines

We analyzed the phenotype of seeds in four rice FOX Arabidopsis lines [k42340 (*LOC_Os08g 41910*), k06835 (*LOC_Os04g43210*), k32722 (*LOC_Os03g49180*), and k15507 (*LOC_Os10g 40934*): name of overexpressing gene] (Table 12.2). Table 12.2 shows that seed size and weight increased in comparison with rice FOX Arabidopsis lines and wild-type plants. But yield decreased over all rice FOX Arabidopsis lines. In the case of Jatropha (and unlike Arabidopsis), its yield can be improved by fertilizer application—and for this study, we prioritized seed size over yield.

12.3 The Method of Making Transgenic Jatropha Mediated by *Agrobacterium*

12.3.1 An Improved Method of *Agrobacterium*-Mediated Genetic Transformation

Various methods of transformation have been reported for Jatropha. A transformation procedure for *J. curcas* has been established for the first time via *A. tumefaciens* infection of cotyledon disk explants (Li et al. 2008a).

Here we introduce our method to make transgenic Jatropha mediated by *Agrobacterium*.

(a)

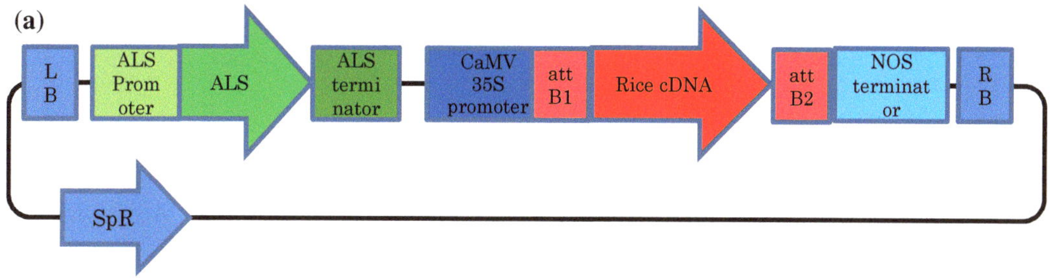

(b) *LOC_Os10g40934.3* (*LOC_Os10g40934.11*)

```
[1]ATGGCGGGAG  CAAGATCCGT  TGGTTCTCTC  CCAGTGCCTA  ATGTGCAGGC
GCTTGCAGAA  ATTTGCAATG  ATCCAGATGA  ACACATACCT  GAAAGGTACA
TCAGACCAGA  GGCTAGTTCT  GAGGAGGTCA  TCAACAACTA  CCAAGGTGAC
ATGGCGATTC  CGATCATCGA  TCTCAAGAAA  TTGCTTTGTC  CGCAATCATC
AGAAGAGGAG  TGTGTGAAGC  TGAGATCTGC  CTGCCAGTAT  TGGGGGTTCT
TTCTGCTCAT  TAACCATGGA  GTGCCAGATG  AAGTGATTGC  AAACCTGAAG
AGAGACATTG  TTGATTTCTT  TAGCCAGCCA  CTGGACACCA  AGAAGGAGTA
CACACAGCTA  CCGAATAGCC  TAGAAGGCTA  TGGACAGAGC  TTTGTTTTCT
CTGAGGACCA  GAAACTTGAC  TGGGCAGAC[2]A  TGCTGTATCT  TCATGTCCAT
CCTAGCGATT  CAAGGGACCT  AAGGTTCTGG  CCCACTTCTC  CAGCATCTTT
CAGGCAATCC  ATTGATGCAT  ACTCATCAGA  GACTAAAAGC  CTGGCACTCT
GCTTGTTTGA  GTTCATGGCT  AAGGCTGTGG  GCGCTAAGCC  AGAGTCACTT
TTAGATTTAT  TTGAAGAGCA  GCCTCGAGGC  TTAAGGATGG  CCTATTACCC
ACCATGCCGG  CAAGCTGACA  AGGTGATGGG  CCTTTCGCCA  CACTCTGATG
CGGGTGGCCT  GACACTGCTG  CTCGAGATTA  ACAATGTGCA  GGGCCTGCAG
ATCAAGAAAG  ATGGCAAGTG  GTTCTCCATA  GATGCTCCAA  ATGGTGCACT
TATTGCTAAC  ATTGGCGACA  CACTTGAGAT  CCTGAGTAAC  GGAAAGTTCA
GAAGTGTTGA  ACACAGGGCC  GTGATAAACC  CAAATAAAGA  GCGAATTTCA
GCAGCACTCT  TCCACTATCC  AAGTGAAAAT  ATGGTGATCA  GCCCTCTGCC
GGAGTTTGTG  AAAGATGGTA  AAGTGAAGTA  CAGATCAATA  AGTTACCTTG
ATTTCATGAA  ACAAATCTTC  ACACAACAGC  TTGATGGAAA  AAACCGAGTG
GAGGT[3]ATTAA  AGCTGGACCA  GTAG
```

Fig. 12.1 Overexpressing construct of candidate larger seed genes (from rice FOX Arabidopsis). **a** Overexpressing construct, *ALS Arabidopsis thaliana* W574L/S653I acetolactate synthetase gene, *LB* T-DNA *left* border, *RB* T-DNA *right* border, *SpR* spectinomycin resistance gene for selection maker in *Escherichia coli* and *Agrobacterium*. **b** Coding sequence of *LOC_Os10g40934* and two sizes of transcripts using overexpressing construct

Table 12.2 Results for rice FOX Arabidopsis lines [where wild-type plant (Colombia ecotype) results = 1]

Rice FOX Arabidopsis line	Gene name	Number of seeds in silique (seeds)	Seed weight (mg/seed)	Seed length (*X* axis) (mm)	Seed length (*Y* axis) (mm)	Seed yield per plant (mg)
k42340	*LOC_Os08g41934*	0.49	1.35	1.19	1.07	0.78
k06835	*LOC_Os04g43210*	0.99	1.11	1.10	1.06	0.96
k32722	*LOC_Os03g49180*	0.84	1.12	1.09	1.05	0.89
k15507	*LOC_Os10g40934*	0.76	1.32	1.13	1.05	0.77

Examples are included to illustrate the method of root induction.

The method to make transgenic Jatropha is outlined as follows:

Culture *Agrobacterium* (EHA101 strain) in LB medium containing antibiotic 80 μg/ml spectinomycin, 50 μg/ml rifampicin, 50 μg/ml kanamycin for 2 days at 28°C.

↓

Centrifuge culture solution at 5000 rpm for 5 min to remove LB medium and collect *Agrobacterium*.

↓

Infection solution: Add 100 μM Acetosyringone to MS medium (3% sucrose pH 5.8) and elute *Agrobacterium* to concentration of $OD_{600} = 0.2$.

↓

Remove husk of air-dried Jatropha seeds and soak seeds in 22–26 °C tap water overnight (Fig. 12.2).

↓

Remove cotyledons and endosperm from soaked seeds (Fig. 12.2).

↓

Sterilize cotyledons with 4% sodium hypochlorite including one drop of Tween20 for 2 min (Fig. 12.3).

↓

Wash cotyledons with sterilized water three to five times (Fig. 12.3).

↓

Cut cotyledons into 3–5 mm^2 pieces by knife (Fig. 12.3).

↓

Infect 3–5 mm^2 pieces in *Agrobacterium* infection solution at 50 kPa for 5 min using an airtight vacuum container (Fig. 12.3).

↓

Remove infection solution from the square pieces using filter paper.

↓

Co-cultivate infected cotyledon pieces on co-cultivation medium [MS medium, 3% sucrose, pH 5.8, 100 μM Acetosyringone, 4.4 μM (1.0 mg/l) BA, 2.46 μM (0.5 mg/l) IBA, 0.3% Gelrite] lined with filter paper, containing infected cotyledon pieces. Co-cultivation continues for 4 days (Fig. 12.3).

↓

Transfer cotyledon pieces into callus induction medium [MS medium, 3% sucrose, pH 5.8, 1 tablet/l Augmentin250RS, 4.4 μM (1.0 mg/l) BA, 2.46 μM (0.5 mg/l) IBA, 0.3% Gelrite] at 26 °C under 16-h light conditions.

↓

Transfer calluses to shoot induction medium [MS medium, 3% sucrose, pH 5.8, 1 tablet/l Augmentin250RS, 13.3 μM (3.0 mg/l) BA, 0.5 μM (0.1 mg/l) IBA, 0.3% Gelrite] at 26 degrees under 16-h light conditions.

↓

After 2 weeks, transfer calluses and shoots into selection medium [MS medium, 3% sucrose, pH 5.8, 1 tablet/l Augmentin250RS, 13.3 μM (3.0 mg/l) BA, 0.5 μM (0.1 mg/l) IBA, 20 nM Bispyribac Na, 0.3% Gelrite] at 26 °C under 16-h light conditions.

↓

Transfer elongated shoots into root induction medium (Gamborg'B5 medium, 2% sucrose, pH 5.8, 1 tablet/l Augmentin250RS, 0.25 g/l Active Chacol, 20 nM Bispyribac Na, 0.2–0.3% Gelrite) (Fig. 12.4a–c).

↓

Once roots have fully developed, transfer transformed plants to soil and acclimatize them in a growth room.

Fig. 12.2 Removing the husks from Jatropha seed. **a** Air-dried Jatropha seeds. **b** Seeds prior to husk removal, and post-husk removal. **c** After husks are removed, seeds are soaked in 22–26 °C tap water overnight

12.3.2 Examination of Infection Using *Agrobacterium*

The efficiency of transformation using the *Agrobacterium* strain LBA4404 and phosphinothricin (herbicide) for selection showed an improvement over using EHA105 and hygromycin. About 55% of the cotyledon explants produced phosphinothricin-resistant calluses. Finally, 13 transgenic plants were obtained from 100 cotyledon explants of Jatropha, representing an overall transformation efficiency of 13% (Li et al. 2008a).

A system was developed in which the herbicide bispyribac Na, which inhibits acetolactate synthase, was used as the selection agent, and a two-point-mutated acetolactate synthase gene (mALS) was used to confer resistance upon transformants. This system significantly improved the efficiency of *Agrobacterium*-mediated transformation (Kajikawa et al. 2012). We compared green fluorescence in pieces of

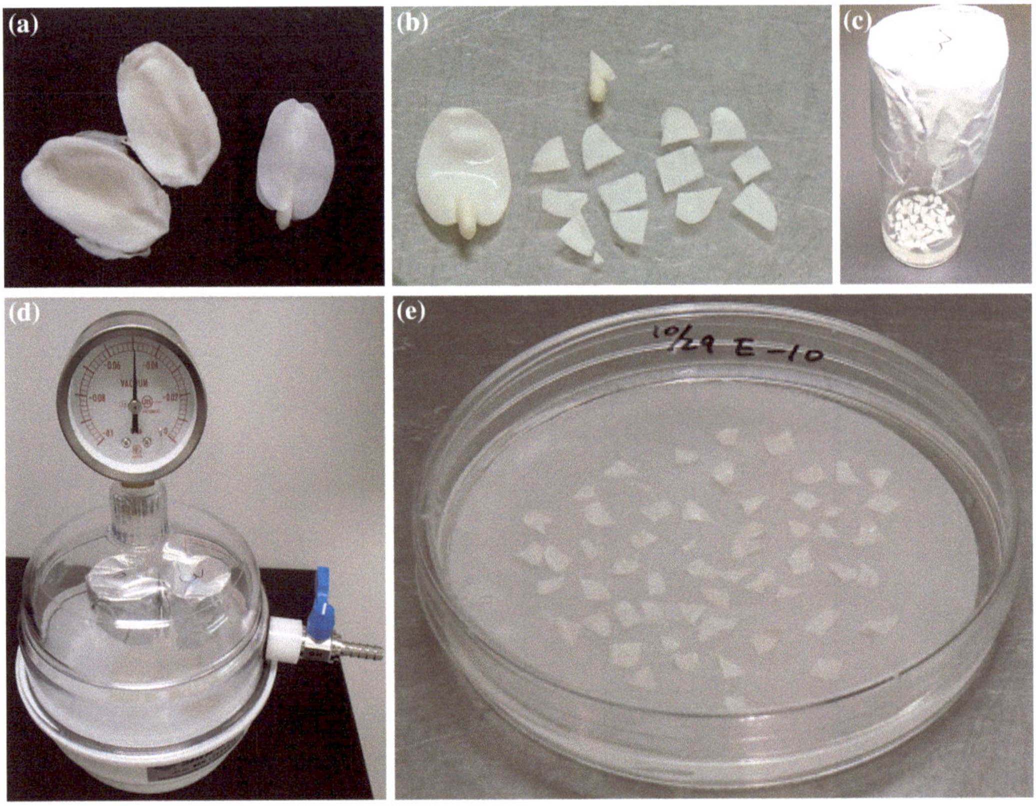

Fig. 12.3 Infecting cotyledons with *Agrobacterium*. **a** Cotyledons (*right*) and endosperm (*left*) from Jatropha seeds. **b** Cotyledons cut into 5 mm^2 pieces. **c** Cotyledon pieces being soaked in infecting solution. **d** Using an airtight vacuum container, the cotyledon pieces are depressurized at 50 kPa for 5 min (to ease *Agrobacterium* infection of cells). **e** Dish lined with filter paper, containing infected cotyledon pieces. Co-cultivation continues for 4 days

cotyledon using various conditions of infection and observed high efficiency of transformation using EHA101 strain and bispyribac Na.

12.3.3 Conditions for Regeneration

High efficiency of callus regeneration is required when performing *Agrobacterium*-mediated transformation. We investigated the conditions needed to make calluses and shoots from cotyledon explants.

When we treated explants with Nos. 2 and 3 (Table 12.3), a higher efficiency of callus inductions was observed. The most shoots were observed after treatment with No. 3 for 73 days. When we treated explants with No. 5, multiple shoots were observed, but these multiple shoots could not develop leaves and did not grow. Explants treated with condition Nos. 2 and 5 had higher callus induction, but lost the ability to regenerate shoots if their callus and explants are cultivated for more than 1 month. BA 1.0 + IBA 0.5 mg/l (Table 12.3, No. 2) was chosen as a callus induction medium, and BA 3.0 + IBA 0.1 mg/l (Table 12.3, No. 3) was chosen as the shoot induction medium. Explants were cultivated on callus induction medium for 2 weeks and then transferred to shoot induction medium.

Fig. 12.4 Transgenic Jatropha. **a** *LOC_Os03g49180* on root induction medium. **b** *LOC_Os10g40934* on root induction medium. **c** Long version construct of *LOC_Os10g40934* on root induction medium. **d** *LOC_Os04g43210* in soil. **e** *LOC_0s08g41934* in soil

12.3.4 Conditions for Root Induction from Shoots

Only 17 of 332 transferred shoots induced roots in callus induction medium and shoot induction medium. We need to improve the conditions for root induction, because the rate of root formation was only 5.1% in these mediums.

Root stimulators, used to induce root growth in cuttings and bulbs, contain phytohormone IBA. When cultivating chrysanthemum by direct cutting, root growth is induced by soaking in

Table 12.3 Number of calli and shoots under various treatment conditions

Treatments of		Number of treated explants	Number of calli	Number of shoots
Cytokinin (mg/l)	Auxin (mg/l)			
BA 0.5	IBA 0.05	24	10 (42%)	3
BA 1.0	IBA 0.5	24	16 (67%)	2
BA 3.0	IBA 0.1	28	5 (18%)	12
Kn 0.5	IBA 0.05	27	8 (30%)	0
TDZ 0.5	IBA 0.05	30	17 (56%)	0

Calli are shown as a number and a percentage
Calli were observed 19 days after treatment, while shoots were observed 73 days after treatment
BA benzyl adenine, *Kn* kinetin, *TDZ* thidiazuron, *IBA* 3-indolebutyric acid

IBA solution. The effects of IBA concentrations on root performance of Jatropha cuttings were investigated (Camellia et al. 2009). Healthy and uniform stem cuttings of 25 cm long were prepared and dipped in fungicide solution. The basal of cuttings (1–2 cm length) were dipped in each concentration of IBA for 5–10 s. After that, the cuttings were allowed to air dry for 30 s before being inserted into sand 1/3 proportion of the cutting inside a propagating tray. As a result, it was observed that the application of IBA (10,000 or 20,000 mg/l) increased number of roots.

We tested auxins in root induction. As auxins, we used both 2.0 mg/l IBA and 0.1 mg/l Naphthaleneacetic acid (NAA) in an attempt to induce roots in medium, but the roots of each shoot were not observed for 2 months following transfer and the shoot surface (in contact with the medium) became callused. IBA and NAA treatments did not yield satisfactory results. Next, we considered liquid (active component: IBA 0.4%) or powder (active component: IBA 0.5%) OXYBERON (Bayer Cropscience Co Ltd) which is known to induce root growth in cuttings. Liquid OXYBERON induced roots in 4/20 shoots treated (20% efficiency), while powder OXYBERON did not induce any roots.

Low levels of macronutrients in medium are beneficial for root initiation. Total nitrogen content of cuttings was negatively correlated with the number and weight of roots in woody plants. Percentages of root induction without NH_4NO_3 were as high as 100% in apple cultivars when KNO_3 was provided at full strength (Sriskandarajah et al. 1990). Moreover, acidity and concentration of sucrose in medium affected root induction of fruit rootstocks in vitro (Orlikowska et al. 1992).

In particular, the use of NH_4^+ in tissue culture medium affected the pH of the medium because H^+ ions are released as this form of nitrogen is used. On the other hand for the same reason, NO_3^- induced alkalization. The MS formulation contains inorganic nitrogen as both ammonium and nitrate in the ratio of $2NO_3^-:1NH_4^+$. Removal of NH_4^+ had a significant effect on medium pH, its presence caused a decrease in pH as the culture period proceeded (Bennett et al. 2003; Woodward et al. 2006). Because of this, we considered using Gamborg'B5 medium which has a lower concentration of nitrogen instead of MS medium and also decreased sugar density.

Gamborg'B5 medium contains KNO_3 as the main source of nitrogen and supplies only 1/10 the amount of NH_4^+ compared to MS medium (Table 12.4). Moreover, because it was found that some plants secrete rooting inhibitor into the medium during in vitro cultivation, we considered adding active charcoal to absorb it (Table 12.4). Treatment with MS medium at full or 1/2 concentrations with 1 or 0% sucrose did not induce roots. Gamborg'B5 induced roots at 28% efficiency within 1 month of treatment. It follows that we should use Gamborg'B5 medium containing 2% sucrose and 250 mg/l active charcoal as a root induction medium. After transferring shoots into root induction medium

Table 12.4 Concentration of nitrogen and sucrose in root induction medium

Medium		NO_3–N (mg/l)	NH_4–N (mg/l)	Sucrose (%)	Active charcoal (mg/l)
MS1/2	Sucrose 1%	1130	185	1	–
MS1/2	Sucrose 0%			0	
MS	Sucrose 1%	2261	371	1	
MS	Sucrose 0%	2261	371	0	
Gamborg'B5	Sucrose 2%	1535	18	2	250

from shoot induction medium, shoots started making roots after about 40 days. Some plants induced roots only 3 weeks after transfer. Shoots that do not induce roots prior to 50 days after transfer are generally not going to induce any roots.

12.4 Conclusions

We succeeded in making transgenic Jatropha with the candidate genes of larger seeds obtained from rice FOX Arabidopsis lines (Fig. 12.4). However, these transgenic Jatropha need ample time to grow their seeds and as such we must wait a long time to observe seed size in transgenic Jatropha.

FT (*Flowering locus T*) encodes a florigen that acts a key regulator in the flowering pathway (Turck et al. 2008). The transgenic plants in which *Citrus FT* was expressed constitutively showed early flowering and fruiting, which reduces its generation time (Endo et al. 2005). An *FT* homolog was isolated and designated as *JcFT* (Li et al. 2014). We should attempt to overexpress *JcFT* to reduce the generation time in Jatropha when we observe phenotype of seeds in transgenic Jatropha. If we can develop a system for measuring seed efficiency, we can transform candidate genes of larger seed phenotypes obtained from rice FOX Arabidopsis lines. Reporter gene (*GUS*) expression was examined by histochemical staining of GUS activity in leaves of kanamycin-resistance regenerated shoots that were transformed via *Agrobacterium* LBA4404 strain with kanamycin selection marker gene. Of the total 120 kanamycin-resistant shoots, 37 showed *GUS* expression in leaves. A total of 120 kanamycin-resistant shoots contained a lot of non-transformed plants (escape) (Pan et al. 2010).

Kajikawa et al. obtained 13 putative transgenic shoots from 300 cotyledon explants, but of the 13 putative transgenic shoots, only six were determined to be transgenic shoots including the transgene by PCR analysis. Of the six transgenic shoots, three produced roots on the root induction medium (Kajikawa et al. 2012). The remaining problem is to improve the decreasing number of non-transformed plants (escape) and root induction.

Acknowledgments We would like to thank Prof. Kiriiwa (Department of Agriculture, Shizuoka University) for plant care.

References

An S, Park S, Jeong DH, Lee DY, Kang HG, Yu JH et al (2003) Generation and analysis of end sequence database for T-DNA tagging lines in rice. Plant Physiol 133:2040–2047

Bauer MJ, Birchler JA (2006) Organization of endoreduplicated chromosomes in the endosperm of *Zea mays* L. Chromosoma 115:383–394

Bennett IJ, McDavid DAJ, McComb JA (2003) The influence of ammonium nitrate, pH and indole butyric acid on root induction and survival in soil of micropropagated *Eucalyptus globulus*. Biol Plant 47:355–360

Camellia NA, Thohirah LA, Abdullah NAP, Khidir OM (2009) Improvement on rooting quality of *Jatropha curcas* using Indole Butyric Acid (IBA). Res J Agric Biol Sci 5:338–343

Chen LY, Shi DQ, Zhang WJ, Tang ZS, Liu J, Yang WC (2015) The Arabidopsis alkaline ceramidase TOD1 is a key turgor pressure regulator in plant cells. Nat Commun 6:6030

Cohen JD, Bandurski RS (1982) Chemistry and physiology of the bound auxins. Annu Rev Plant Physiol 33:403–430

Elliott RC, Betzner AS, Huttner E, Oakes MP, Tucker WQ, Gerentes D et al (1996) *AINTEGUMENTA*, an *APETALA2*-like gene of Arabidopsis with pleiotropic roles in ovule development and floral organ growth. Plant Cell 8:155–168

Endo T, Shimada T, Fujii H, Kobayashi Y, Araki T, Omura M (2005) Ectopic expression of an FT homolog from citrus confers an early flowering phenotype on trifoliate orange (*Poncirus trifoliata* L. Raf.). Transgenic Res 14:703–712

Forson FK, Oduro EK, Hammond-Donkoh E (2004) Performance of Jatropha oil blends in a diesel engine. Renew Energy 29:1135–1145

Francis D (2007) The plant cell cycle—15 years on. New Phytol 174:261–278

Gendreau E, Traas J, Desnos T, Grandjean O, Caboche M, Höfte H (1997) Cellular basis of hypocotyl growth in *Arabidopsis thaliana*. Plant Physiol 114:295–305

Gonzalez N, Gévaudant F, Hernould M, Chevalier C, Mouras A (2007) The cell cycle-associated protein kinase WEE1 regulates cell size in relation to endoreduplication in developing tomato fruit. Plant J 51:642–655

Hayashi H, Czaja I, Lubenow H, Schell J, Walden R (1992) Activation of a plant gene by T-DNA tagging: auxin-independent growth *in vitro*. Science 258:1350–1354

Heidelberg JF, Paulsen LT, Nelson KE, Gaidos EJ, Nelson WC, Read TD et al (2002) Genome sequence of the dissimilatory metal ion-reducing bacterium *Shewanella oneidensis*. Nat Biotechnol 20:1118–1123

Hirochika H, Guiderdoni E, An G, Hsing YI, Eun MY, Han C et al (2004) Rice mutant resources for gene discovery. Plant Mol Biol 54:325–334

Ichikawa T, Nakawaza M, Kawashima M, Iizumi H, Kuroda H, Kondou Y et al (2006) The FOX hunting system: an alternative gain-of-function gene hunting technique. Plant J 48:974–985

Kajikawa M, Morikawa K, Inoue M, Widyastuti U, Suharsono S, Yokota A et al (2012) Establishment of bispyribac selection protocols for *Agrobacterium tumefaciens*- and *Agrobacterium rhizogenes*-mediated transformation of the oil seed plant *Jatropha curcas* L. Plant Biotechnol 29:145–153

Kandler O, Hopf H (1982) Oligosaccharides based on sucrose (sucrosyl oligosaccharides). In: Loewus AF, Tanner W (eds) Plant carbohydrates 1. Intracellular carbohydrates, encyclopedia of plant physiology. Springer, Berlin, pp 348–383

Kanter U, Usadel B, Guerineau F, Li Y, Pauly M, Tenhaken R (2005) The inositol oxygenase gene family of Arabidopsis is involved in the biosynthesis of nucleotide sugar precursors for cell-wall matrix polysaccharides. Planta 221:243–254

Klucher KM, Chow H, Reiser L, Fischer RL (1996) The *AINTEGUMENTA* gene of Arabidopsis required for ovule and female gametophyte development is related to the floral homeotic gene *APETALA2*. Plant Cell 18:137–153

Li C, Luo L, Fu Q, Niu L, Xu ZF (2014) Isolation and functional characterization of *JcFT*, a *FLOWERING LOCUS T* (*FT*) homologous gene from the biofuel plant *Jatropha curcas*. BMC Plant Biol 14:125

Li M, Li H, Jiang H, Pan X, Wu G (2008a) Establishment of an *Agrobacteriuim*-mediated cotyledon disk transformation method for *Jatropha curcas*. Plant Cell Tiss Org Cult 92:173–181

Li S, Xiao X, Sun P, Wang F (2008b) Screening of genes regulated by cold shock in *Shewanella piezotolerans* WP3 and time course expression of cold-regulated genes. Arch Microbiol 189:549–556

Loewus FA, Murthy PPN (2000) *myo*-Inositol metabolism in plants. Plant Sci 150:1–19

Martínez-Andújar C, Martin RC, Nonogaki H (2012) Seed traits and genes important for translational biology—highlights from recent discoveries. Plant Cell Physiol 53:5–15

Methe BA, Nelson KE, Deming JW, Monen B, Melamud E, Zhang X et al (2005) The psychrophilic lifestyle as revealed by the genome sequence of *Colwellia psychrerythraea* 34H through genomic and proteomic analyses. Proc Natl Acad Sci USA 102:10913–10918

Miyao A, Iwasaki Y, Kitano H, Itoh J, Maekawa M, Murata K et al (2007) A large-scale collection of phenotypic data describing an insertional mutant population to facilitate functional analysis of rice genes. Plant Mol Biol 63:625–635

Mizukami Y, Fischer RL (2000) Plant organ size control: AINTEGUMENTA regulates growth and cell numbers during organogenesis. Proc Natl Acad Sci USA 97:942–947

Nakazawa M, Ichikawa T, Ishikawa A, Kobayashi H, Tsuhara Y, Kawashima M et al (2003) Activation tagging, a novel tool to dissect the functions of a gene family. Plant J 34:741–750

Nagel W (1978) Endopolyploidy and polyteny in differentiation and evolution. Elsevier, Amsterdam

Orlikowska T (1992) Effects of mineral composition and acidity of media, saccharose level, brand and quantity of agar on rooting of fruit rootstocks *in vitro*. Biol Plant 34:45–52

Pan J, Fu Q, Xu Z-F (2010) *Agrobacterium tumefaciens*-mediated transformation of biofuel plant *Jatropha curcas* using kanamycin. Afr J Biotechnol 9:6477–6481

Ral JP, Bowerman AF, Li Z, Sirault X, Furbank R, Pritchard J et al (2012) Down-regulation of Glucan, Water-Dikinase activity in wheat endosperm increases vegetative biomass and yield. Plant Biotechnol J 10:871–882

Rice Annotation Project (2007) Curated genome annotation of *Oryza sativa* ssp. Japonica and comparative genome analysis with *Arabidopsis thaliana*. Genome Res 17:175–183

Rice Annotation Project (2008) The rice annotation project database (RAP-DB): 2008 update. Nucleic Acids Res 36:D1028–D1033

Rice Full-Length cDNA Consortium (2003) Collection, mapping, and annotation of over 28,000 cDNA clones from japonica rice. Science 301:376–379

Ruan YL, Patrick JW, Bouzayen M, Osorio S, Fernie AR (2012) Molecular regulation of seed and fruit set. Trends Plant Sci 17:656–665

Sakurai T, Kondou Y, Akiyama K, Kurotani A, Higuchi M, Ichikawa T et al (2011) RiceFOX: a database of Arabidopsis mutant lines overexpressing rice full-length cDNA that contains a wide range of trait information to facilitate analysis of gene function. Plant Cell Physiol 52:265–273

Salas JJ, Markham JE, Martínez-Force E, Garc'es R (2011) Characterization of sphingolipids from sunflower seeds with altered fatty acid composition. J Agric Food Chem 59:12486–12492

Schneider S, Schneidereit A, Udvardi P, Hammes U, Gramann M, Dietrich P et al (2007) Arabidopsis INOSITOL TRANSPORTER2 mediates H+ symport of different inositol epimers and derivatives across the plasma membrane. Plant Physiol 145:1395–1407

Sriskandarajah S, Skirvin RM, Abu-Qaoud H (1990) The effect of some macronutrients on adventitious root development on scion apple cultivars *in vitro*. Plant Cell Tiss Org Cult 21:185–189

Sun F, Suen PK, Zhang Y, Liang C, Carrie C, Whelan J et al (2012) A dual-targeted purple acid phosphatase in *Arabidopsis thaliana* moderates carbon metabolism and its overexpression leads to faster plant growth and higher seed yield. New Phytol 194:206–219

Turck F, Fornara F, Coupland G (2008) Regulation and identity of florigen: FLOWERING LOCUS T moves center stage. Annu Rev Plant Biol 59:573–594

Zhang Y, Yu L, Yung K, Leung DYC, Sun F, Lim BL (2012) Over-expression of *AtPAP2* in *Camelina sativa* leads to faster plant growth and higher seed yield. Biotechnol Biofuels 5:19–28

Woodward AJ, Bennet IJ, Pusswonge S (2006) The effect of nitrogen source and concentration, medium pH and buffering on *in vitro* shoot growth and rooting in *Eucalyptus marginata*. Sci Hort 110:208–213

Wu JX, Li J, Liu Z, Yin J, Chang ZY, Rong C et al (2015) The Arabidopsis *ceramidase* AtACER functions in disease resistance and salt tolerance. Plant J 81:767–780

Ye J, Liu P, Zhu C, Qu J, Wang X, Sun Y et al (2014) Identification of candidate genes *JcARF19* and *JcIAA9* associated with seed size traits in Jatropha. Funct Integr Genom 14:757–766

Germplasm Establishment and Selection of Drought-Tolerant Lines of Jatropha in the Philippines

13

Irish E. Bagsic, Primitivo Jose A. Santos and Maria Lea H. Villavicencio

Abstract

The preliminary evaluation of the Philippine Jatropha germplasm resulted in the identification of 13 accessions from divergent local and foreign provenances as potential materials for hybridization. Selection criteria for potential varieties include medium-to-high oil content, very low to almost zero free fatty acid content, abundance of fruits in a cluster, high branching density, and heavy seed weight. Six of these selections plus two other accessions from Tanzania and Thailand were further evaluated for drought stress. Based on the morphological and physiological data generated, GB57540 appears to be superior as it continuously hits the highest mean value for most of the parameters used. Mean values for GB 57262 on the other hand appeared to be not significantly different from the former thus is also proposed for selection. This study confirms that the main mechanism of the crop to tolerate drought stress is through strict stomatal control to reduce transpirational water loss. Congruent with stomatal closure, a reduction in leaf area was apparent, thus decreasing the transpirational surface area. The succulent nature of the crop's stem also adds to its adaptive mechanism as this serves as a water reservoir that supplies the crop with water during periods of severe drought.

13.1 Introduction

Countries all over the world have been exploring possible alternative sources of energy to compensate the continuous depletion and price increase in fossil fuel. One of the feedstocks gaining high potential as substitute or additives for diesel fuel is Jatropha, *Jatropha curcas* L. (Fairless 2007; Kumar and Sharma 2008; Berchmans and Hirata 2008).

I.E. Bagsic (✉)
International Rice Research Institute, Pili Drive, University of the Philippines Los Baños, College, 4031 Laguna, Philippines
e-mail: irish_up1123@yahoo.com

I.E. Bagsic · P.J.A. Santos · M.L.H. Villavicencio
Institute of Plant Breeding, Crop Science Cluster, College of Agriculture, University of the Philippines Los Baños, College, 4031, Laguna, Philippines

S. Tsuchimoto (ed.), *The Jatropha Genome*, Compendium of Plant Genomes,
DOI 10.1007/978-3-319-49653-5_13

This crop described as drought tolerant and capable to thrive on marginal and poor soils (Heller 1996; Fairless 2007; Parawira 2010), has been preferred as biodiesel source as it does not compete with areas intended for food crops (DOE 2007; Fairless 2007). It can grow without irrigation in areas with varying amount of rainfall, from 250 up to 3000 mm year^{-1} (Foidl et al. 1996; Katwal and Soni 2003). This characteristic of Jatropha is viewed to aid in reclaiming degraded land, preventing soil erosion, improving soil microenvironment and avoiding desertification (DOE 2007; Chaudharry et al. 2007; Kumar and Sharma 2008; Parawira 2010; Chavan et al. 2014).

Jatropha curcas is reported to originate from Central and South America and was distributed to Asia and Africa by Portuguese ships (Parawira 2010). Traditionally, the crop was introduced as hedge crops by traders and was used for medicinal purposes (Heller 1996; Gubitz et al. 1999). Apart from these uses, the hype of Jatropha production was further driven by the high amount of oil content in its seeds that can be readily converted to biodiesel (Maes et al. 2009; Parawira 2010). The seed kernels were estimated to contain 40–60% (w/w) oil—indicative of a good feedstock for biodiesel production (Makkar et al. 1997). In addition, the quality of crude oil produce from Jatropha is reported to be comparable to that of rapeseed and meets the US and the European standards for biodiesel (Azam et al. 2005; Parawira 2010). These characteristics of the crop resulted to the establishment of hectares of Jatropha plantations globally (DOE 2007; Achten et al. 2010a, b; Chavan et al. 2014). In 2008, cultivated Jatropha covered an estimated area of 900,000 ha worldwide, where 85% (760, 000 ha) are situated in Asia, 13% (120,000 ha) in Africa, and 2% (20,000 ha) in Latin America (Chavan et al. 2014). Large-scale production in Asia is centered in Indonesia, Myanmar, India, and China (Chavan et al. 2014). In terms of global absolute biodiesel production potential however, Malaysia ranks 1st, predicted to produce around 14.5 billion liters of biodiesel at $0.53 l^{-1}; Indonesia as 2nd with an approximate production of 7.5 billion liters at $0.42 l^{-1}, while the Philippines ranks 8th with a predicted biodiesel production of around 1.2 billion liters at $0.53 l^{-1} (Johnston and Holloway 2007).

The Philippine's high potential for biodiesel production can be, in part, attributed to the climatic suitability of the country for Jatropha cultivation and the availability of considerable area for production. Based on a study by the Philippine Department of Science and Technology (DOST) in 2006, around 5 million hectares can be utilized for Jatropha production given that farmers will take the initiative to plant in field boundaries and practice intercropping (DOE 2007). This area can yield an estimate of 25 million metric tons (MMT) of biodiesel feedstock (DOE 2007).

The enactment of the Philippine Republic Act 9367, known as the Biofuel Act of 2006, had significantly influenced the country's efforts in exploring non-food and underutilized crop species to reduce the dependence on imported crude. In addition, the cultivation of these fuel crops is viewed to increase the economic activity in rural areas and decrease the level of greenhouse gas emission. Pilot plantations had been established in Negros Occidental, Negros Oriental, Leyte, Sorsogon, and North Cotabato to assess the agronomic requirements and site conditions of Jatropha (DOE 2007). Joint efforts of several government offices including the Department of Energy (DOE), Department of Agriculture (DA), University of the Philippines Los Baños (UPLB), Philippine National Oil Corporation (PNOC), and DOST have been in place to look at the existing agriculture and production technologies in other countries utilizing Jatropha biodiesel (DOE 2007).

The combination of the above-mentioned economic and environmental claims for this multipurpose crop had received substantial attention of many researchers worldwide resulting in a production hype. However, most of these claims are either yet to be realized or poorly supported by scientific evidence (Jongschaap et al. 2007). In particular, claims in mobilizing the economic activity in areas with poor soils and minimum rainfall are lacking verifiable data, thus the productivity of Jatropha under sub-optimal conditions is in question. Selection of superior or elite accessions for important traits such as oil yield and tolerance to abiotic stresses like drought is lacking.

13.2 Establishment of Germplasm Collection

The conservation of genetic biodiversity of underutilized crops like Jatropha is of utmost importance because these are potential materials for future genetic improvement. It is therefore important not to lose these genes through extinction of species or their genotypes, as they cannot be recreated (Kiew 1996). Initial efforts on the conservation of the diversity of Philippine *Jatropha* spp. were conducted in 2007 (IPB-UPLB unpublished). Exploration trips from 11 regions resulted to a total of 775 collected accessions (Fig. 13.1). Majority of the materials were *J. curcas* (83%) comprising of local accessions and few introductions, while the remaining 17% of the assembled gene pool were other species within the genus *Jatropha*. The collection of other *Jatropha* species was done to capture the maximum diversity within the genus.

Stem cuttings from representative trees were established ex situ at the National Plant Genetic Resources Laboratory (NPGRL), UPLB. Whenever available, fruits and seeds were also collected for conservation since this type of conservation (i.e., through seeds) was proven to be more cost-efficient for many plant species having orthodox seeds (Ford-Lloyd and Jackson 1986; Chin 1996). Jatropha seeds exhibit an orthodox seed behavior (Heller 1996) which can be dried to a low moisture content (MC) and retain their viability over a long period of time under low temperature. The optimum MC for seed storage as reported (Parreño-de Guzman and Aquino 2009) is at 4–5%. Seeds stored at this MC level were found to retain high germination rate regardless of the storage temperature, i.e., at 0 or 18–20 °C (Figs. 13.2, 13.3). Obtained results conform to earlier reports that MC of oily seeds and those that have poor storage characteristics should be maintained at around 3–8% (Rao et al. 2006). The moisture level of seeds is considered as the most critical factor in storage (Bonner 1991). Hence, knowledge on the appropriate MC of seeds in combination with other factors such as initial seed quality and temperature is of key importance for genetic conservation.

13.2.1 Characterization and Morphological Diversity of Collected Germplasm

Determination of diversity on the collected accessions requires their characterization. The process of characterization refers to the description of the germplasm material in the collection. The information generated from this activity is not only used to assess the extent of genetic diversity present within the collection but also facilitate the utilization of the germplasm collection. Specifically, the characterization of the germplasm is important for the following reasons: (1) It looks at the value of the germplasm for utilization wherein desirable or unique traits can be identified; (2) it differentiates one accession from another and identifies duplicates among the accessions; (3) it can be used for gap filling; (4) it determines taxonomic relationships among and within the germplasm accessions; and (5) it can identify germplasm for core collections (NPGRL 1998).

The diversity of Philippine *J. curcas* collections in terms of morphology was examined through the devised Descriptors' List by the National Plant Genetic Resources Laboratory (IPB-UPLB unpublished). Phenotypic variations were reported for both qualitative and quantitative characters assessed. Displayed Shannon–Weaver diversity index (H') for 9 of the 21 characters evaluated ranged from 0.7608 to 0.9527 which indicates high variation for these characters on the gene pool. Percent oil content, branching pattern, and branching density are among those characters displaying high degree of variation. Such morphological variations in some of the seed, leaf, and general tree architecture were found in a number of published reports

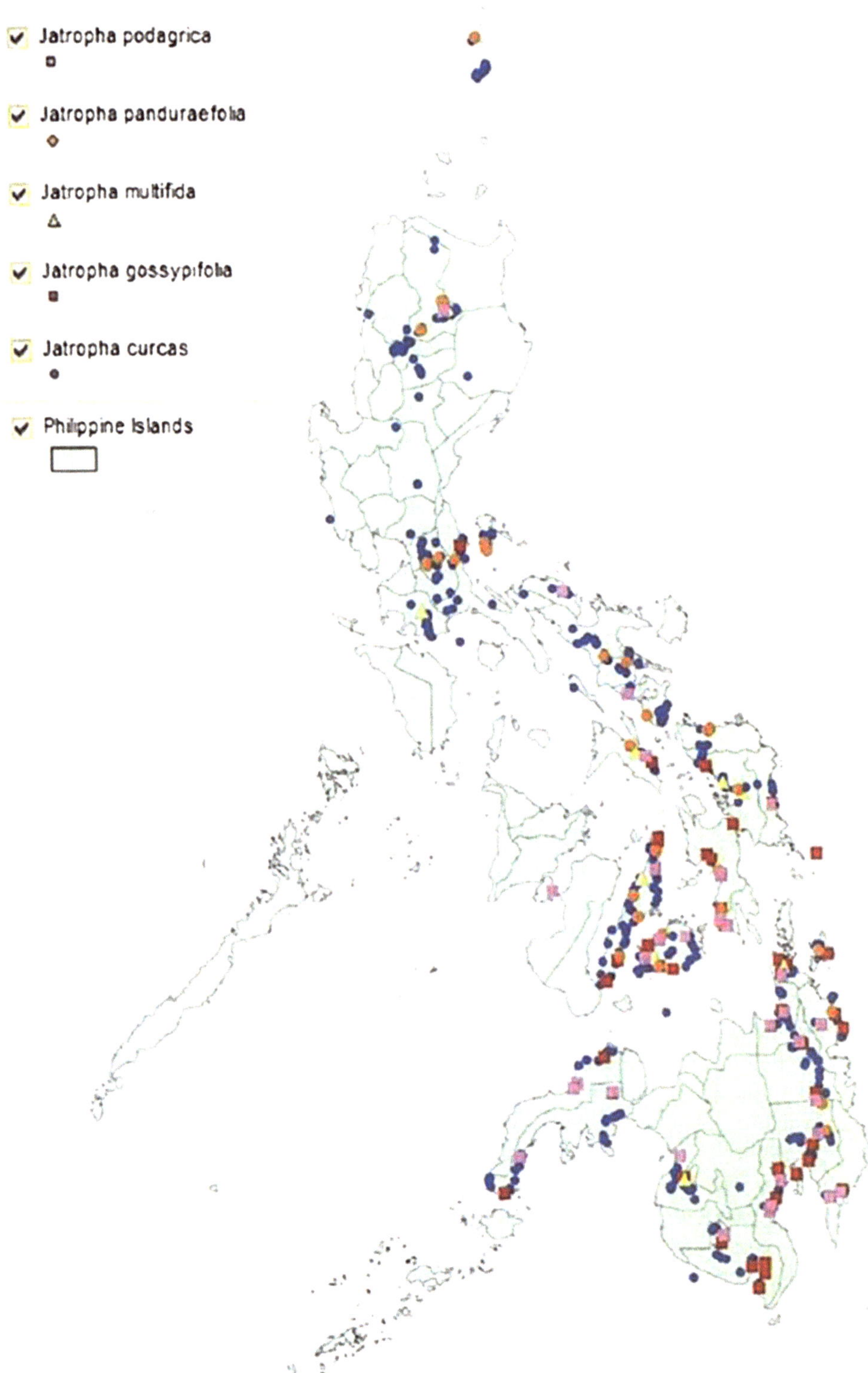

Fig. 13.1 Geographic provenances of *Jatropha* spp. germplasm in the Philippines. The geographic distribution was mapped using the diversity analysis (DIVA) software

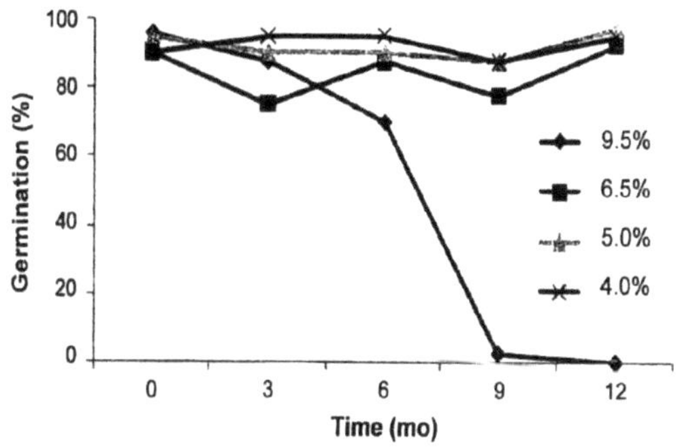

Fig. 13.2 Percent germination of stored *J. curcas* seeds with varying MC level. Seeds were kept in sealed aluminum foil packs at 18–20 °C for 12 months. *Source* Parreño-de Guzman and Aquino 2009

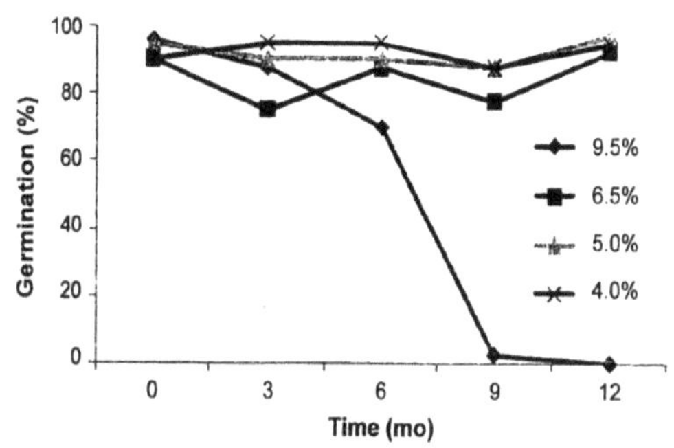

Fig. 13.3 Percent germination of stored *J. curcas* seeds with varying MC level. Seeds were kept in sealed aluminum foil packs at 0 °C for 12 months. *Source* Parreño-de Guzman and Aquino 2009

(Ginwal et al. 2005; Pant et al. 2006; Kaushik et al. 2007; Rao et al. 2008; Das et al. 2010). Contrary to these findings, exploration of genotypic variability through the employment of various molecular markers including random amplified polymorphic DNA (RAPD), inter-simple sequence repeat (ISSR), and amplified fragment length polymorphism (AFLP) revealed low genetic diversity of collected accessions from different provenances (Basha and Sujatha 2007; Ranade et al. 2008; Sun et al. 2008; Pamidimarri et al. 2010; Grativol et al. 2011). Often, results from these studies showed very low diversity in accessions collected within a country. These results suggest a narrow genetic base of *J. curcas* and its high elasticity to different environmental conditions. In the Philippines, however, the evaluation of *J. curcas* accessions is limited only to morphological characterization, and no studies were conducted in characterizing the collections on the molecular level. Nonetheless, data on morphological and oil quality characteristics of acquired accessions have a significant implication on the development and identification of elite individuals exhibiting high potential for oil content production. Hence, the potential of Jatropha as a commercially viable crop necessitates the establishment of an appropriate breeding program in developing elite varieties with sustained production in a range of environmental conditions.

13.2.2 Selection of Parental Materials for Crop Improvement

The germplasm characterization is a key in the selection of potential parental materials for crop improvement. Following establishment and evaluation of Philippine *J. curcas* accessions, preliminary works were conducted to select parental genotypes for varietal improvement. Combination of tree architecture and oil characters was considered for selection (IPB-UPLB unpublished). In general, plants with high vigor, medium-to-high oil content, very low to almost zero free fatty acid content of seeds, abundant fruits in a cluster, high branching density, and heavy seed weight were principally selected as parental genotypes. On top of these, all introduced alien accessions were included in the selection to widen the genetic base since limited diversity in species within countries of origin was earlier reported (Basha and Sujatha 2007; Ranade et al. 2008; Sun et al. 2008; Pamidimarri et al. 2010; Grativol et al. 2011). The inclusion of these accessions from foreign provenances is viewed to maximize gains in varietal improvement and hybridization.

Based on the criteria mentioned, 13 accessions have been selected as potential materials for hybridization (IPB-UPLB unpublished). Hybridization has been initiated among selected parental germplasm. Cross-pollination in the species was found to be quite simple and has proven to produce adequate viable progeny seeds. Observations of pollination traits under local conditions resulted on the detection of period of anther dehiscence and stigma, receptivity, i.e., in the morning, a few hours after floral opening. F_1 plants had been established and were found to exhibit high variation in terms of general vigor, plant development, and pest and disease incidence. These investigations, however, are still in their infancy. Considering the long duration to achieve optimum production ($\sim$5 years), the evaluation of progenies resulting from these varietal development works will require an ample period of time for the efficient selection of improved cultivar. Estimated period for both conventional breeding and induced mutagenesis for *J. curcas* is 8–10 years (Datta and Pandey 2013). This relatively long period for breeding could be a reason for the limited number of released *J. curcas* varieties for commercial production to date. Based on available data, only five varieties with specific characteristics including high yield (IP-1 and IP-2), high oil content (Chuan R-Sc-Jc-002-2005), high toxin #1 (Chuan R-Sc-Jc-001-2005), and high adaptability to semi-arid and arid conditions (SDAUJ1-Chatrapathi) were released in Indonesia, China, and India, respectively (Chen 2007; Hasnam 2007; Prastowo 2007; Datta and Pandey 2013). This information implies the need to develop long-term projects toward the development of improved *J. curcas* cultivars for sustainable production. Abiotic stress aspects of the crop should also be considered to allow selection of lines highly adapted to stress conditions with acceptable productivity.

13.3 Preliminary Study on Responses of *Jatropha curcas* to Drought Condition

Knowledge on the physiological processes associated with abiotic stress tolerance like drought is indispensable for the implementation of approach for sustained productivity of jatropha in marginal areas. *J. curcas* is reported to be hardy on drought and capable of growing in areas where the amount of rainfall is limited (Heller 1996; Katwal and Soni 2003; Fairless 2007; Parawira 2010).

The morphological and physiological responses of pooled *J. curcas* accessions were evaluated to gain initial understanding of the crop's mechanism for tolerance. For this purpose, 2-week-old seedlings with relatively similar plant height were transplanted in sandponics beds developed by Sumitomo Electric Industries (SEI). Prior to drought imposition, seedlings were allowed to stabilize and recover from transplanting stress for 42 days. Initial cycle of drought imposition which lasted 10 days showed a rapid decrease in stomatal conductance (gs) (Fig. 13.4) and subsequent reduction in the total leaf area (data not shown). This conforms to previous findings that the main mechanism of the crop to tolerate drought stress is through strict stomatal control to reduce transpirational water loss (Thompson et al. 2007; Pompelli et al. 2010; Silva et al. 2010; Díaz-López et al. 2012; Sapeta et al. 2013). Stomatal conductance reading on the 7th day of drought period dropped to almost zero (0.003 mol/m^{-2} s^{-1}). At this point, an almost nil (0% field capacity) moisture level in the substrate (sand) was recorded. This finding indicates that stomatal conductance is highly responsive on the available moisture level on the substrate. Earlier studies provided strong evidences on

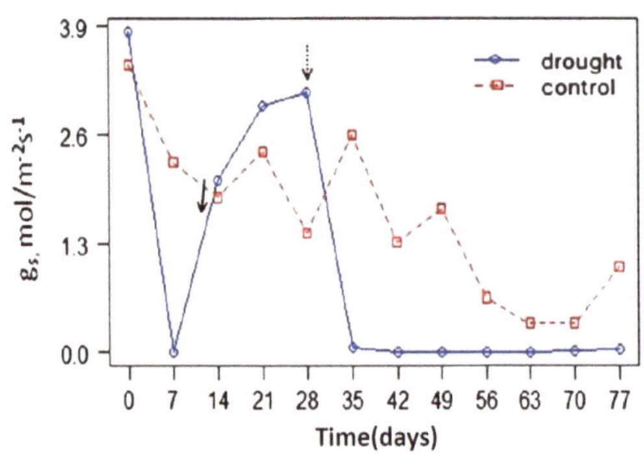

Fig. 13.4 Changes in the stomatal conductance (gs) values under stressed and controlled conditions. *Arrow* on day 10 indicates re-watering after the first cycle of drought, while the *arrow* on day 28 indicates the start of 2nd drought cycle

stomatal regulation in *J. curcas* under low soil water availability (Jones 1985; Díaz-López et al. 2012; Sapeta et al. 2013). Such mechanism for drought tolerance was also reported for other species within the family Euphorbiaceae including *Ricinus communis* (Sausen and Rosa 2010), *Hevea brasiliensis*, and *Manihot esculenta* (El-Sharkawy 2007).

Initial manifestation of drought stress symptoms including partial browning and withering of leaves occurred as early as 4th day of drought imposition, simultaneous to substantial soil moisture lost. Rapid recovery was observed on the stressed plants 3 days after re-watering (Fig. 13.5). During the recovery period of the 1st drought cycle, a 20% difference in plant height between the controlled and stressed plants was apparent. The abrupt decrease in stomatal conductance under high level of water deficit could have possibly affected the net photosynthesis of the crop. Although measurement of net photosynthesis was not performed, this hypothesis is supported by experimental results of Sapeta et al. (2013), where plants subjected to severe stress (gs of almost zero) showed an abrupt drop in net photosynthesis. Depending on

Fig. 13.5 Visual characteristic of pooled *J. curcas* accessions **a** before drought treatment; **b** 4 days after drought imposition; **c** 9 days after drought im-position; and **d** 3 days after re-watering

the level of stress imposed, the recovery of photosynthetic rates can vary. For moderate stress imposition, photosynthetic rates recovery can be fast and complete. In the case of severe water deficit, however, it can be from days to weeks and in some instances, is never complete (Miyashita et al. 2005; Flexas et al. 2006). The same trend was observed in the recuperation of gs (Pompelli et al. 2010). Observations on the accessions that had been subjected to drought for this study showed their ability to recover. They, however, were not able to catch up with the growth of the well-watered plants and continued to be left behind as the second cycle of drought treatment, which lasted for 7 weeks, was imposed (data not shown). The difference in plant height became more apparent at the end of the second cycle of drought. Stressed plants were observed to be 60% smaller than its well-watered counterpart. On the 42nd day of recovery, plant height of stressed plants was still 30% reduced compared to the control.

Majority of the stressed plants had completely shed-off their leaves at the 7th week of drought imposition. Indeed, these plants survived although without leaves while growing in bone-dry (approximately 1.11% MC) sand substrate. The succulent nature of its stem or trunk could possibly serves as a water reservoir where it could draw water little by little while the growing condition was still unfavorable; this could also enable them to recover from the ill effects of drought if re-watered. Thus, changes in the girth diameter could be an important

Fig. 13.6 *J. curcas* accession at 210th day of drought treatment

parameter in characterizing *J. curcas* for drought tolerance. Upon termination of this experiment, accessions planted on one of the water-stressed sandponics bed were left to stand to identify the maximum number of days that the crop can withstand such low amount of moisture level. Observations in this side experiment revealed the ability of *J. curcas* to survive under relatively dry (approximately 0% MC) conditions for seven months. Not only did the plants survive but they were also observed to develop leaves although with a noticeable difference in anatomy (Fig. 13.6).

13.4 Evaluation of Selected *J. curcas* Lines for Drought Tolerance

Following initial findings from the preliminary study on the responses of *J. curcas* to drought, six of the 13 selections (Table 13.1) from varietal improvement works were evaluated for drought stress with the objective of identifying accessions more adaptable to dry conditions. Additionally, two foreign accessions (GB 58959 originating from Tanzania and GB 58960 from Thailand) were also screened for comparison.

Although materials for evaluation from these accessions were grown under the same growth condition and the same time point during establishment, differences in visual plant vigor and plant height were noticeable. This could indicate the presence of genetic differences between accessions. Assessment in terms of different physiological and morphological characters including plant height (cm), total leaf area (m^2), stem diameter (cm), root dry matter (g), stem dry matter (g), and leaf dry matter (g) revealed minimal differences among the accessions tested (Table 13.2; Fig. 13.7). This result conforms to earlier studies indicating that the relative performance of the crop is highly affected by growth conditions than the individual genetic background (Maes et al. 2009; Achten et al. 2010a, b; Sapeta et al. 2013). At the end of the drought period which lasted for 93 days, GB 57540 was observed to be superior compared to the other accessions based on the parameters tested (Table 13.2). Data on gs showed the same pattern as the previous experiment conducted. The drop in gs was observed 7 days after withholding irrigation and was found not to differ among accessions. At this point, all eight accessions exhibited gs reading between 0.016 and 0.025 mol m^{-2} s^{-1}. Indeed, this result supports

Table 13.1 Promising *J. curcas* germplasm selected as potential parental genotypes for varietal improvement. *Source* UPLB 2010

Genebank number	Origin	Desirable traits
57,147	South Cotabato (local)	High oil content; medium % steryl ester and % sterol; good fruiting potential
57,262	Davao City (local)	High oil content; medium % steryl ester; 0% free fatty acids; large seeds
57,259	South Cotabato (local)	High oil content; high % steryl ester; large seeds
57,328	Eastern Samar (local)	High oil content; medium-to-high % sterol
57,540	Palawan (local)	Medium-to-high oil content; medium-to-low % free fatty acid; high % sterol
57,258	Surala (local)	High oil content; large seeds; good fruiting potential

Table 13.2 Estimated mean values of parameters used for drought evaluation of 8 *J. curcas* accessions

Accession	Plant height (cm)	Total leaf area (m^2)	Stem diameter (cm)	Root dry matter (g)	Stem dry matter (g)	Leaf dry matter (g)
GB 57259	21.5bc	0.42ab	1.102bc	0.35c	3.43b	2.21bcd
GB 57258	13.0c	0.22b	0.78c	0.50c	2.47b	1.41d
GB 57328	20.0bc	0.39ab	1.01bc	0.38c	2.67b	2.74abc
GB 58959	21.9bc	0.4ab	1.083bc	0.59c	3.32b	1.73cd
GB 57540	31.4a	0.5a	1.467a	1.37ab	6.69a	3.72a
GB 57147	19.7bc	0.49a	1.098bc	0.47c	3.75b	1.52d
GB 58960	18.8bc	0.39ab	1.032bc	0.91bc	5.2ab	3.28ab
GB 57262	25.1ab	0.42ab	1.248ab	1.61a	7.18a	3.33ab

Duration of drought period is 93 days. Root, stem, and leaf dry matter were measured at maximum stress, while plant height, total leaf area, and stem diameter were recorded in 7-day interval ($n = 6$). Means followed by the same letter are not significantly different at $\alpha = 0.05$ according to Duncan's multiple range test (DMRT)

the claim that the main mechanism of *J. curcas* to withstand drought stress is through stomatal control (Thompson et al. 2007; Pompelli et al. 2010; Silva et al. 2010; Díaz-López et al. 2012; Sapeta et al. 2013). Persistence of stomatal closure over a long period would result to the reduction in plant growth and yield. This is because such closure of stomates limits the intake of CO_2, thus limiting photosynthesis. In our study, evidence of delayed plant growth was observed on the recorded weekly plant height of the accessions evaluated. On the average, a 1.4 cm increase was observed on the eight accessions subjected to 93 days of drought treatment. Although this value appears as negligible, this suggests that *J. curcas* can grow even at severe stress. Findings of Achten et al. (2010a, b) detected that withholding irrigation arrest growth, but at low soil water availability, plants would continue to grow although at a reduce rate than the well-watered plants. The same observation was generated by Sapeta et al. (2013). In relation to the displayed drought mechanism of *J. curcas* through stomatal closure, an apparent decrease in the total leaf area was also observed. This reduction in canopy size is one mechanism for water conservation allowing plants to minimize the available area of transpiration (Boyer 1970) therefore maintaining good water status (Díaz-López et al. 2012). The displayed total leaf area during the period of drought treatment was highest for GB57540 although was not significantly different from five other accessions. This suggests the ability of *J. curcas* to maintain a minimal amount of its leaves enough to possibly sustain limited photosynthesis which can explain the observed increase in plant height discussed above. Another interesting aspect of the crop's mechanism to drought is its ability to store water in its stem, thereby providing moisture to the plant whenever conditions are unfavorable. *J. curcas* is a deciduous shrub exhibiting a succulent stem (Foidl et al. 1996) with soft wood (Henning 2006). These characteristics of *J. curcas* further categorized it as deciduous lightwood tree with large stem water storage or deciduous stem-succulent tree (Maes et al. 2009).

The observed trend on stem diameter for most accessions is that it decreases starting the 14th day of drought treatment. Similar result was reported by Achten et al. (2010a, b). As it is, it appears that the succulent nature of *J. curcas* stem possibly serves as a water reservoir where it could draw water little by little to survive during periods of drought. In addition, Maes et al. (2009) concluded that the succulent stem of the crop balances the minimal water losses of the leaves during drought.

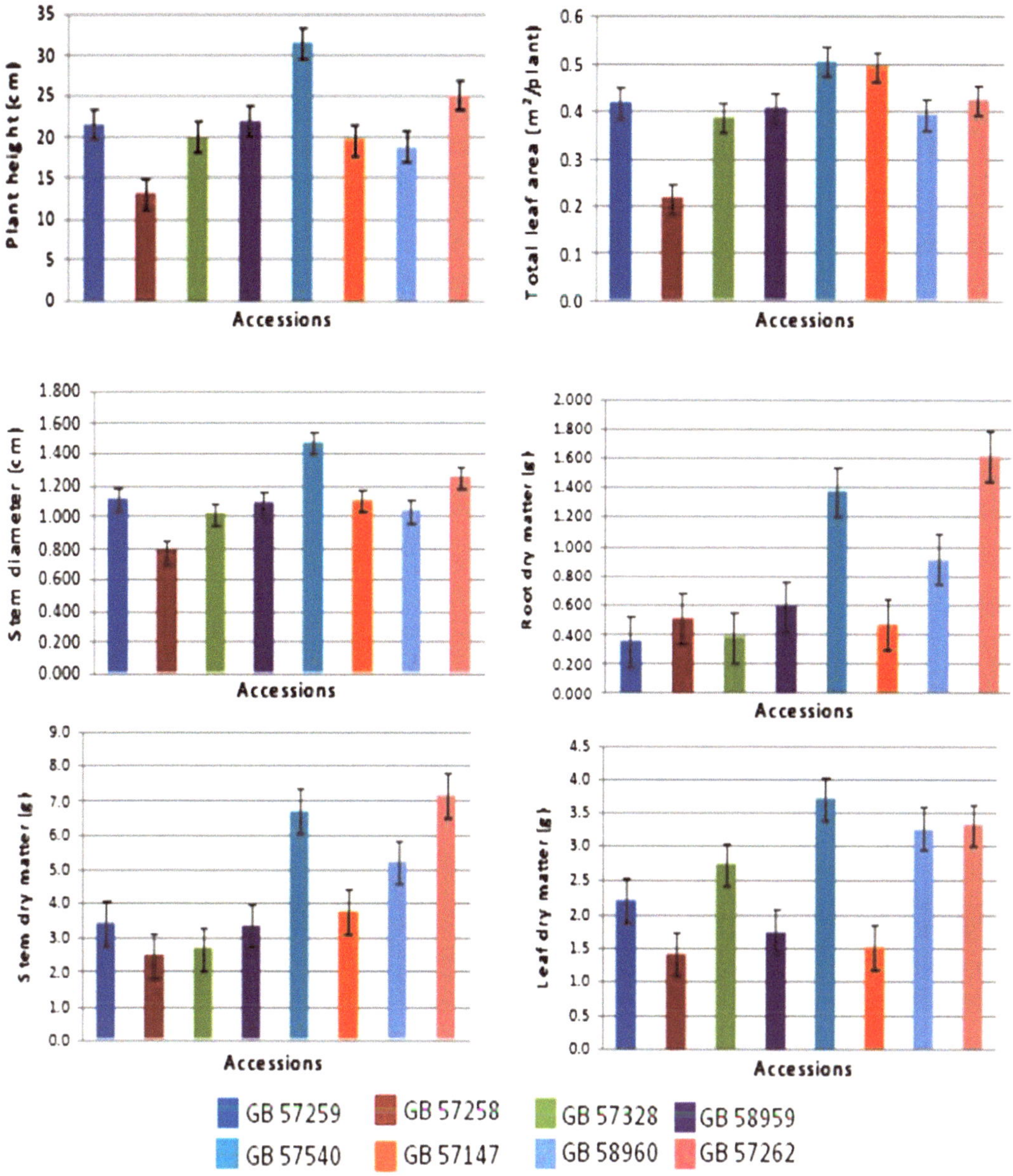

Fig. 13.7 Effect of drought stress on the different *J. curcas* accessions subjected to 93 days of drought period on sand culture. *Bars* represent means ± SE (n = 6)

Data on root, stem, and leaf dry weight were also measured at the end of drought period. The highest dry weight for stem and leaf dry weight were observed in GB57540 while for root dry weight, the highest mean value was recorded for GB57262. Comparing the means of the two accessions, however, showed no significant difference in root dry weight. Although not statistically analyzed, the computed root/shoot ratio was highest in GB57262 (0.15) but did not show much difference with GB57540. Dry weight data of the different tissues are determined in order to know the extent of biomass reduction under drought condition as opposed to controlled condition. In general, during periods of extreme water deficit, higher reduction in the leaves than in roots was observed. This significant reduction in leaf biomass usually translates to an increase in root-to-shoot ratio under drought (Achten et al. 2010a, b; Díaz-López et al. 2012). In here, it can be suggested that in response to severe water deficit, photoassimilates are repartitioned to the roots, therefore resulting in the reduction in the evaporative surface area (Liu and Stuetzel 2004). In the present experiment, however, our focus is to determine differences in the biomass produced among accessions tested therefore, the assumption by Díaz-López et al. (2012) and Achten et al. (2010a, b) was not confirmed. As it is, the identification of accessions with significant biomass value under drought condition could facilitate in the selection of planting materials suitable in areas with limited water availability.

13.5 Concluding Remarks

The exploration of genetic diversity in plants facilitates in the utilization of germplasm collections. Plants with unique traits can be identified, thus allowing their effective use in downstream programs. The information generated from the characterization of *J.curcas* collections at NPGRL-IPB, UPLB resulted to the identification of potential accessions for breeding and subsequent drought screening.

Knowledge on the physiological processes associated with abiotic stress tolerance such as drought is indispensable for the implementation of approach for sustained productivity of Jatropha in marginal areas. In our study, we confirm that the main mechanism of the crop to tolerate drought stress is through strict stomatal control to reduce transpirational water loss. Congruent with stomatal closure, a reduction in leaf area was apparent, thus decreasing the evaporative surface area leading to a decrease in net photosynthesis. The succulent nature of the crop's stem also adds to its adaptive mechanism as this serves as a water reservoir that supplies the crop with water during periods of severe drought. Biomass produced in terms of root, leaf, and stem dry weight under drought treatment could also assist in the selection of planting materials suitable in areas with limited water availability. From this set of parameters, GB 57540 was selected as it continuously hits the highest mean value except for root dry matter. GB57262, which exhibit the highest mean value for root dry matter, appeared not to significantly differ with GB57540 in other parameters tested.

While the preliminary screening resulted to selection of *J. curcas* accessions that performed well under drought at early vegetative stage, the need to assess the performance of these accessions in terms of yield should also be considered. Thus, in order to draw a more reliable conclusion in selecting the most promising *Jatropha* accessions under drought, data on yield should be included in the set of parameters to be gathered. In addition, the performance of these accessions in the field should also be evaluated.

Acknowledgments The authors wish to acknowledge Osaka University (OU) and Sumitomo Electric Industries (SEI) for funding our research and providing the sandponics system for drought screening. We also acknowledge Mr. Naoki Ikeguchi for the technical support provided for operating the sandponics system.

References

Achten WMJ, Almeida J, Fobelets V, Bolle E, Mathijs E et al (2010a) Life cycle assessment of Jatropha biodiesel as a transportation fuel in rural India. Appl Energy 87:3652–3660

Achten WMJ, Maes WH, Reubens B, Mathijs E, Singh VP et al (2010b) Biomass production and allocation in *Jatropha curcas* L. seedlings under different levels of drought stress. Biomass Bioenergy 34:667–676

Azam MM, Waris A, Nahar NM (2005) Prospects and potential of fatty acid methyl esters of some non-traditional seed oils for use as biodiesel in India. Biomass Bioenergy 29:293–302

Basha SD, Sujatha M (2007) Inter and intra-population variability of *Jatropha curcas* L. characterized by RAPD and ISSR markers and development of population-specific SCAR markers. Euphytica 156:375–386

Berchmans HJ, Hirata S (2008) Biodiesel production from crude *Jatropha curcas* L. seed oil with a high content of free fatty acids. Bioresour Technol 99:1716–1721

Bonner FT (1991) Measurement of moisture content. In: Gordon AG, Gosling P, Wang BP (eds) GORTree and shrub seed handbook. International Seed Testing Association (ISTA), Zurich, pp 12–1 to 12–7

Boyer JS (1970) Leaf enlargement and metabolic rates in corn, soybean, and sunflower at various leaf water potentials. Plant Physiol 46:233–235

Chaudharry DR, Patolia JS, Ghosh A, Chikara J, Boricha GN et al (2007) Changes in soil characteristics and foliage nutrient content in *Jatropha curcas* plantations in relation to stand. FACT Foundation, Wageningen

Chavan A, Gour VK, Basha H (2014) *Jatropha curcas* L.: a predominant panacea for energy security and climate change. Curr World Environ 9(1):130–136

Chen F (2007) Advances in Jatropha industry research and development. International workshop on the development of the JCL industry, China, 29–31 October 2007, pp 7–8

Chin HF (1996) Ex situ conservation of tropical tree species. In: Quah SC, Kiew R, Bujang I, Kusnan M, Haq N et al (eds) Underutilized tropical plant genetic resources: conservation and utilization. Universiti Pertanian Malaysia Press, London, pp 143–152

Das S, Misra RC, Mahapatra AK, Gantayat BP, Pattnaik RK (2010) Genetic variability, character association and path analysis in *Jatropha curcas*. World Appl Sci J 8(11):1304–1308

Datta SK, Pandey RK (2013) Studies on *Jatropha curcas* L. and its improvement through induced mutagenesis. In: Bahadur B, Sujatha M, Carels N (eds) Jatropha, challenges for a new energy crop. Springer, New York, pp 312–334

Díaz-López L, Gimeno V, Simón I, Martínez V, Rodríguez-Ortega WM et al (2012) *Jatropha curcas* seedlings show a water conservation strategy under drought conditions based on decreasing leaf growth and stomatal conductance. Agric Water Manag 105:48–56

DOE (2007) Philippine Department of Energy. https://www.doe.ov.ph/energy-resources-alternative-fuels/biofuels/biodiesel/323-jatropha-methyl-ester. Accessed 12 July 2015

El-Sharkawy MA (2007) Physiological characteristics of cassava tolerance to prolonged drought in the tropics; implications for breeding cultivars adapted to seasonally dry and semiarid environments. Braz J Plant Physiol 19:257–286

Fairless D (2007) Biofuel: the little shrub that could-maybe. Nature 449:652–655

Flexas J, Bota J, Galmés J, Medrano H, Ribas-Carbó M (2006) Keeping a positive carbon balance under adverse conditions: responses of photosynthesis and respiration to water stress. Physiol Plant 127:343–352

Foidl N, Foidl G, Sanchez M, Mittelbach M, Hackel S (1996) *Jatropha curcas* L. as a source for the production of biofuel in Nicaragua. Bioresour Technol 58:77–82

Ford-Lloyd BV, Jackson M (1986) Plant genetic resources: an introduction to their conservation and use. Cambridge University Press, Cambridge

Ginwal HS, Phartyal SS, Rawat PS, Srivastava RL (2005) Seed source variation in morphology, germination and seedling growth of *Jatropha curcas* Linn. in Central India. Silv Genet 54(2):76–80

Grativol C, Lira Medeiros CF, Hemerly AS, Ferreira P (2011) High efficiency and reliability of inter-simple sequence repeats (ISSR) markers for evaluation of genetic diversity in Brazilian cultivated *Jatropha curcas* L. accessions. Mol Biol Rep 38:4245–4256

Gubitz GM, Mittelbach M, Trabi M (1999) Exploitation of the tropical oil seed plant *Jatropha curcas* L. Bioresour Technol 67:73–82

Hasnam (2007) Improvement of *Jatropha curcas* L. in Indonesia: promise and performance. International workshop on the development of the JCL industry, China, 29–31 October 2007, pp 22–27

Heller J (1996) Promoting the conservation and use of underutilized and neglected crops. In: *Physic nut. Jatropha curcas* L. Institute of Plant Genetics and Crop Plant Research, Gatersleben/International Plant Genetic Resources Institute, Rome, Italy

Henning (2006) The Jatropha system. http://www.jatropha.de/. Accessed 2 Aug 2015

Johnston M, Holloway T (2007) A global comparison of national biodiesel production potentials. Environ Sci Technol 41(23):7967–7973

Jones HG (1985) Partitioning stomatal and non-stomatal limitations to photosynthesis. Plant Cell Environ 8:95–104

Jongschaap REE, Corré WJ, Bindraban PS, Brandenburg, WA (2007) Claims and facts on *Jatropha curcas* L., Plant Research International. BV Wageningen, Netherlands

Katwal RPS, Soni PL (2003) Biofuels: an opportunity for socioeconomic development and cleaner environment. Indian For 129:939–949

Kaushik N, Kumar K, Kumar S, Kaushik N, Roy S (2007) Genetic variability and divergence studies in seed traits and oil content of Jatropha (*Jatropha curcas* L.) accessions. Biomass Bioenergy 31:497–502

Kiew R (1996) Biodiversity and its implications for genetic conservation. In: Quah SC, Kiew R, Bujang I, Kusnan M, Haq N et al (eds) Underutilized tropical plant genetic resources: conservation and utilization. Universiti Pertanian Malaysia Press, London, pp 1–10

Kumar A, Sharma S (2008) An evaluation of multipurpose oil seed crop for industrial uses (*Jatropha curcas* L.): a review. Ind Crops Prod 28:1–10

Liu F, Stuetzel H (2004) Biomass partitioning, specific leaf area, and water use efficiency of vegetable amaranth (*Amaranthus* spp.) in response to drought stress. Sci Hortic 102:15–27

Maes WH, Achten WMJ, Reubens B, Raes D, Samson R et al (2009) Plant-water relationships and growth strategies of *Jatropha curcas* L. seedlings under different levels of drought stress. J Arid Environ 73:877–884

Makkar HS, Becker K, Sporen F, Wink M (1997) Studies on nutritive potential and toxic constituents of different provenances of *Jatropha curcas* L. J Agric Food Chem 45:3152–3157

Miyashita K, Tanakamaru S, Maitani T, Kimura K (2005) Recovery responses of photosynthesis, transpiration, and stomatal conductance in kidney bean following drought stress. Environ Exp Bot 53:205–214

NPGRL (1998) Protocols on genebank operations. National Plant Genetic Resources Laboratory, Institute of Plant Breeding, UPLB

Pamidimarri D, Mastan SG, Rahman H, Reddy MP (2010) Molecular characterization and genetic diversity analysis of *Jatropha curcas* L. in India using RAPD and AFLP analysis. Mol Biol Rep 37:2249–2257

Pant KS, Khosla V, Kumar D, Gairola S (2006) Seed oil content variation in *Jatropha curcas* Linn. in different altitudinal ranges and site condition in HP India. Lyonia 11:31–34

Parawira W (2010) Biodiesel production from *Jatropha curcas*: a review. Sci Res Essays 5:1796–1808

Parreño-de Guzman L, Aquino AL (2009) Seed characteristics and storage behavior of *Jatropha curcas* L. Philipp J Crop Sci 34(1):13–21

Pompelli MF, Barata-Luís R, Vitorino HS, Goncalves ER, Rolim EV et al (2010) Photosynthesis, photoprotection and antioxidant activity of purging nut under drought deficit and recovery. Biomass Bioenergy 34:1207–1215

Prastowo B (2007) Biofuel development in Indonesia. International Workshop on the Development, China, 29–31 October 2007, pp 13–20

Ranade SA, Srivastava AP, Rana TS, Srivastava J, Tuli R et al (2008) Easy assessment of diversity in *Jatropha curcas* L. plants using two single-primer amplification reaction (SPAR) methods. Biomass Bioenergy 32:533–540

Rao GR, Korwar GR, Shanker AL, Ramakrishna YS (2008) Genetic associations, variability and diversity in seed characters, growth, reproductive phenology and yield in *Jatropha curcas* L. accessions. Trees 22:697–709

Rao NK, Hanson J, Dulloo ME, Ghosh K, Nowell D et al (2006) Manual of seed handling in genebanks. Handbooks for genebanks No. 8. Bioversity International, Rome, Italy

Sapeta H, Costa J, Lourenco T, Maroco J, van der Linde P et al (2013) Drought stress response in *Jatropha curcas*: growth and physiology. Environ Exp Bot 85:76–84

Sausen TL, Rosa L (2010) Growth and carbon assimilation limitations in *Ricinus communis* (Euphorbiaceae) under soil water stress conditions. Acta Bot Bras 24:648–654

Silva EN, Ferreira-Silva SL, Fontenele AD, Ribeiro RV, Viégas RA et al (2010) Photosynthetic changes and protective mechanisms against oxidative damage subjected to isolated and combined drought and heat stresses in *Jatropha curcas* plants. J Plant Physiol 167:1157–1164

Sun Y, Christensen B, Liu F, Wang H, Müller R (2008) SSR and AFLP markers reveal low genetic diversity in the biofuel plant *Jatropha curcas* in China. Crop Sci 48:1865–1871

Thompson AJ, Andrew J, Mulholland BJ, McKee JT, Hilton HW et al (2007) Overproduction of abscisic acid in tomato increases transpirational efficiency and root hydraulic conductivity and influences leaf expansion. Plant Physiol 143:1905–1917

Utilization of Wastewater for Cultivation of Jatropha in Egypt

14

Adel Hegazy

Abstract

The world demand for energy has hugely increased in the recent decades. Traditional energy sources are limited and for the long term are not sufficient for the forthcoming generations. Utilization of biofuel plants to fulfill this demand is an objective of great interest. *Jatropha curcas* was considered worldwide as one of the most promising biofuel crops for the next several decades. Biodiesel production from jatropha seed oil has a high content of free fatty acid. However, the expansion was faced by obstacles, such as the lack of elite lines with high productivity as well as insufficient good water for agriculture in semiarid region. Plant tissue culture was considered as the technique for rapid propagation and production of sufficient plant materials with high quality. Direct organogenesis presents an advantage to permit maximum fidelity in terms of genetic stability of the produced plants. Actually, there are very few laboratories worldwide that can use the micropropagation technique to produce in vitro jatropha plants at the commercial level, and JOil laboratory in Singapore is considered one of the best producers of them. Egypt has a strong potential for large-scale plantations, to be one of the biggest inedible oil production countries promoting the renewable energy development. Geographically, it has strong sunlight, high temperature in summer, warm in winter season, large-scale nonarable land (about 94%), and availability of wastewater. According to Holding Company for Water and Wastewater (HCWW) report, the total amount of wastewater in Egypt is estimated as 7.3 billion m^3/year, and approximately, 3.6 billion m^3/year is currently treated. It is predicted that by the year 2037, the discharge of wastewater may be about 32 million m^3/day, equal to 11.7 billion m^3/year. In the present study, an overview of our successful research results in

A. Hegazy (✉)
Genetic Engineering and Biotechnology Research Institute (GEBRI), University of Sadat City, 79/, Sadat City, Egypt
e-mail: adel.heazy@gebri.usc.edu.eg

S. Tsuchimoto (ed.), *The Jatropha Genome*, Compendium of Plant Genomes,
DOI 10.1007/978-3-319-49653-5_14

utilizing treated wastewater of residential city in the irrigation and the productivity of jatropha plantations. This will play an important role and contribute to the provision of freshwater for food crops. Evaluate jatropha growth characters of elite local and imported jatropha varieties were evaluated in greenhouse and open field under the complete drip irrigation program of treated wastewater.

Stakeholders

Abbreviation: Governmental bodies, organizations, and ministries

ARC	Agriculture Research Center
CJO	Crude jatropha oil
EEAA	Egyptian Environmental Affairs Agency
ECHEM	Egyptian Petrochemicals Holding Company
EU	European Union
FAO	Food and Agriculture Organization of the United Nations
FDI	Foreign direct investment
GHG	Greenhouse gas
GMO	Genetically modified organism
HCWW	Holding Company for Water and Wastewater
JICA	Japan International Cooperation Agency
JDI	Japan Development Institute
LE	Egyptian pound
MALR	Ministry of Agriculture and Land Reclamation, Egypt
MWRI	Ministry of Water Resource and Irrigation
NWRP	National Water Resources Plan
UAE	Under-secretariat for Afforestation and Environment
USAID	United States Agency for International Development

14.1 Introduction

The genus *Jatropha* belonging to the *Euphorbiaceae* family contains approximately 170 known species. The name *Jatropha* was derived from the Greek words jatrós (doctor) and trophé (food), which imply medicinal uses. *Jatropha cuscas* (jatropha) is a small tree or a large shrub, which can reach a height of 3–5 m, while under favorable conditions, it can attain a height of 8 or 10 m (Ashwani and Sharma 2008). The trees are deciduous, and flowering occurs during the wet season, and two flowering peaks are often seen, i.e., during summer and autumn. In permanently humid regions, flowering occurs throughout the year in the hot seasons (Dehgan and Webster 1979). Bioenergy produced from plants, solar or wind power, and others are called "renewable energy." Using new energy sources eventually become "energy self-sufficiency" especially in developing countries. (Metwally 2009). The jatropha flowers are pollinated by insects especially honey bees. Each inflorescence yields a bunch of approximately 10 or more ovoid fruits. The black seeds mature about 3–4 months after flowering. The weight per 1000 seeds is about 727 g, 1375 seeds/kg in average. At optimal conditions, about 5 tons/ha of seeds are produced per year (Li et al. 2007). The seeds contain 30–40% oil with a fatty acid pattern similar to that of edible oils (Gubitz et al. 1999). Jatropha has attracted global attention as an important source of biodiesel (Annarao et al. 2008), because the seed oil can be combusted as fuel without being refined (Chhetri et al. 2008), and burns with clear smoke-free flame, tested successfully as fuel for the simple diesel engine (Forson 2004). Micropropagation is becoming an important technique for increasing shoot proliferation rates and producing new varieties through genetic transformation, it considered as an alternative to the conventional vegetative propagation methods (Li et al. 2007). However, conventional propagation of jatropha is limited by poor seed germination, scanty, and delayed rooting of seedlings and vegetative cuttings (Purkayasha et al. 2010). Recently, Aldoori (2014) established an applicable micropropagation protocol of jatropha by overcoming oxidative browning and hyperhydricity, which are in vitro physiological disorders as results of oxidation of phenolic compounds, excessive hydration, and low lignification in tissue.

There are three kinds of water to utilize for irrigating jatropha plantations. 1) Primary or secondary treated wastewater (TWW) produced by residential city usage. Its effect on agricultural plantation and on the biofuel productivity in jatropha plantations were studied (Kotb et al. 2000; El-Tohamy et al. 2012; Rajaona et al. 2012; Silva et al. 2014), 2) Agricultural drainage water (ADW) that is recommended for jatropha irrigation. It increased the leaf area when mixed with the freshwater in a ratio of 25% (Hussein and Abdellrauf 2013), and 3) Industrial drainage water (IDW). The possibility to use this poor-quality water in jatropha production was studied (Achten 2010; Abd El-Baky et al. 2010, 2013).

The aim of our research work is mainly to evaluate the growth of local and imported elite jatropha trees under Luxor governorate climate conditions, selecting the appropriate irrigation system, a suitable one under the prevailing conditions in such an arid and hot region to ensure its sustainable agricultural development. It protects the environment through using a treated wastewater instead of disposing into water stream. We finally find a sustainable and environmentally friendly alternative to today's fossil fuel, which reduces the CO_2 emission that is considered to be the first culprit of the global warming. Generally, such study may be helpful for identifying the best agricultural management process to subsequently achieve the highest productivity of biodiesel.

14.2 Wastewater Use for Jatropha Cultivation

Nowadays, freshwater resources in Egypt are limited and insufficient either to cope with designed agricultural development projects or to other water uses. Therefore, such forefront issue represents the main constraints for any future agricultural expansion (FAO 1992). Water resources management including unconventional resources, i.e., groundwater drainage and treated sewage effluent, is considered as an urgent issue for the expansion of irrigated agriculture. That means that the reuse of such alternative water resources becomes part of the official policy of actual and future extension programs for the agricultural development in Egypt, particularly, by using a high technology for maximizing the efficiency in usage of these limited water resources. Ministry of Public Works and Water Resources recycles about 5 billion m^3 of wastewater officially and its goal is to increase that volume to 7 billion m^3 (Kotb et al. 2000). Jatropha has a high efficiency on soil stabilization, withstands extreme weather conditions, and has the possibility of being irrigated by sewage, producing high amount of oil used as biofuel in various industries (El-Tohamy et al. 2012). The uncontrolled application of such waters must have many effects on both soil properties and plant growth; in particular, the long-term use must have the hazardous effects which are mainly dependent on the soil nature and water quality, as well as the kinds of crops grown and irrigation systems applied. To formulate a responsible schedule for the proper use of the wastewater resources and the protection of the cultivated lands and crops, several attempts must be taken to recommend the superiority of these uses as well as the suitable irrigation system. One strategy to increase the horizontal expansion, as well as available wastewater resources for the reclaimed marginally desert soils on both sides of the Nile Valley, is to change the traditional farming systems in turn. Such soils would be used for agriculture under the safe limits of water hazard for each soil type or plant. Hence, one of the best agricultural developments, which have the maximized desert soil potentiality as well as the sustainability through reclamation process, is to control the positive interaction between soil varieties and plant characteristics as affected by different water qualities under different irrigation methods (Nessrien Ahmed 2010). Agricultural systems are increasingly under pressure from competition for water. Most affected by this are arid and semiarid regions. The reuse of treated sewage effluents (TSE) should be taken into account as an alternative improved water management for irrigation. However, the use of TSE in agricultural production can lead to the build-up of soil salinity, as well as leaching of nutrients into the groundwater. Thus, potential crops need to be salt tolerant, adapted to arid areas, and ideally, used for nonfood products to avoid potential health risk. Jatropha, claimed to be suitable for growth under adverse conditions, may be an option to produce the biofuel from plantations irrigated with wastewater. The size of irrigated surface will depend on leaching fraction (LF). Soil salinity has to be considered, since jatropha has been reported to be salt sensitive. N (nitrogen) supply from TSE is not sufficient to produce a moderate seed yield, while P and K demands can be satisfied. Suitability of the irrigation system depends on environmental factors such as climate, soils, and the overall water availability, and thus, the applicability of the model results to other cases needs further study (Rajaona et al. 2012). The behavior of macronutrients and micronutrients in the soil was evaluated when swine wastewater (SW) was supplied successively in growing jatropha. The doses of SW were set in proportions of 0, 40, 80, 120, 160, and 200 m^3/ha. The soil attributes related to acidity did not suffer influence from successive applications of SW. In the first application, none of the evaluated parameters suffered interference of the doses. In macronutrients, K and Ca presented quadratic behavior in the second application. An increase in micronutrient Cu by the dose of SW was recorded in the second and third applications. Fe increased only in the third application of SW (Silva et al. 2014). This is mainly due to direct usage of sewage effluent that represents one of the main pollution sources for the environment. It contains a lot of solid, semisolid, and liquid wastes that generate substantial amount of toxic organic and inorganic pollutants, which will lead to serious environmental consequences or problems if dumped in the soil ecosystem without treatment. So, the management practices for this wastewater before discharging on land represent an urgent process in order to prevent the serious problems of soil ecosystem, i.e., salinity and sodicity stresses as well as soil biosphere. In optimizing for reducing the possible adverse effects of the marginal desert sandy soil, the suitable irrigation system is drip irrigation, which is partially capable of retaining enough available sort moisture range for grown plants and biological activity, and cause more pronounced increments in plant growth, seed yield and seed oil yield with high quality than others (El-Tohamy et al. 2012). Jatropha plants can grow on the poor-quality land unsuitable for food crops, and need little water or fertilizers. They also thrive in rock crevices and in a variety of climates (i.e., tropical, subtropical, and slightly cool with light frost) and soils (i.e., gravelly, sandy, saline, stony, semiarid, and arid). There are other advantages i.e. pest-resistance and high absorbance rate of carbon dioxide (CO_2) from the atmosphere (Mkoka and Shanahan 2005). The phyto-protective action against pests and pathogens provides additional protection to intercropped plants (Jones 2004).

Hussein and Abdellrauf (2013) revealed that the leaf area increased when freshwater was mixed with 25% agricultural drainage water (ADW) and used for the irrigation of jatropha. The leaf area decreased when the water contains 50% of ADW compared to the case of 25%, but it is still more than the control (0% of ADW). When ADW percentage increased over 50%, the area decreased to less than the control. Dry weight of leaves or the stem negatively responded to the ratio of ADW in the irrigation water. The plant fresh weight or dry weight responded positively to 100 ppm of plant fertilizers in the sprayed solution, but there was no difference between 100 ppm and 200 ppm. On the other hand, several researches have been done for the possibility of use of poor-quality water in jatropha irrigation (Achten 2010). The effect of irrigation by mixed drainage water on enzymes of the oxidative defense of jatropha plants was studied. Under the unflavored condition, reactive oxygen species such as superoxide radical, H_2O_2, and OH radicals have a role in lipid peroxidation that leads to the membrane damage and degradation of biomolecules such as proteins, phospholipids, and pigments. Antioxidant protection involves compounds such as carotenoids, ascorbic acid, tocopherol, phenolics, and flavonoids (Abd El-Baky et al. 2010) and a number of enzymes including superoxide dismutase (SOD), ascorbate peroxidase (APX), catalase (CAT), and glutathione reductase (GR). They are believed to play crucial roles in scavenge of different free radicals (Abd El-Baky et al. 2008). Recently, the effect of NPK foliar fertilizer and varied percentages of industrial drainage water (IDW) in irrigation water on lipid peroxidation (LP) and antioxidant enzyme activities (SOD, APX, and CAT) of jatropha plants was investigated. They found that concentration of LP increased as percentage of IDW increased up to 75%. A positive relationship was recorded between the enzyme activities and the level of IDW. The maximum values of the three enzymes activities were obtained with application of NPK with the ratio of 2:2:2. The highest lipid peroxidation was detected in leaves of non-fertilized plants irrigated by IDW, whereas the lowest values were detected in leaves of plants fertilized by NPK (2:2:2) irrigated by the freshwater (Abd El Baky et al. 2013).

14.3 Advantage of Jatropha Plant and Its Biofuel

Jatropha is an important nonedible oil plant growing in the wasteland. Its biofuel is of high quality, and clean, nontoxic, eco-friendly, and economic due to its low production cost. It can be a good plantation material for ecorestoration in all types of wasteland, and also is an important medicinal plant. George et al. (2005) reported that jatropha biodiesel could be used in today's vehicle fleets worldwide, and might offer a viable means to sustainable transportation; lower greenhouse gas emissions and enhanced mobility even in remote areas. Mitigation of global warming and creation of new regional employment opportunities could be important cornerstones of any forward-looking transportation system for emerging economies. Misra and Misra (2010) revealed that jatropha could prevent soil erosion, and could be grown as the live fence and used as an alternate commercial crop. It could be used for biofuel feedstock, soap, glue, or dye industry. As described above, researchers confirmed the beneficial usage of treated wastewater to irrigate jatropha plantations (El-Tohamy et al. 2012; Hussein and Abdellrauf 2013).

Both developed and developing countries are keen to secure the biofuel supply capacity (energy). They are highly conservative to set national policy on biofuel promotion, due to the severe food price rise caused by historically highest records of crude oil price and expectation of biofuel industry development, which led to imbalance of food supply especially in developing countries, which depend on low cost imported grains and vegetable oils. As a result of such misleading for biofuel policies, not only in biofuel-producing countries but also in consuming countries consider "sustainability" of biofuel production for a long term, and have revised their ambitious biofuel policies to improvise present technical and physical infrastructure development for biofuel feed production without negative impacts on food supply (Hegazy et al. 2012).

Now that the "jatropha fever" is over, public/private stakeholders are seriously looking for the reasonable and dependable solutions for jatropha development engaging in activities as below. Through series of scientific studies, the world is now confident to use jatropha as biofuel feedstock.

1. High yielding jatropha variety development has been started since 2005.
2. Best available agronomical method has been tested in many countries, and many practical cultivation techniques/technologies have already been developed.
3. Many jatropha-related researchers have been active for high-yield varieties, agronomy, seed cake utilization, and oil extraction.
4. In addition to the yield improvement, the use of seed cake has seriously been considered. A few countries have started to use seed cake for organic fertilizer and for biofuel for boiler and electric power.

In the updated jatropha market situation, there are strong demands in biojet fuel and other industrial usage such as bioplastic other than primary application for biodiesel. The jatropha biofuel supply is very limited at this moment with less than 2000 tons of crude jatropha oil (CJO) marketed in 2010. However, in 2011, CJO is sold around US$1000 to $1100/ton mainly for experimental purposes.

Another demand is identified in jatropha seed cake, which is used as 1) organic fertilizer, 2) biofuel substitute for coal, or 3) possible animal feed with detoxification in the future. Based on the experts' experiences in Myanmar, jatropha seed cake could be sold with $200–300/tons as an alternative of oil seed cake of sesame or castor oil for organic manure production. The price of coal substitute purpose is about $80–100/ton in Myanmar (market price).

14.4 Egyptian Potentiality for Biofuel Industry Development

One of the unique advantages of the development of the Egyptian jatropha biofuel industry is its renewable nature produced from the sun, wastewater, and nonarable land (Fig. 14.1). Another advantage is its inedible nature having no competition with food production and supply, which raised a worldwide concern after the food price hyped in 2008. The promising bioenergy market was identified in Europe as big market. It has been increasing in demand and price of biodiesel feedstock by a set of policies encouraging renewable energy use. Also setting strict Greenhouse Gas (GHG) cap regulations within the region. Egypt has a competitive access to the European market compared to other biofuel-producing countries in Africa and Asia. It has geographical closeness and relatively well-developed logistics networks. Economic impacts from the bioenergy industry in Egypt are multifaceted. The industry contributes to environmental sustainability by ensuring freshwater and energy security with a smart use of treated wastewater, and generates vast employment opportunities, especially, jatropha being suitable under constant sunlight and high temperature. Cultivation of jatropha encourages the regional development of Upper Egypt where it was being neglected long time by industrial activities. This addresses the issue and contributes to the mitigation of development disparity between the regions. Although a significant potentiality and benefits were identified, the main challenge for a commercially viable Egyptian jatropha biofuel development model is the low productivity of the existing varieties. A definite need for the use of new high-yield varieties and technologies requires inviting private investors with the appropriate technology. In specific, investigation of the potentiality of new technology in Egypt, confirmation of investor-friendly policy environment, and re-examination of the most viable

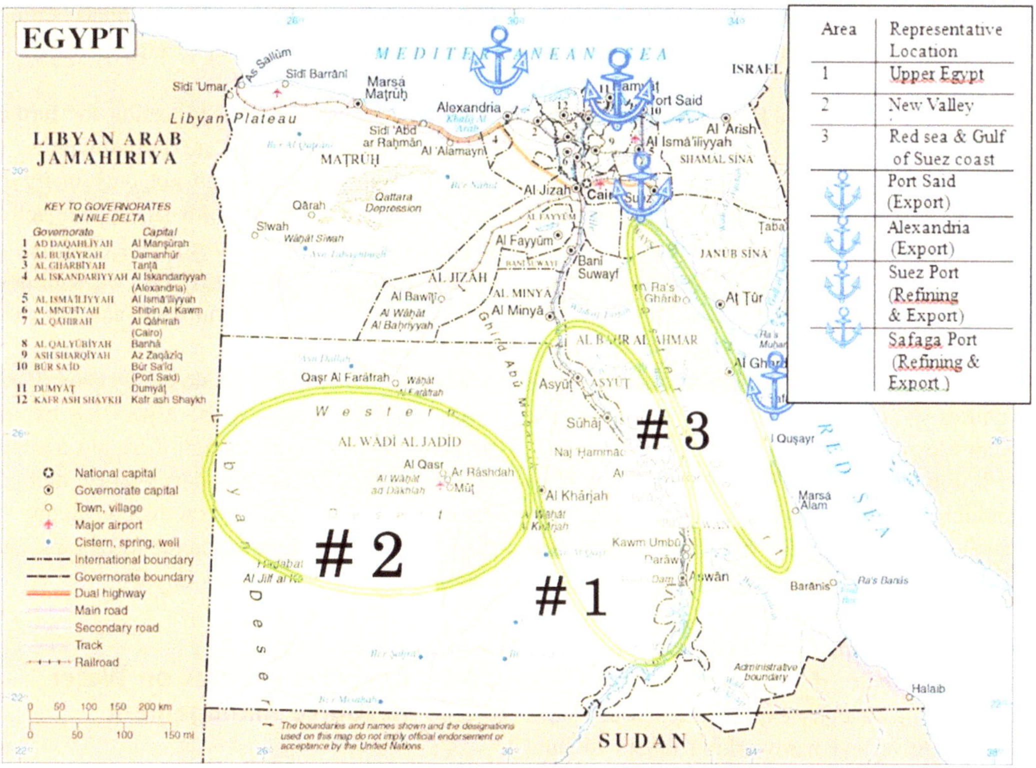

Fig. 14.1 Potential focused areas for jatropha plantations and ports

business model with the Egyptian governmental support are required (Hegazy et al. 2012).

14.4.1 Job Creation in Egypt

Egyptian economy is unable to absorb increasing number of job seekers entering into the job market every year. The mismatch of demand and supply within the labor market for skills and qualification has also hampered the employment opportunities for the well-educated young and future generation. The population of Egypt has been growing over 2.0% per year until now. The total population has reached 81 million, and the population growth rate was 2.2% for 2010/2011. As the population growth will remain average of 2.2% for the coming decades, the population of Egypt is expected to reach 100 million by 2020, facing with problems of how to provide necessary housing, education, and infrastructure (Ministry of Economic Development 2012).

Egyptian government is a center of world's attention expecting to see changes in the existing political, economic, and social structure. Employment of Egypt has been depending on 1) services (51%), 2) agriculture (32%), and 3) industry (17%). Service sector provides 51% of employment including 1) tourism, 2) wholesale/retail trade, and 3) other services. Agriculture is still very important provider of employments (32%) with one of three workers engaged in agriculture. However, the agriculture sector has remained underdeveloped and is keeping the productivity level at 19.4% of that of the industry. Egyptian Government should be encouraged to modernize this sector in the future. Considering the abundant sunshine, land, water, and labor, Egypt has a potentiality to expand the agriculture sector as a major provider of employment by attracting FDI/DDI as is demonstrated by several modern large-scale agrobusiness companies which are already operating (World Bank and Central Bank of Egypt 2012).

14.4.2 Creation of New Employment in Rural Communities

Until the Arab Spring Revolution, the promising workplaces were abroad in Gulf countries and Libya. However, job availability in the Arab countries is declining due to the revolution and/or slowdown of the global economy. As a result, Egypt is facing a very challenging task of providing jobs for rapidly growing young generation. However, unlike other developing and least developed countries in arid areas in Africa and Middle East regions, Egypt still has tremendous potentialities of its sustainable and strong economic development by using its unused or inefficiently exploited resources, such as competitive and hardworking labors in rural communities, Nile river, groundwater, treated wastewater, and arable land with irrigation.

Due to the lack of job opportunities in rural communities, many young people had to go to big cities in Egypt or overseas for survival. However, under present unfavorable circumstances, such as few investments in Egypt and prolonged Eurozone financial crisis, it has gotten harder for young generation day by day.

In general, it is not easy to create new employment in rural communities under globally competitive capitalism. However, unlike other countries, Egypt may have good potentiality to develop agribusinesses by

1. leveraging and improving Nile river water application for higher value commodity production in existing farm land,
2. developing unused groundwater resources for higher value commodity production with new farm land development in unused land,
3. Applying "Treated wastewater" for inedible high value or strategic commodity (ex. biofuel feeds or industrial materials) production in Non-Arable Land where there is little potentiality to utilize conventional farming activities with Nile river or ground water.

In any water resource application for further economic development in rural communities, it must be competitive enough not only in Egypt but also in global markets for sustainable activities led by private sectors. Conventional agribusiness sector shall likely take off with new government's strong commitments and privates' active participation.

On the contrary, there is no clear vision of the water resource exploitation, especially treated wastewater, for Egypt, though there are tremendous potentialities for rural development and job creation. Our study focuses on the potentiality of the water resource application for sustainable and meaningful resource use for Egypt.

14.5 Egyptian Policies on Water Resource Management

14.5.1 National Water Resource Plan-2017

Due to the lack of renewable water source except the Nile River, water resource management is one of the highest priorities and bottleneck of sustainable development for Egypt. In order to develop the long-term strategy for water use in Egypt, Ministry of Water Resource and Irrigation (MWRI) initiated and took a lead of National Water Resources Plan-2017 (NWRP) with the collaboration of all relevant authorities in Egypt (http://www.mwri.gov.eg/En/plan46.html).

NWRP presented the past status (up to 2004) of the nationwide water usage, problems and issues, and the proposal for solutions. The NWRP includes relevant policy decisions and measures that are to be implemented. All stakeholders were involved in NWRP implementation, and careful planning and coordination were conducted.

An implementation framework of NWRP-2017 was developed to assign clear responsibilities for the implementation of the varied plans. It also included the budgetary requirements for the implementation, including

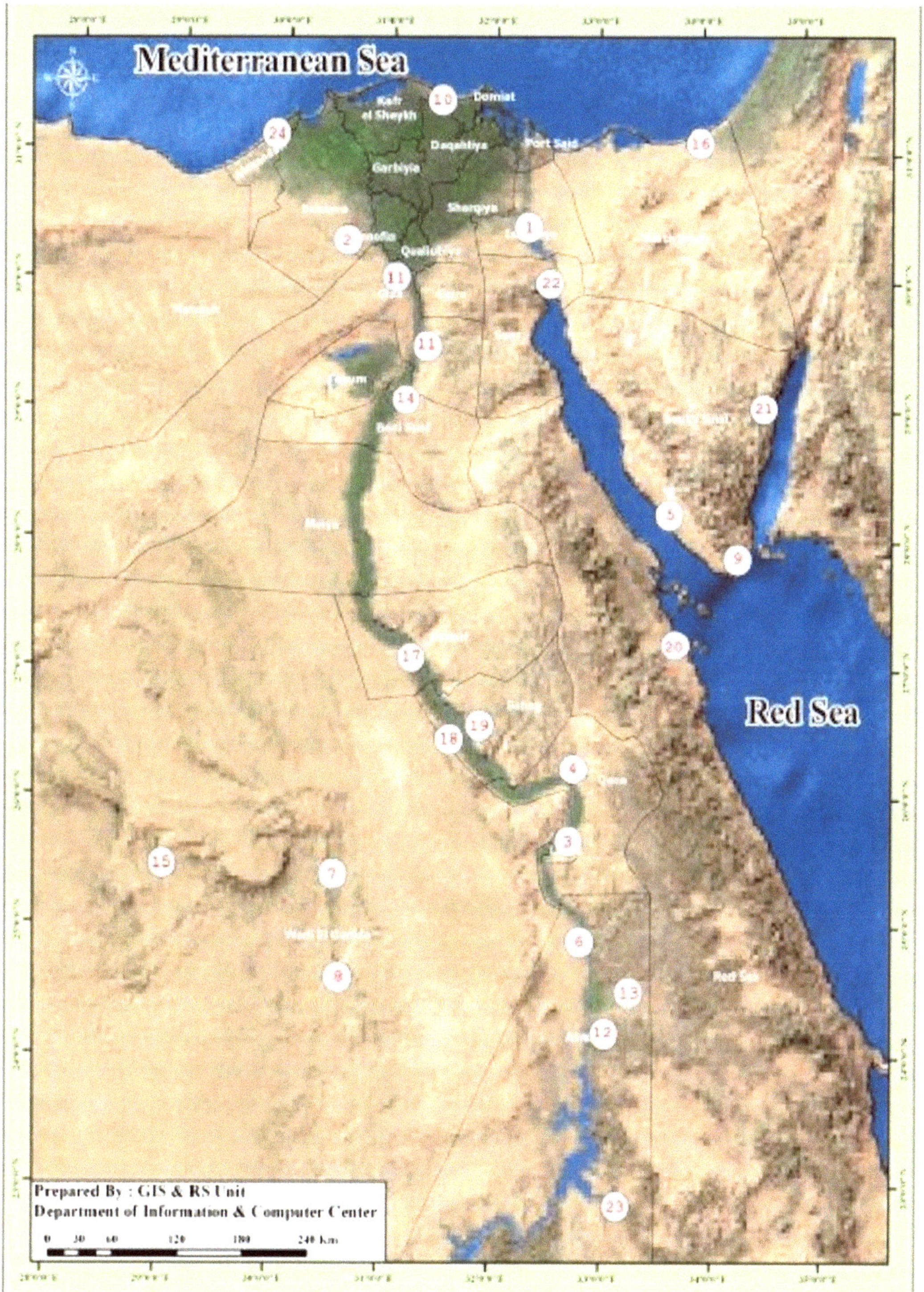

Fig. 14.2 Forests under safe use of treated sewage water for afforestation program. *Source* EEAA Annual Report 2007

investments and recurrent costs. However, due to the development stage of the new government after the revolution in 2011, few activities seemed taken into proceeding. As a part of national water resources, treated wastewater is accounted for safety discharge in afforestation (Fig. 14.2) or inedible crop production to avoid the Nile River contamination.

14.5.2 Egyptian Code for Using TWW in the Field of Agriculture ECP501

For the purpose of the safety wastewater management, Ministry of Housing, Utilities and Urban Development set the Egyptian code (ECP501) for using the treated wastewater in the field of agriculture. ECP501 comprises of main legislation and comprehensive guidelines for actual application of TWW. It only allows woods and inedible oil cultivation. According to this code, the agricultural sector is divided into three groups, which in turn are subdivided into 11 groups. Table 14.1 indicates the referred classification.

14.5.3 Potential Use of Wastewater for Inedible Oil Production

Presently, there are 38 forests developed by government of Egypt and mostly irrigated by treated wastewater (TWW). However, none of them have been operated as commercialized plantations. On the contrary, some existing oil crops and forests shall be converted to

Table 14.1 Classification of plants and crops irrigable with treated wastewater

Grade	Agricultural group	Description
A	G 1.1 Plants and trees grown for greenery at tourist villages and hotels	Grass, Saint Augustine grass, kinds of cactus, ornamental palm trees, climbing plants, fencing bushes and tree, wood trees, and shade trees
	G 1.2 Plants and trees grown for greenery inside residential areas at the new cities	Grass, Saint Augustine grass, kinds of cactus, ornamental palm trees, climbing plants, fencing bushes and tree, wood trees, and shade trees
B	G 2.1 Fodder/feed crops	Sorghum
	G 2.2 Trees producing fruits with peel	On conditions, they are produced for canning and manufacturing purposes such as lemon, mango, olive trees, date palms, and nuts such as almond and pecan
	G 2.3 Trees used for forestation of highways and green belts around cities	Casuarinas, camphor, Athol tamarix, oleander, fruit producing trees, date palms, and olive trees
	G 2.4 Nursery plants	Nursery plants of wood trees, ornamental plants, and fruit trees
	G 2.5 Roses and cut flowers	Local roses, eagle roses, bulbs (e.g. gladiolus and bird-of-paradise)
	G 2.6 Fiber crops	Flax, jute, hibiscus, and sisal
	G 2.7 Mulberry for producing silk	Japanese mulberry
C	G 3.1 Industrial oil crops	Jojoba, castor, and jatropha
	G3.2 Wood trees	Kaya, camphor, and other wood trees

commercially viable practices to justify the public expenditure in such activities. The potential supply capacity of jatropha biofuel depends on Holding Company for Water and Wastewater (HCWW)'s afforestation plan throughout Egypt and availability of treated wastewater and potentiality of jatropha plantation development in Upper Egypt (Table 14.2).

The national wastewater treatment capacity of Egypt was 12 million m^3/day (4.4 billion m^3/year) in 2008 and is estimated to increase up to 20 million m^3/day (7.3 billion m^3/year) in 2012 (Fig. 14.3). On the other hand, the present national discharge of wastewater (the amount of collected wastewater) of Egypt is estimated at 6.5 billion m^3/year of wastewater in 2010

Table 14.2 Potentiality of jatropha plantation development in Upper Egypt based on the TWW availability

Location	Unit	Actual	Design capacity	Condition
Availability of treated wastewater				
Fayum	1000 m^3/day	NA	68	
Mniya	1000 m^3/day	NA	110	
Asyut	1000 m^3/day	57	472	
Suhaj	1000 m^3/day	76	292	
Quina	1000 m^3/day	158	325	
Luxor	1000 m^3/day	35	56	
New Valley	1000 m^3/day	15	78	
Aswan	1000 m^3/day	106	153	
Total	Million m^3/y	163	567	Annual wastewater availability
Potentiality of jatropha plantation development				
Wastewater available for jatropha cultivation	Million m^3	82	284	50% of wastewater
Jatropha watering	litter/year	365	365	
Maximum jatropha trees	million trees	224	777	
Maximum jatropha plantation	Feddan (Ha)	487,500 (195,000)	1,690,000 (676,000)	2.5 feddan/ha

Source HCWW 2008 (actual/design capacity), JDI (the rest of the calculation wastewater supply potential for inedible oil corp cultivation)

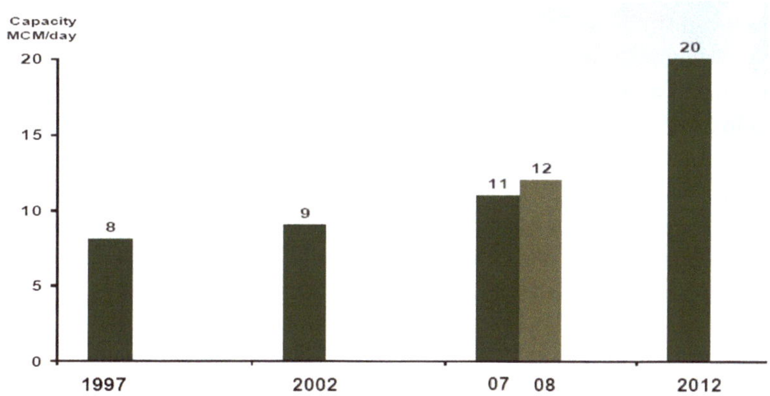

Fig. 14.3 National wastewater treatment capacity of Egypt. *Source* HCWW (2011) "Wastewater Reuse in Egypt: Opportunities and Challenges"

Fig. 14.4 Wastewater supply and treated wastewater for agroforestry (2010). *Source* HCWW (2011) "Achieving the MDGs for Water and Sanitation Sector in Egypt"

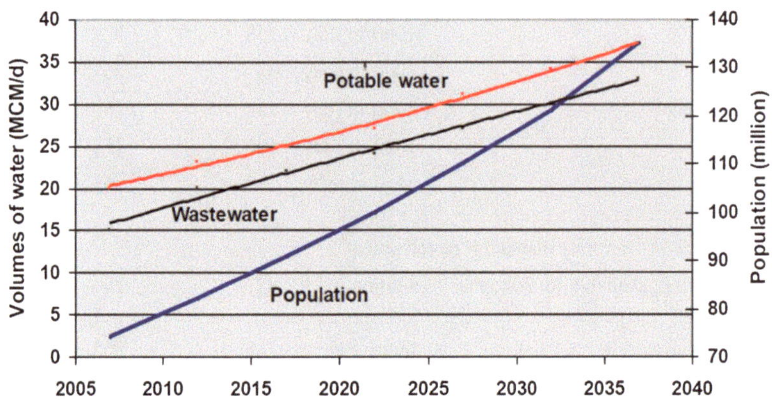

Fig. 14.5 Wastewater forecast based on HCWW master plan original data from HCWW master plan up to 2037. *Source* CEDARE (2011) "Water use efficiency and economic approach"

(Fig. 14.4). Of that amount, about 3.6 billion m^3/year is currently treated, and 0.7 billion m^3/year of TWW is allocated for afforestation (0.26 billion m^3/year is undergoing secondary treatment and 0.44 billion m^3/year undergoing primary treatment).

Therefore, the current use of TWW (696,000 m^3/day or 0.25 billion m^3/year) for afforestation discussed above still represents only 36% of the capacity allocated for afforestation, 7% of the total TWW (3.6 billion m^3/year), and 4% of the total wastewater discharge (6.5 billion m^3/year). In HCWW's 2009 Master Plan, the total wastewater discharge in Egypt is predicted up to 2037 (Fig. 14.5). By 2037, wastewater will increase up to about 32 million m^3/day (11.7 billion m^3/year).

The real potential of jatropha inedible oil production with TWW can be estimated with the full use of the total available wastewater in Upper Egypt and the entire country. The share of the Upper Egypt in the national wastewater discharge is 8% as is shown in Table 14.3. Assuming the 8% share of Upper Egypt in the national treated wastewater in 2037, we estimated the full potential of wastewater supply of both Upper Egypt and the whole country from 2010 to 2037 and summarized in Table 14.4. In this estimation, the water requirement of jatropha is assumed to be at 5000 m^3/feddan/year. The planting density of 2.5 m × 2.5 m is assumed.

Based on the above estimation, the potential of jatropha oil production in Upper Egypt is estimated in Table 14.5. In order to estimate the

Table 14.3 National wastewater discharge and treatment capacity

	Governorate	Installed capacity (CM/day)	Actual discharge (CM/day)	Treated effluent discharge (CM/day)
1	Alexandria	777,000	1,191,254	585,000
2	Aswan	53,500	87,654	46,594
3	Asyut	50,000	182,355	52,680
4	Beheria	184,592	295,055	71,236
5	Beni Suef	25,920	139,108	35,000
6	Cairo	1,711,880	2,250,365	1,378,880
7	Dakahlia	150,446	422,034	143,853
8	Damietta	143,800	137,965	89,945
9	Faiyum	108,500	188,704	71,868
10	Gharbia	294,000	287,443	156,920
11	Giza	1,030,000	995,676	794,000
12	Ismailia	100,000	150,227	79,500
13	Kafr el-Sheikh	18,500	250,906	18,500
14	Matruh	25,000	14,086	3900
15	Minya	88,000	171,462	43,000
16	Monufia	143,250	187,258	111,150
17	New Valley	25,610	50,285	21,870
18	North Sinai	51,000	37,157	29,000
19	Port Said	190,000	136,231	124,190
20	Qalyubia	625,000	354,751	373,624
21	Qena (incl. Luxor)	38,000	213,271	48,000
22	Red Sea	0	46,286	0
23	Sharqia	81,000	430,991	68,900
24	Sohag	22,000	164,383	18,000
25	South Sinai	20,330	28,557	11,830
26	Suez	130,095	140,326	120,050
Total		6,087,423	8,953,792	4,497,490
Total (upper Egypt)		189,110	744,234	187,144
Share (upper Egypt)		3%	8%	4%

Source USAID (2010) "Integrated water resource management 2"

economic impact to the national economy, the nationwide potential of inedible oil production is also estimated in the Table 14.6.

14.5.4 Irrigation System Cost

The existing irrigation systems for the afforestation under MALR use minimal amount of water for growing forest trees in Egypt. However, this is not enough for the commercial crop production. An average of 10–40 L/tree/day of irrigation water should be needed to maximize the production. The cost of irrigation system development depends on the design of irrigation system including specifications of pump and pipes because there are many different grades of irrigation equipments available in the local market

Table 14.4 Potential treated wastewater supply for inedible oil production

	2010	2030	2037
National wastewater discharge (billion m^3/year)	6.5	10.6	11.7
National treated wastewater (billion m^3/year)	4.4	9.5	10.5
National treated wastewater (billion m^3/year)	3.6	5.8	6.4
Upper Egypt wastewater discharge (billion m^3/year)	0.52	0.85	0.9
Upper Egypt treated wastewater (billion m^3/year)	0.47	0.76	0.8
Non-Upper Egypt area wastewater discharge (billion m^3/ year	5.98	9.75	10.76
Non-Upper Egypt treated wastewater (billion m^3/year)	3.93	8.78	9.69

Source CEDARE (2011) "Water use efficiency and economic approach," Hegazy et al. (2012)

Table 14.5 Potential production of jatropha oil in Upper Egypt in 2037

	Wastewater discharge (billion m3/year)	Planation area (feddan)	Oil production (ton)
Upper Egypt	0.4	84,240	105,300

Source Hegazy et al. (2012)

Table 14.6 National wide potential of inedible oil production in 2037

	Wastewater discharge (billion m3/year)	Planation area (feddan)	Oil production (ton)
Nationwide total	5.3	1,174,095	638,118

Source Hegazy et al. (2012)

with a wide price range. Proper selection of pumps and pipes is important for cost minimization and a long-term success of the plantation business.

14.5.5 Potential Improvement of Jatropha for Commercial Plantations in Egypt

Assuming the condition below, Egypt is expected to reach commercially viable jatropha biofuel production during the Pilot Project period (2013–2015).

- Successful outcome of high-yield varieties
- Crude jatropha oil (CJO) price will remain around US$1000/ton (Implying the crude petroleum oil price will remain around $100/barrel or higher)
- The use of jatropha seed cake either for organic fertilizer or biofuel for boiler or power generation as a substitute for coal

The Pilot Project (2013–2015) is targeting to achieve 3.5–4.0 ton/ha by 2015 and 5.0–6.0 ton/ha by 2020, which can be achievable based on the past improvement experiences of key crops. It is highly recommendable to hire local consultants/experts to get the import permit, but it may take a few months to clear all paper work, which is up to a few weeks work in most countries. In addition to the Egyptian side processes, seed suppliers shall also consider the time to acquire "strict" pest risk assessment report from authorities of plants' origin, National Plant Protection Organization (NPPO)

14.6 Objective and Profile of Proposed Pilot Project

14.6.1 Objective

The objective of the proposed Pilot Project is the testing of commercial viability of high-yield varieties of jatropha introduced from Asia as a main crop.

14.6.2 Selection of Pilot Project Site

The selection of the Pilot Project site is one of the most critical factors to achieve the commercial viability of the Egyptian biofuel model. The JICA Study team examined several candidate locations for selecting one site for the proposed Pilot Project (4 hectare) based on critical criteria, in order to prove the commercial viability not only for the 4-hectare Pilot Project but also for the following commercial scale supply chain development. Due to the limitation in time and information, the evaluation was carefully conducted and confirmed among the JICA study team, relevant authorities, and technical experts. Several candidate locations for the future commercial projects were proposed after the collection of the most updated information such as land availability (physical and jurisdiction), wastewater supply (supply capacity, cost and long-term reliability), logistic costs, and natural disaster (flooding and storm). The following map shows the four candidate sites (Asyut, Qena, Luxor, and Aswan) in the Upper Egypt.

The details of the selection criteria are as follows:

1. Plant Quarantine Availability:
   Existence of the plant quarantine is a favorable condition to introducing new varieties to the Pilot Project.
2. Greenhouse Facility Availability:
   Existence of the greenhouse facility is required for the quarantine procedure.
3. International Airport Availability:
   Existence of international airport (frequency of the flight and distance from the site) is evaluated.
4. Distance from Sea Port:
   Distance and closeness to a sea port are criteria for possible export of CJO to EU and other overseas markets.
5. Land Availability:
   Land requirement for the Pilot Project is 4 hectare.
6. Soil Conditions:
   Soil samples were analyzed in the laboratory with chemical and physical properties in February and March 2012.
7. Existing jatropha plantation with irrigation for renovation:
   After the implementation of the proposed Pilot Project, the next target should be the renovation of existing plantations. Therefore, the existence of already developed jatropha plantations with irrigation is a factor for identifying candidate sites.
8. Wastewater Availability:
   The wastewater requirement for the 4-hectare Pilot Project is 20,000 m^3/year (=5000 m^3/hectare × 4 hectare).
9. Wastewater Quality:
   According to HCWW, the quality of wastewater is very similar for all sites due to guidelines set by HCWW.
10. Project Building Availability:
    Existence of a building (within 1 km), which can be used for the proposed Pilot Project as the project office, is favorable for economical and speedy implementation of Pilot Project in terms of avoiding extra construction cost of building and basic infrastructure (water, electricity, etc.).

14.6.3 Cost of TWW and Land Lease

The price of treated wastewater (to be set by HCWW) and the government's land lease fee for jatropha and inedible oil development projects should be set at low levels in order to attract private investors into this sector. At present, investors would be reluctant to invest in this new sector due to uncertainties of the new industry such as future productivity. In this situation, the investment will not go smoothly if the Egyptian government wishes to enjoy upfront profits from the beginning. In the petroleum and gas industry in the world, the concept of "profit sharing" can be applicable. Likewise this concept can be applied in this new inedible oil industry development so that the land and wastewater will not be charged until the profit of a new project comes up and the profit will be shared between private investors and the local government. Under the assumption of the profit sharing scheme, we set the cost of land and water at zero. For example, the price of treated wastewater is currently set at zero LE/ton for MALR's "Man Made Forest" Project.

14.6.4 Conditions for Proposed Pilot Project Implementation in Luxor

- Ministry of Agriculture, Forestry Department provided 4 hectare (10 feddan) land at identified and agreed site.
- Wastewater will be supplied from the existing facilities of HCWW using the Ministry of Agriculture owned pumping/piping system.
- Project site office & factory/laboratory building shall be provided by the Ministry of Agriculture (Existing 5 room building of about 150 m^2).
- Assisting Institutions/Agencies shall provide all other irrigation, equipment, cars, and office building improvement, oil extraction machine and necessary experts and manpower for the Pilot Project.

14.6.5 Afforestation and Irrigation System in Luxor

The present situation of the afforestation in Luxor (El Hebiel) is known as the majority of the existing forest is *Khaya senegalensis* out of the total space (1741 feddan). Jatropha is planted in 121 feddan.

There are three wastewater plants in Luxor. El Hebiel plant is the by far the largest. Therefore, study team examined the afforestation site adja-

Fig. 14.6 Stages of wastewater on oxidation ponds

Fig. 14.7 Sand filters for the crude wastewater

Fig. 14.8 Pipe outlet providing TWW

cent to the treatment plant. This wastewater treatment plant in Luxor is on the eastern side of Nile River. It is operated by Luxor Company of water and wastewater. Oxidation pond for TWW is shown in figures (Fig. 14.6), sand filters (Fig. 14.7) and pipe outlet (Fig. 14.8).

14.6.6 Supply of TWW

TWW is conveyed from the existing treatment plant to the reservoir by HCWW in Luxor. TWW is conveyed through the conveyance pipeline (800 mm diameter, approximately 5 km in length) to the reservoir in total 43,000 m^3 per day. The study team collected a TWW sample from the site and conducted chemical analyses at ARC of MALR. The analysis result is presented in Table 14.7. The most influential water quality criterion on crop productivity is the water salinity as measured by electrical conductivity (EC). The EC results of 1.0 mmho/cm indicate slight higher salinity compared to the safety level (0.7 mmho/cm), but is lower than the level of moderate salinity hazardous level (1.5–3.0 mmho/cm) and much lower than the severe level (>3.0). It is the level usually not harmful to crops. The sodium imbalance (sodicity) is indicated by sodium absorption ratio (SAR). It was 3.5 in the sample water. It suggests no risk of water infiltration problem unless the EC is lower than 0.4 mmho/cm. The pH of the water is in the normal range (6.5–8.4). No heavy metal pollution was detected. As a conclusion, the treated wastewater from the Luxor treatment plant can be used for the crop cultivation (Table 14.7).

The suitability criteria of water source for irrigation purpose in Luxor indicated that it lies in the first category C1S1, i.e., no problems for salinity and sodicity are expected. An elemental composition analysis of N, P, K, Fe, Mn, Zn, Cu, Cd, Co, Pb, Ni, and Cr as well as biological criteria (i.e., COD, BOD, Fecal coliforms, Salmonella, and Shigella) was executed on each of the studied irrigation water and experimental soil, and their contents were found to be within the permissible limits. Since their soluble values in the irrigation water source are more than the freshwater, a field experiment was conducted on the chosen soil sites. The obtained results showed a beneficial effect of the applied irrigation water source on the grown plants, more pronounced increments in plant growth, seed yield and seed oil yield with high quality (El-Tohamy et al. 2012).

14.6.7 Distribution of TWW

The most irrigation systems applied for are surface or drip irrigation (Fig. 14.9). A part of TWW is supplied without filter to the irrigation area by the pipe (12 inch diameter) in total 5000 m^3/day and irrigated 60 feddan of afforestation area (Fig. 14.10). Remaining part of TWW is overflowed from the reservoir and flowing down to the pond adjacent to the highway in total 28,000 m^3/day through the natural

Table 14.7 Analysis result of treated wastewater from proposed project site

		Concentration (mg/l)	
EC (mmhos/cm)	1.02	(NH_3^+)	31.8
ppm	653	(NO_3^-)	0.000
pH	6.71	(P)	4.060
Soluble anions (meq/L)		(Fe)	0.098
CO_3^-	–	(Mn)	0.105
HCO_3^-	0.90	(Zn)	0.046
Cl	4.11	(Cu)	0.000
SO_4^-	4.58	(Co)	0.000
Soluble cations (meq/L)		(Ni)	0.000
Ca^{++}	1.50	(Pb)	0.000
Mg^{++}	2.50	(Cr)	0.000
Na^+	5.03	(Cd)	0.000
K^+	0.56	(Mo)	0.000
RSC	–	(Se)	0.000
SAR	3.57		

Source Hegazy et al. (2012)

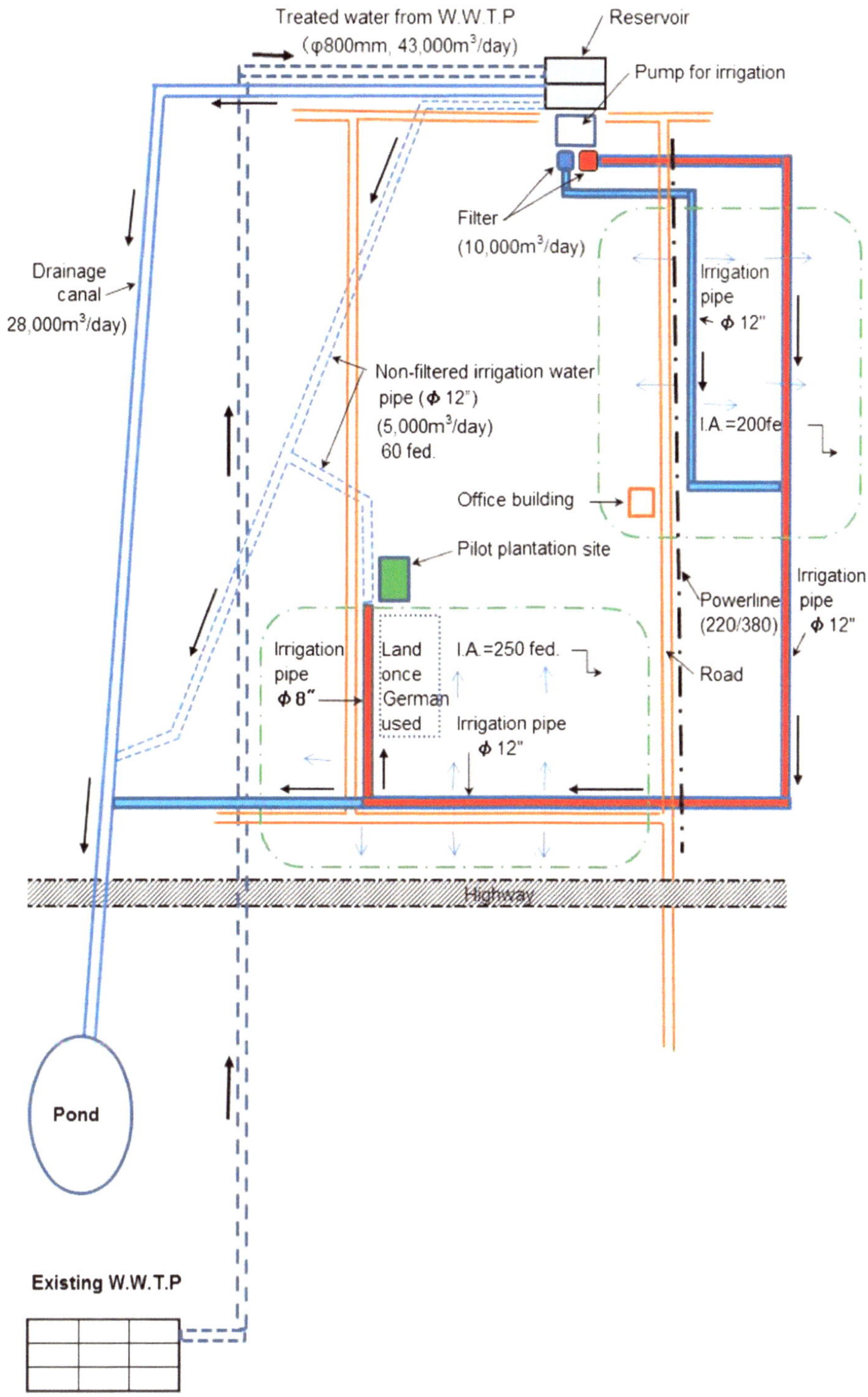

Fig. 14.9 Location of the facilities for afforestation and irrigation (Luxor station)

Fig. 14.10 Drip and surface irrigation systems of jatropha plantations in Egypt **a** different locations, **b** small plastic pipe connected to dripper called Macaroni system, and **c** Macaroni connected to 18-mm pipe (special for treated wastewater)

canal of 7 km in length. A special irrigation system was superior, in which the small pipe of TWW connects to the dripper called Macaroni (Fig. 14.9). Vegetative growth using TWW was superior to that using freshwater (Fig. 14.11). As for the supply of filtered TWW, pump station is located adjoining to the reservoir, in which five pumps (three pumps: diameter is 8/10 in., discharge capacity is 400 m^3 per hour each, and 2 pumps: diameter is 6 inch/8 inch, discharge capacity is 200 m^3 per hour each) are installed, and is supplying the water to the irrigation area. These pumps once send TWW to the sand filter plant, and filtered water is delivered to the irrigation area. There are two sand filter systems (indicated in blue color and red color each in Fig. 14.9). The water filtered by blue colored system is supplied to the 200-feddan irrigation area, and filtered water by red color system is supplied to the 250-feddan irrigation area. At present, only 15% of TWW (43,000 m^3/day) is filtered and sent to the irrigation area. Therefore, the big amount of water is available to use for the irrigation of jatropha or afforestation purpose.

Selection of the land for the pilot plantation project is carried out considering spread of the area, topographical condition, land ownership, possibility of water supply, and existence of building to minimize the cost to arrange the plantation. JICA study team selected three candidates of the area, which are shown in the Fig. 14.12. Each candidate site conditions are as follows.

- Site-1
  - Topographic condition: Very wide and flat
  - Land allocation: Governorate of Luxor for land fill site
  - Water supply: Pipe from the pump station locates nearby

Fig. 14.11 Jatropha growth under freshwater and TWW **a** Jatropha seedlings grown using freshwater, **b** Jatropha seedlings grown using TWW, **c** Jatropha trees grown using freshwater, and **d** Jatropha trees grown using TWW

  - Existence of building: MALR's building is located nearby
- Site-2
  - Topographic condition: Land is long shape, but fairly flat
  - Land allocation: Luxor water and wastewater company under HCWW
  - Water supply: Pipe end from the pump station is located beside the site
  - Existence of building: MALR's building is located nearby
- Site-3
  - Topographic condition: Land is undulating and soil seems to be hard
  - Land allocation: Luxor water and wastewater company under HCWW
  - Water supply: WWTP locates adjacent to the site
  - Existence of building: No building to use easily

Although each site has merit and demerit, the land ownership should be mainly considered. Because it is essential to obtain the land in early time to implement this pilot plantation project. Site-2 seems to be predominant land comparing to others in this point of view. Additionally, water source and building are available nearby, so that Site-2 should be selected as the pilot plantation project site.

14.6.8 Irrigation Network on the Pilot Plantation

The selected land (site-2) has a rectangular shape (Fig. 14.12). The optimum length of the drip pipe for water supply at the end portion of irrigation network is 25 m. Considering these conditions, the layout of the irrigation network is drawn as the general design (Fig. 14.13).

Fig. 14.12 Location of three candidate sites for pilot plantation site. *Source* Hegazy et al. (2012)

14.7 Climatic and Soil Condition of Project Site

14.7.1 Soil Conditions

The dominant soil types on the east side of Nile River in Luxor are lithosols or calcaric fluvisols according to the soil map of Egypt (Fig. 14.14). The soil in the 10 feddan site is sandy soil and the land is fairly flat. The study team collected 10 soil samples (10–30 cm depth and 60 cm depth) from the site. The pH of the soil is slightly alkaline (8.2–8.6). Contents of heavy metals are in the safe range. The organic matter in the soil is quite low. However, nitrogen and potassium contents in the soil are relatively high, while phosphorus is low. According to the mechanical analysis of the soil samples, sand constitutes 87% of the soil, rendering sandy texture. The soil in the site does not have soil layers, being typical desert soil (Fig. 14.15).

14.7.2 Climate Conditions

The climatic conditions of Luxor are summarized in Table 14.8. The average temperature is about 32 degrees Celsius and can reach more than 40 degrees Celsius in summer time. During the noontime in summer, agricultural workers cannot normally work due to high temperature. The rainfall is practically zero. Therefore, continuous irrigation water must be supplied to the Pilot Project field throughout a year.

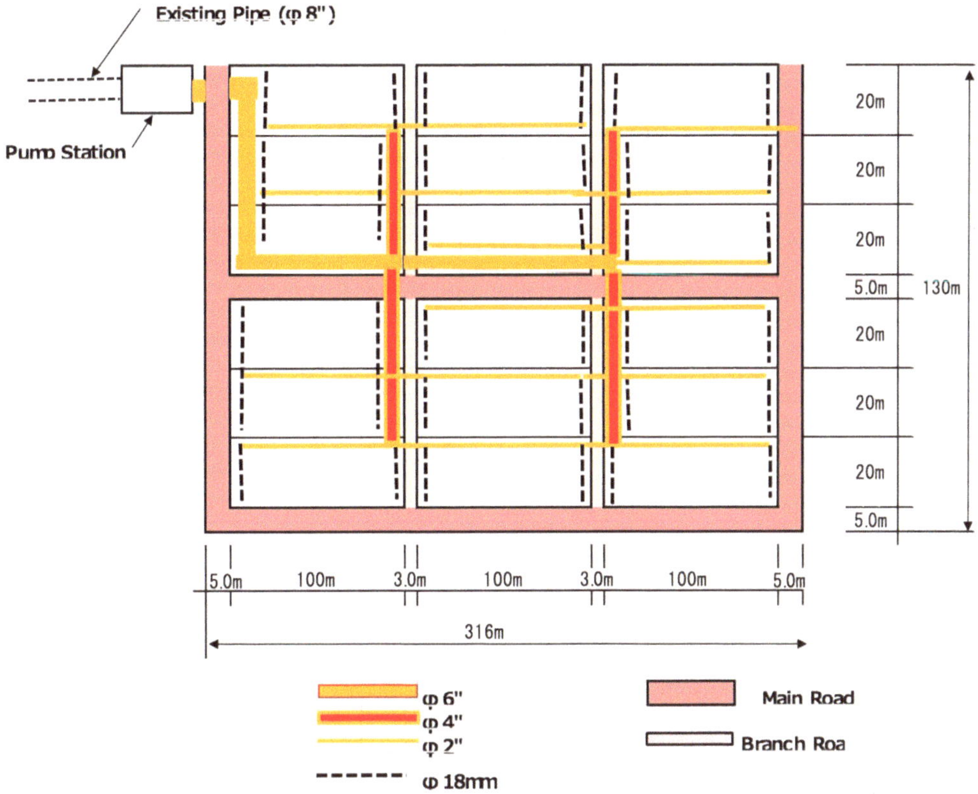

Fig. 14.13 Layout of the irrigation network. *Source* Hegazy *et al.* (2012)

14.8 Profile of Pilot Project

The proposed Pilot Project is expected to start in 2013 and last 3.5 years to 2016. The first component of the Pilot Project is the 10 feddan pilot plantation with new jatropha varieties (Fig. 14.16). The Pilot Project shall test all available new jatropha varieties. The second component of the Pilot Project is the pilot oil factory (Figs. 14.17 and 14.18). Other alternative oil crops may also be in the scope of Pilot Project based on the further investigation of their viability. In order to demonstrate the use of jatropha oil for practical use in Egypt, the Pilot Project introduces an oil extraction degumming facility (Fig. 14.19).

14.9 Evaluation of Commercial Feasibility as Biofuel and Organic Fertilizer

The jatropha oil (Strait Vegetable Oil: SVO) will be analyzed for quality evaluation at laboratories of local and international inspection organizations. The project team will check whether the product(s) meet the market needs or not. Once satisfactory quality of SVO is produced, it will be sold or offered as a sample to biojet fuel producers for quality evaluation. As a by-product from oil extraction, seed cake (75% by weight of the processed seeds) will be produced, too. The project team will check whether an organic fertilizer made from the seed cake is feasible in Upper Egypt or not.

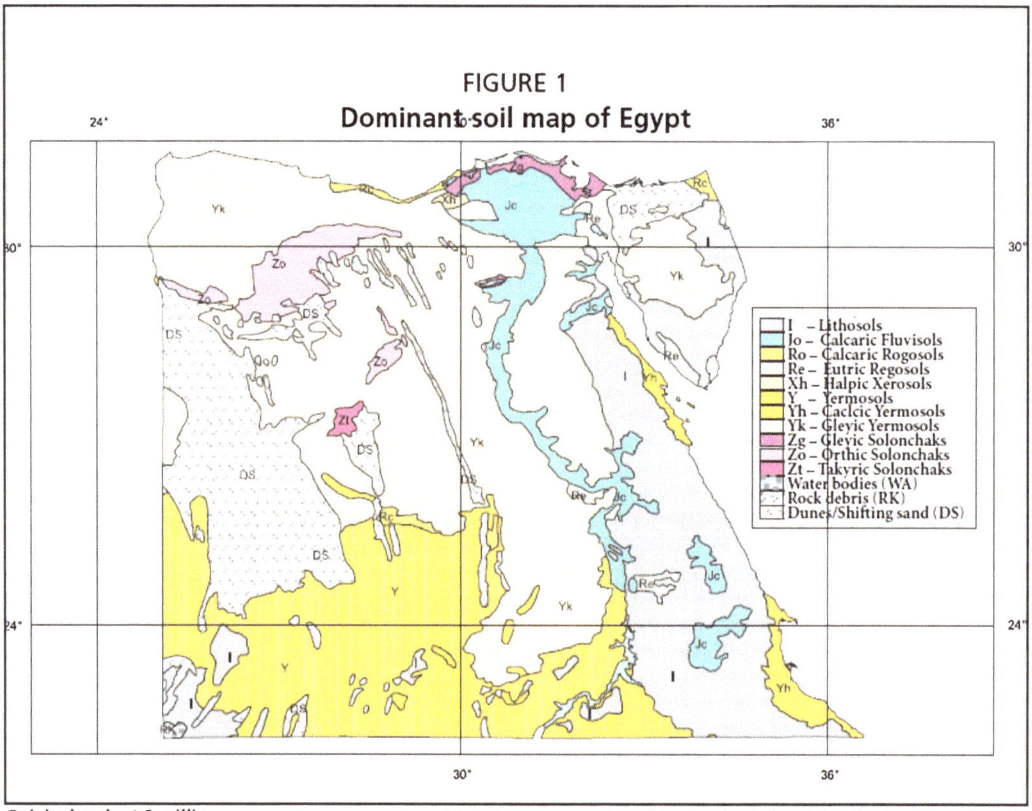

Original scale: 1:5 million
Source: DSMW-FAO-UNESCO.

Fig. 14.14 Soil map of Egypt

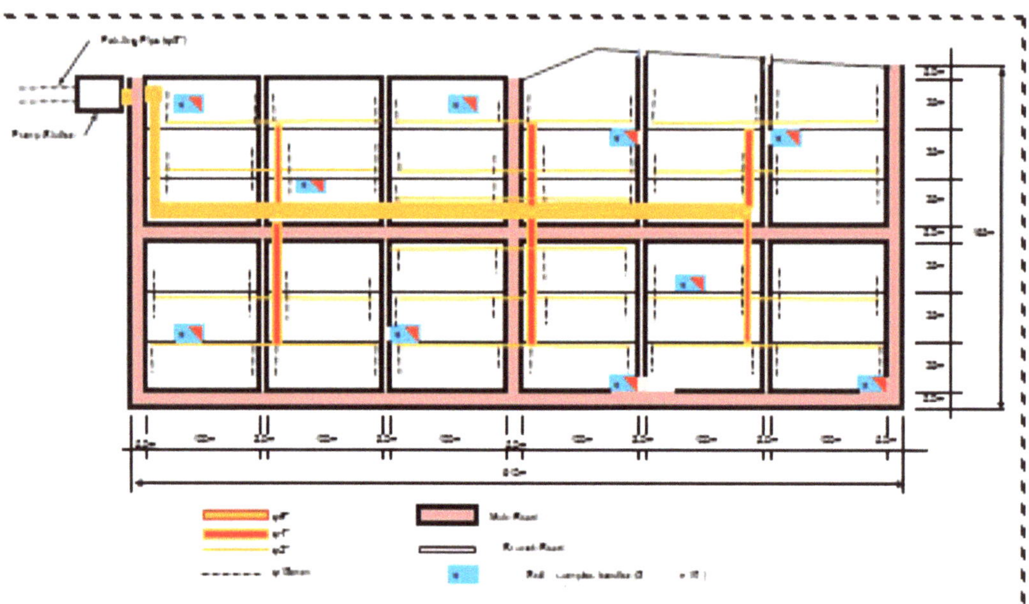

Fig. 14.15 Locations of soil sampling for analysis

Table 14.8 Climatic data of Luxor

Month no.	Avg. temperature (°C)	Max. temperature (°C)	Min. temperature (°C)	Relative humidity (%)	Wind speed (knot)	Avg. sun shine (h)	Avg. radiation (MJ/M^2)	Total rain (mm)	Avg. Et_0 (mm)
1	14	23	5.4	52	3.2	9.1	16	0.1	2.6
2	16	25	7	42	3.6	9.7	19.1	0.2	3.5
3	20.2	29.5	10.6	34	4.3	10.1	22.4	0	5
4	26	34.8	15.7	26	4	10.8	25.3	0	6.4
5	30	38.7	20	22	3.7	11.6	27.4	0.3	7.3
6	32.4	41.1	23	23	3.5	13.1	29.7	0	7.9
7	32.9	40.6	23.6	26	3.2	13	29.3	0	7.7
8	32.5	40.7	23.4	27	3	12.2	27.7	0	7.1
9	30	38.6	21.5	32	2.6	11.8	25.4	0	6.1
10	25.4	35.3	17.5	40	2.8	10.8	21.3	0	4.8
11	20	29.6	12.1	47	3	9.6	17.2	0	3.5
12	15	24.6	7.2	53	3	9	15.2	0	2.6

Source Central laboratory for agricultural climate (2004)

Fig. 14.16 New jatropha plantation

14.10 Cultivation Design

The productivity of jatropha largely depends upon planting density, irrigation water, fertilizer application, and variety. The agronomy of jatropha cultivation in desert environment has not been studied well not only in Egypt but also in the world. Therefore, it is advisable to have cultivation experiments in the proposed pilot plantation in order to establish the best cultivation methods. The following experimental design is suggested by the study team in consultation with Egyptian local experts that are specialized in agriculture in desert environment. A general location map of the afforestation and the

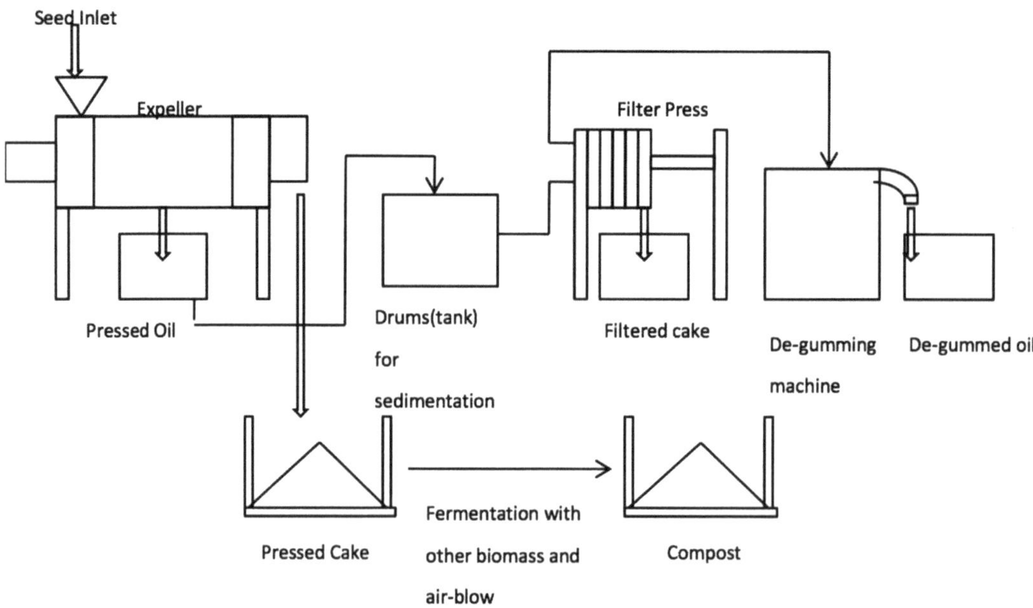

Fig. 14.17 Normal jatropha oil extracting system flowchart. *Source* Hegazy *et al.* (2012)

irrigation system under MALR in Luxor was drawn (Fig. 14.20). Using plants produced by micropropagation (Fig. 14.21) through direct organogenesis pathway (true-to-type to the mother plant) was superior to seedlings.

14.11 Brief Report About Data Analysis—Monitoring of Jatropha Plants Under Plant Quarantine Shade House Conditions in Luxor Site

14.11.1 Introduction

Based on the consensus among stakeholders, jatropha Biofuel Industry Development Study (2011–2012), Under-secretariat for Afforestation and Environment, Ministry of Agriculture and Land Reclamation (UAE/MALR) agreed to proceed a comparison test of imported jatropha varieties with local control varieties (Table 14.9 and Fig. 14.22).

The major objectives of this study are as follows:

1. Monitoring the growth of the imported varieties with the reference,
2. Measuring the seed productivities under Egyptian environment with TWW,
3. Assuming the expected productivities at the matured stage,
4. Evaluating the feasibility of the jatropha production based on the productivities,
5. Giving recommendation to Egyptian authorities for the further needs of actions.

Our experiments aim to compare growth parameters and seed production in the following:

A Shade house:
 Imported saplings
B Open field:

1. Imported saplings.
2. Plants from selected local cuttings.
3. Plants from seeds from Philippines.
4. Plants from seeds from Egypt.

Fig. 14.18 Extracted oil, oil expeller, and oil extraction facility

14.11.2 Stock Plant Materials

This work was carried out under plant quarantine shade house conditions at Luxor afforestation started from May 5, 2012. Jatropha saplings (270 healthy and uniform plants) were imported from JOiL Singapore laboratory and implemented in this study.

14.11.3 Plants Under the Shade House in the Nursery

Transparent Agril sheets (Fig. 14.23a, b) to maintain relative humidity were gradually removed within the first 4 weeks. Saplings were maintained under the shade house covered with black Seran sheets (63% shade), around 35 °C and 70–80% relative humidity (Fig. 14.23c, d). Plants were weekly fertigated with nutrient solution of NPK (Nitrolev 1.0 g/L) at the ratio of 20:20:20. Sprinkler mist system was used to compensate the reduction in relative humidity. Saplings were not ready for open field transfer because of

- High temperature in July month.
- According to the plant quarantine laws, it is not allowed to transplant the saplings to the open field unless it will spend 1 year (4 seasons) under monitoring in the nursery.

14.11.4 Recommended Data to Record

Morphological comparisons of all plantlets were done, and the flowing data were recorded along 15 months:

- Plant height (cm): From the soil surface up to the last emerged leaf edge.
- Canopy width (cm): The distance between the horizontally faraway two leaves
- Plant seed yield number: Seed number/tree/year
- Plant seed yield weight (g): Seed weight/tree/year
- 100 seeds weight (g): Randomly selected seeds regardless of the size

14.11.5 Nursery

Data within 2 months confirmed that good results were almost recorded with the soil mixture containing peat moss alone or in a combination with foam, clay and coarse sawdust (treatments 1, 3, 4, and 5), respectively (Tables 14.10 and 14.11; Fig. 14.23). The absence of peat moss negatively affected vegetative growth of saplings (treatment number 8; Table 14.10). On the other hand, soil mixture

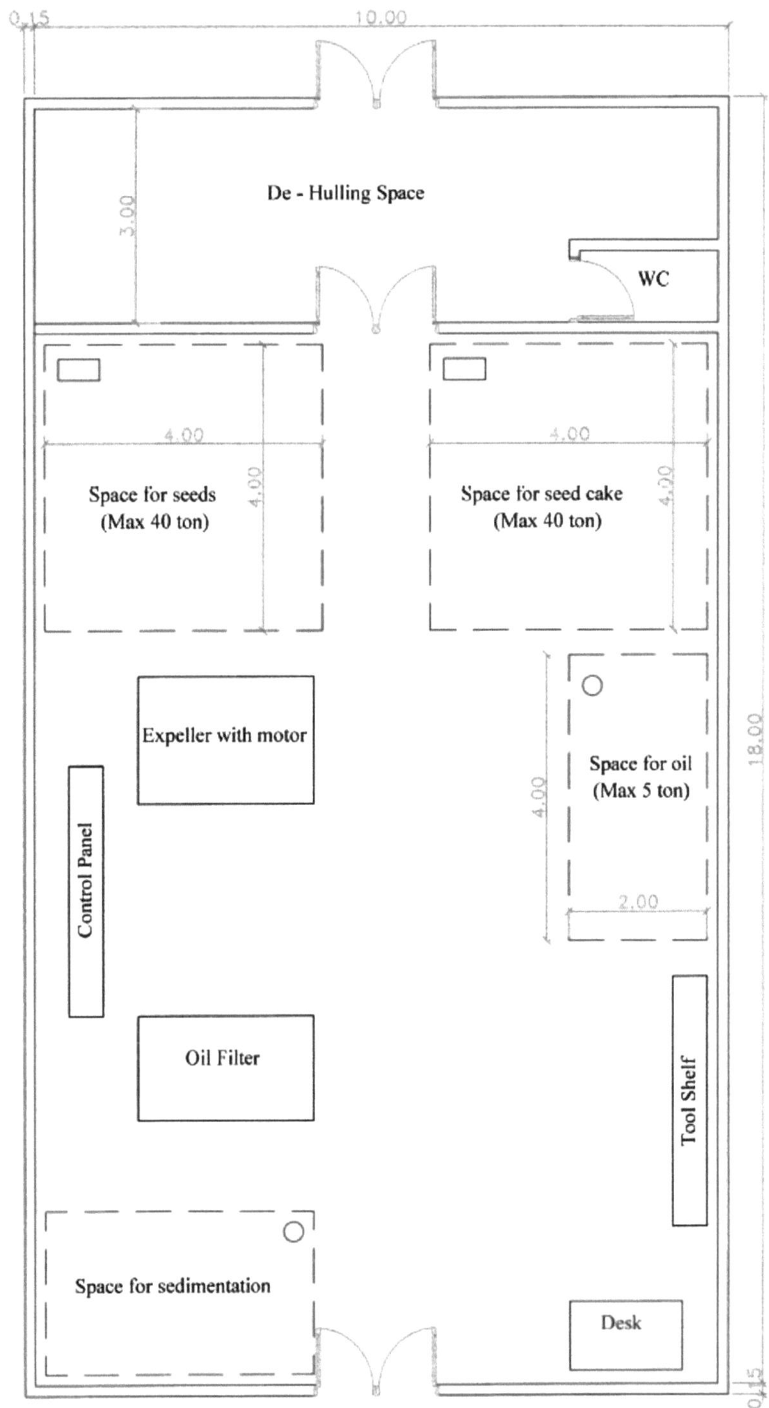

Fig. 14.19 Jatropha oil factory for proposed Pilot Project. *Source* Hegazy et al. (2012)

containing peat moss and sand (treatment number 2) recorded the lowest survival percentage along 2 months as compared with the other treatments.

In general, data indicated that the fast response in vegetative growth was obtained with the imported saplings as compared with the local cuttings (Table 14.11; Fig. 14.23). Also,

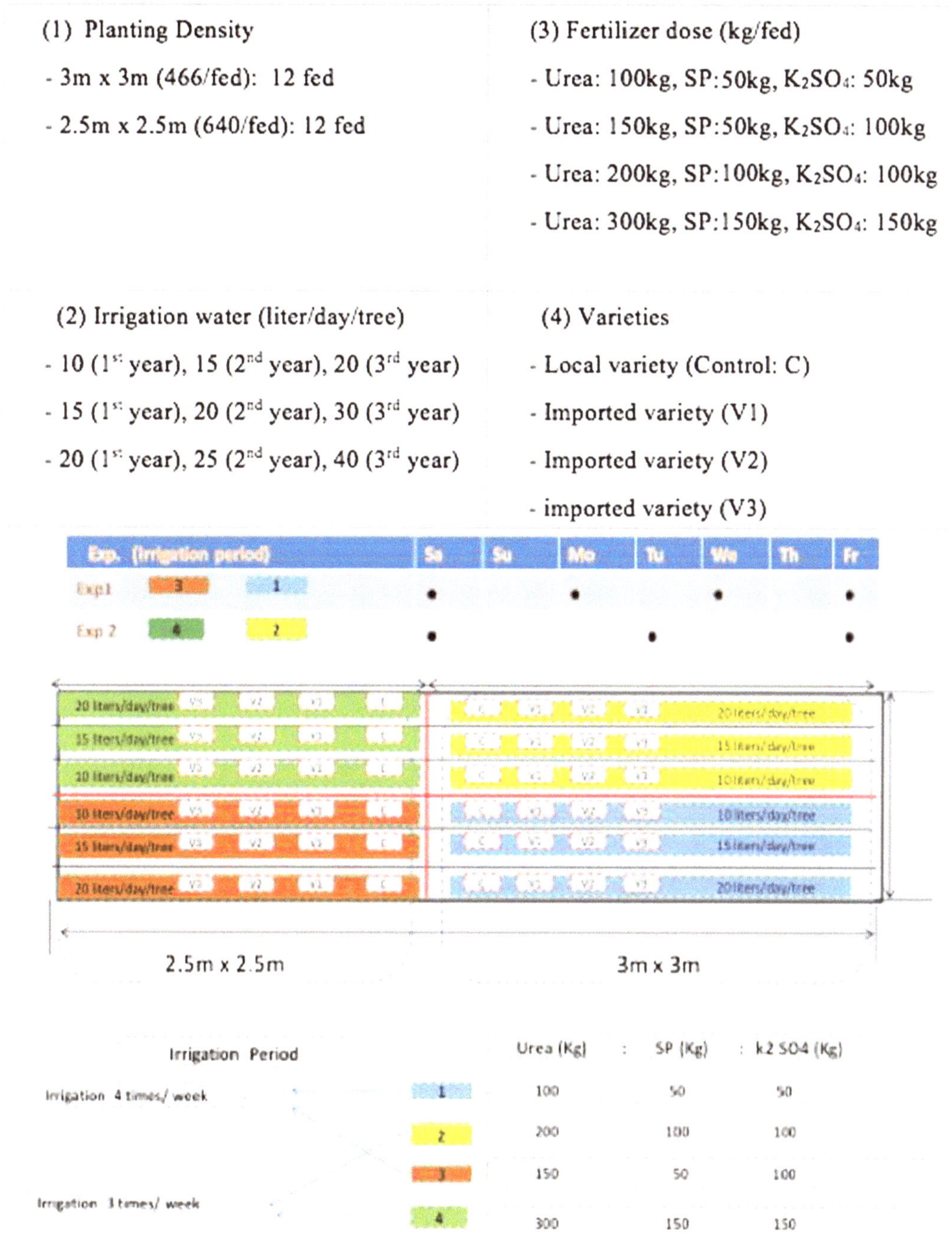

Fig. 14.20 Design of jatropha cultivation. *Source* Hegazy et al. (2012)

local cuttings indicated low survival percentage under all soil mixture types after 2 months in the shade house as compared with the imported plants. Finally, soil mixture containing peat moss in combination with foam (treatment number 3) recorded the highest survival percentage either with in vitro plants or local cuttings.

Fig. 14.21 Micropropagation stages of *Jatropha curcas* **a** Selected local germplasm with high quality and quantity yield, **b** plant tissue culture laboratory, **c** explant surface sterilization procedure, **d** explant culturing under a septic condition in starting stage, **e** shoot cluster in vitro, **f** individual shoots on rooting stage, **g** healthy well-rooted ex-agar plantlets, and **h** successful in vitro plants in acclimatization stage in the greenhouse

Table 14.9 Implementation of action plan

Oct. 2013	Oct. 2012–Sep. 2013	June–Sep. 2012	Particulars	Ser.
			Plants under quarantine inspection in shade house in Luxor site	1-
			Open field comparison assay among: –Imported saplings from JOil Lab growing in Luxor site –Plants from selected local cuttings –Plants from seeds from Philippines –Plants from seeds from Egypt	2-
			Evaluation of the comparison results	3-

Note:

1. It could be concluded that the choice of jatropha apical cuttings is very important and is a limiting factor to raise the survival percentage in the shade house.
2. The insect named *Tuta absoluta* order Lepidoptera attacked a few plants in shade house, Fig. 14.24a indicated leaf infection symptoms and larva appeared inside the yellowish leaf color. Pesticides (Lannate, Coragen, or Vertimic) were used, and obtained good results to overcome the infection and to destroy the insects.
3. During summer time and because of high humidity inside the shade house:

   (a) The insect named *Aphis gossypii* family *Aphidoidea* attacked a few plants that showed leaf infection symptoms (dry leaves and yellowish color as shown in Fig. 14.24b). Insecticide Actellic 50EC was used, which obtained good results to overcome the infection and to destroy the insects.

   (b) The insect named whitefly remarkably attacked a few plants (Fig. 14.24c), and the same insecticide (Actellic 50EC) was used.

14.11.6 Brief Report About Data Analysis on Oct. 20

Imported saplings in the shade house and the open field

Fig. 14.22 Luxor nursery local control varieties and imported varieties

Fig. 14.23 Growing performance and inspection of local and imported jatropha in vitro plants in the greenhouse **a** Agril sheet used to keep high humidity through cover tunnels in the greenhouse, **b** plant tissue culture laboratory, **c** sprinkler used to maintain humidity around tunnels, **d** mist system used to elevate humidity it constricted in the sealing of the greenhouse, **e** preparation of growing mixture and polyethylene pages fell, **f** local cuttings and imported vitro plants cultured in plastic pages, **g** transfer of cultured plastic pots under tunnels, **h** healthy well growing in vitro plants, **i** local cuttings growing in greenhouse, **j** monitoring environmental conditions, **k** monthly data collection, and **l** monthly plant quarantine committee inspection for plant disease

A. Shade house
   - In this part of study and according to the plant quarantine laws, it is not allowed to transplant the saplings to the open field unless it would spend one year (four seasons) under monitoring in the nursery.
   - We transferred three plants from each treatment out of all eight treatments to the open field to evaluate $3 \times 8 = 24$ saplings (Fig. 14.25a, b), compared their behaviors in the open field with their same colleagues still growing in the

Table 14.10 Effect of growing mixture type on the survival percentage and growth parameters of jatropha in vitro plantlets after 2 months from transplantation in the shade house

Rank	Growing mixture types		No. of plants evaluated in						Survival evaluation
			June			July			
			Survived	Dead	Surv. %	Survived	Dead	Surv. %	
1	Peat moss	–	33	2	94.3	32	1	97.0	B
2		+Sand (1:1, v/v)	20	5	80	16	4	80.0	H
3		+Foam (1:1, v/v)	24	1	96	24	–	100	A
4		+Clay (1:1, v/v)	23	2	92	22	1	95.7	C
5		+Coarse sawdust (1:1, v/v)	21	4	84	20	1	95.2	D
6		+Clay + foam (1:1:1, v/v)	27	–	100	24	3	88.9	E
7		+Clay + c. sawdust (1:1:1, v/v)	29	2	93.5	24	5	82.8	G
8	Clay + coarse sawdust + sand (1:1:1, v/v)		36	2	94.7	30	6	83.3	F

Table 14.11 Effect of growing mixture type on the survival percentage and growth parameters of local jatropha cuttings after 2 months from plantation in the shade house

Rank	Growing mixture types		No. of plants evaluated in						Survival evaluation
			June			July			
			Survived	Dead	Surv. %	Survived	Dead	Surv. %	
1	Peat moss	–	15	–	100	3	12	20	E
2		+Sand (1:1, v/v)	–	–	–	–	–	–	–
3		+Foam (1:1, v/v)	5	–	100	3	2	60	A
4		+Clay (1:1, v/v)	6	–	100	3	3	50	B
5		+Coarse sawdust (1:1, v/v)	6	–	100	2	4	33.3	C
6		+Clay + foam (1:1:1, v/v)	6	–	100	2	4	33.3	C
7		+ Clay + c. sawdust (1:1:1, v/v)	12	–	100	4	12	33.3	C
8	Clay + coarse sawdust + sand (1:1:1, v/v)		10	–	100	3	7	30	D

shade house for 15 months, and recorded the data.

- Imported plants grown in the open field (Fig. 14.25d, f) showed difference from those grown in the shade house in number of leaves, stem circumference (cm), plant height (cm), canopy width (cm), number of branches, number of flowers and seeds with high significance.
- Saplings from tissue culture grown in the shade house did not produce any flowers within 15 months and had very slow growth compared with those transplanted in the open field which

Fig. 14.24 Leaf infection symptoms **a** *Tuta absoluta*, **b** *Aphis gossypii*, and **c** white fly

Fig. 14.25 Growing performance and inspection of local and imported jatropha in vitro plants in the open field at Luxor site **a** Land preparation, planting process in open field, **b** plants occurred good flowering after 3 months from culturing, **c** vegetative growth of local variety, **d** vegetative growth of imported variety, **e** flowering of local variety showed plenty of male flowers, and **f** flowering of imported variety showed plenty of female flowers

showed high vegetative growth as well as high number of flowers and fruits (Fig. 14.25).

- We noticed that all imported saplings grown under the shade house condition for 15 months did not have any flowers or seeds. On the other hand, imported plants transferred in the open field flowered 4 times (Oct., Feb., May and

Fig. 14.26 Size of inflorescences and number of fruits in the cluster of local and imported varieties, as well as selected fruits from both **a** Local variety showed small inflorescence, **b** imported variety showed big inflorescence, **c** local variety showed small cluster of fruits (6–10), **d** imported vitro plants obtained big cluster of fruits (20–27 fruits), **e** a sample of good selected fruits from local variety, and **f** a sample of good selected fruits from imported variety

Sep.) during the 15 months study period.

- During the winter season (Nov., Dec., Jan., and Feb.), imported plants grown in the shade house shed plenty of leaves (this negatively affected canopy width), and remaining leaves looked pale as compared with those grown in the open field which kept looking more healthy and did not shed any leaf.
- We recorded three different types of insect infection symptoms:

1. On July 2012, the insect named *Tuta absoluta* attacked a few plants in shade house (see Fig. 14.24a) and the plants recovered in three days after the insecticide treatment.
2. On August 2013 because of high humidity inside the shade house, the insect named *Aphis gossypii* attacked the plants (see Fig. 14.24b) and the plants recovered in two days after the insecticide treatment.
3. On August 2013 because of high humidity inside the shade house, the insect named whitefly attacked the plants (see Fig. 14.24c) and the plants recovered in two days after the insecticide treatment.

B. Open field

- Within 15-month growth period, the imported trees from tissue culture saplings and the local trees from cuttings recorded 4 times flowering in the open field (Oct., Feb., May, and Sep).
- Saplings from tissue culture and local trees from cuttings gave flowers after 4 months and fruits after 8 months from transplanting from the shade house to the open field.
- No difference could be noticed in flowering and fruiting between imported

Fig. 14.27 Special methods of jatropha propagation **a** Pruned trees (*left*) and unpruned trees (*right*), **b** multiple grafting to hard pruning tree, and **c**, **d** air layering

saplings and plants from local cuttings in the open field within the period Sep. 2012—May 2013. Numbers of male flowers were equal, around one-third out of the total flower number.

- In general, imported plants and plants from local cuttings produced the panicle (the cluster; from three to seven fruits) from winter flowers during Oct. and fruit sit in Nov. and it gives dry fruit in Feb. We, however, cannot depend on it, so we have to wait till the main flowering and fruiting season in May.
- Imported plants grown in the open field showed difference from the local variety in number of leaves, stem circumference (cm), canopy width (cm), number of branches, number of flowers and seeds with high significance.
- In general, all the plants increased the height in the first 5 months, while increased the width up to 15 months.
- Imported saplings from tissue culture obtained vigor growth and produced plenty of branches and fruits more than local plants from cuttings (Fig. 14.26).
- Numbers of male flowers were less in the imported saplings than local plants.
- Growth performance under plastic tuneless in the open field in winter was almost was the same between imported plants and local ones. They showed significant increase in height, but low number of flowers and low fruit set.

- Seeds of imported saplings were bigger than those of local cuttings trees.
- Terminal cuttings produced good productivity more than base cuttings.

Plants from Egyptian and Philippine seeds

- Plants originally from Philippine seeds grew better in the fast 5 months than those from Egyptian seeds, while similar vegetative growth was observed at the end of the 15-month growth period.
- Trees from Egyptian seeds and those from Philippine seeds did not produce any flowers and subsequently fruits up to the end of 15-month period.

Trials of special methods of propagation

- We established hard pruning for mature trees (Fig. 14.27a, b) for good productivity of trees that originally had low productivity. Also, values can be added to seedlings through implementing grafting techniques (Fig. 14.27b).
- Results indicated that applied air layering technique to branches was considered very useful methods for multiple jatropha (Fig. 14.27c, d).

14.12 Conclusions

Jatropha aims to be one of the commercial crops but unfortunately there is lack of published practical cultivation data, leading to big gap between experimental and applied research. In fact, the practical research in the desert area should be important for breeders or even basic jatropha researchers. Open field research work in Egypt is considered as representative to North African countries, which is almost located in semidry and dry region in the Great Desert. Our unique obtained data in dry region can be the basis to set up research objectives for researchers, which seems to be still lacking.

The information about temperature, strong sunshine, soil type, irrigation source (treated wastewater), irrigation network, water requirement, agriculture distance, fertilization, growth behavior, flowering season, insect infection, and insect control are included here, which may affect the jatropha genome basic studies. Since genetically modified variety is most suitable application for inedible and isolated plantation like jatropha, possibility of the genetically modified variety registration and protection of such variety are the focal point of jatropha business and production of high-yield varieties. It will not be fruitful without including the wide range variety of the jatropha genome. The systematic and scientific approach to this research will provide researchers and breeders with many variables to enrich the scientific field, will provide them with conclusive results, and will ensure jatropha high productivity.

We identified here a great potentiality and advantages of the Egyptian model as well as bottlenecks for the model to be commercially viable especially in other desert countries. In order to achieve the commercial viability in desert environment, it needs to introduce higher yield varieties in the world that have been institutionally or commercially developed. We imported and cultivated the elite jatropha varieties (JOil lab. in Singapore and Philippines clone) and have improved seed yields and fraction of each tree's productivity by using cutting or tissue culture multiplication technique. Genetic modification and multiplication by tissue culture have been studied in limited number of institutes in the world. Technology-wise, currently, most of yield improvements are based on the selection and crossbreeding of selected varieties to increase number of bunches and number of fruits of each bunch and seeds size as well.

Global world suffers from shortage in freshwater for irrigating food crops. Treated wastewater is considered as promising irrigation source for inedible bio energy crops, which save freshwater for food crops. Importing and cultivating elite jatropha varieties under desert conditions with a smart use of treated wastewater is rather new approach. Since none of elite jatropha varieties has been developed with treated wastewater irrigation, it is necessary measures to compare the performance of the existing jatropha varieties and confirm their productivities that enable the commercial and industrial development for different viability before the implementation of the large-scale plantations. Using our jatropha growth and fruiting characteristics data will help to provide recommendations for researchers to support them in their jatropha varieties development programs.

We share our exclusive data subscribing the manuscript to be available in serving jatropha investors and scientists. There is no doubt that one day it will be one of their main demand.

References

Abd El-Baky HH, El Baz FK, El-Baroty GS (2010) Enhancing antioxidant availability in wheat grains from plants grown under seawater stress in response to microalgae extract treatments. J Sci Food Agric 90:299–303. doi:10.1002/jsfa.3815

Abd El-Baky HH, Hussein MM, El-Baroty GS (2008) Algal extracts improve antioxidant defense abilities and salt tolerance of wheat plant irrigated with sea water. Afr J Biochem Res 2:151–164

Abd El Baky HH, Hussein MM, Ibrahem EA (2013) Potential of industrial wastewater use for jatropha cultivation in arid land. Am J Agric Biol Sci 8(4):350–356

Achten W (2010) Sustainability evaluation of biodiesel from *Jatropha curcas* L. A life cycle oriented study. PhD thesis, KU Leuven

Aldoori SA (2014) Micropropagation of *Jatropha curcas* L. plant. MS.c thesis, Floriculture and Ornamental Horticulture and Landscape Gardening. Dep Fac Of Agri El-Shatby Alexandria Egypt

Annarao S, Sidhu OP, Roy R, Tuli R, Khetrapal CL (2008) Lipid profiling of developing *Jatropha curcas* L. seeds using 1 HNMR spectroscopy. Bioresour Technol 99:9032–9035

Ashwani K, Sharma S (2008) An evaluation of multipurpose oil seed crop for industrial uses (*Jatropha curcas* L.): a review. Ind Crops Prod 28:1–10

Chhetri AB, Martin S, Tango MS, Budge SM, Watts KC, Islam MR (2008) Non-edible plant oils as new sources for biodiesel production. Int J Mol Sci 9:169–180

Dehgan B, Webster GL (1979) Morphology and infrageneric relationships of the genus jatropha (Euphorbiaceae), vol 74. University of California Publications in Botany, Oakland

El-Tohamy SA, El-Kholy MM, El-Maghraby TA, Hassanein SA (2012) Profitable maximizing from the treated sewage effluent reuse for irrigation in newly reclaimed dessert soils to produce biodiesel from the grown jatropha trees. J Soil Sci Agric Eng Mansoura Univ 3(3):335–347

FAO (1992) Wastewater treatment and use in agriculture. Irrigation and drainage paper no. 47, Rome, Italy

Forson FK (2004) Performance of jatropha oil blends in a diesel engine. Renew Energy 29:1135–1145

George F, Edinger R, Beeker K (2005) A concept for simultaneous wasteland reclamation, fuel production, and socioeconomic development in degraded area in India: Need, potential and perspectives of *jatropha* plantations. Nat Resour Forum 29:12–24

Gubitz GM, Mittelbach M, Trabi M (1999) Exploitation of the tropical oil seed plant *Jatropha curcas* L. Bioresour Technol 67:73–82

Hegazy A, Nagaoka S, Kobayashi S, Hatsukade T, Okamoto S, Takeda Y, Takeya H, Matsumoto T, Hattori M (2012) Survey on policy and regulation environment for Egyptian biofuel industry development. Final report of the titled project (2011–2012) funded by Japan International Cooperation Agency (JICA), pp 1–227

Hussein MM, Abdelraouf RE (2013) Effect of irrigation by mixed agricultural drainage water and potassium foliar fertilizer on growth and photosynthetic pigments of *jatropha* plants. Middle East J Agric Res 4:116–122

Jones C (2004) Europe adopts biodiesel: can an African bean crack Europe's biodiesel blockage? EcoWorld. http://www.ecoworld.com/home/artic!es2.cfm?tid=356. Accessed 29 Dec 2006

Kotb T, Watanabe T, Ogino Y, Tanji K (2000) Soil salinization in the Nile Delta and related policy issue in Egypt. Agric Water Manag 43:239

Li M, Li H, Jiang H, Pan X, Wu G (2007) Establishment of an Agrobacteriuim-mediated cotyledon disc transformation method for *Jatropha curcas*. Plant Cell Tiss Org Cult 92:173–181

Metwally ZA (2009) The prospects of alternative energy. Egyptian general book, Cairo, pp 12–13

Mkoka C, Shanahan M (2005) The bumpy road to clean. green fuel. SciDev Net. http://www.scidev.net/Features/Index,Cfm?fuseaction=readFeatures&temid=477&language=1. Accessed 29 Dec 2006

Misra M, Misra AN (2010) *Jatropha*: the biodiesel plant, biology, tissue culture and genetic transformation—a review. Int J Pure Appl Sci Technol 1:11–24

Nessrien Ahmed SA (2010) Impact of low quality water used under some modern irrigation systems on soil and plants in Sahl El-Tena region. Ph. D. thesis, Institute of Environmental studies and Research, Ain Shams University, Egypt

Purkayastha J, Sugla T, Paul A, Solleti SK, Mazumdar P, Basu A, Mohommad A, Ahmed Z, Sahoo L (2010) Efficient in vitro plant regeneration from shoot apices and gene transfer by particle bombardment in Jatroph curcas. Biologica Plantarum 54(1):13–20

Rajaona A, Sutterer N, Asch F (2012) Potential of waste water use for jatropha cultivation in arid environments. Agriculture 2:376–392. ISSN 2077-0472

Silva AA, Santos RF, Frigo EP, Bassegio D, Frigo JP, Dieter J, Deonir Secco D, Samuel de Sousa NM (2014) Soil chemical properties after surface application of swine wastewater in the cultivation of Jatropha. Afr J Agric Res 9(1):154–159

Tracing the Dispersal Routes by Local Names of Jatropha

15

Takayuki Ando

Abstract

Jatropha, *Jatropha curcas* L., is a perennial shrub widely grown in the tropical and subtropical regions and has great importance as a source of renewable energy. However, the knowledge of its geographical origin, dispersal and extent of diversity is still scanty. This study was conducted to elucidate the dispersal routes of Jatropha from its postulated center of origin (Mexico) using local/indigenous names of the plant in different locations. Two principal routes of Jatropha dispersal exist: from Mexico to Central and South America, and from Mexico or the Caribbean via Cape Verde to Africa and Asia. The plant is thought to have been passed on along with its local name from one ethnic group to another as a medicinal plant. Jatropha was likely dispersed from Veracruz in Mexico or the Caribbean Islands to Cape Verde by slave ships. Vegetative propagation through cuttings is the common means of Jatropha multiplication, which further limits genetic diversification. Genetically, diverse Jatropha exists in places such as center of origin, center of domestication, and other centers of diversity. Dispersal of seeds by animals or pigeons possibly is involved in high genetic diversity in places other than the origin, which may help Jatropha to adapt to different environmental conditions over a long period of time. In order to facilitate viable and efficient breeding programs, it is essential to identify and collect Jatropha germplasm from different places of high genetic diversity.

15.1 Introduction

Jatropha, *Jatropha curcas* L., has many advantages such as drought tolerance and can grow in semi-arid areas or marginal lands that are not suitable for food production without needing large amounts of farm inputs (Heller 1996; Jongschaap et al. 2007; Robbins 2011). Jatropha cultivation

T. Ando (✉)
Center for International Affairs, Tottori University, Tottori, Japan, 4-101, Koyama Minami, Tottori 680-8550, Japan
e-mail: andota@ciatu.tottori-u.ac.jp

S. Tsuchimoto (ed.), *The Jatropha Genome*, Compendium of Plant Genomes,
DOI 10.1007/978-3-319-49653-5_15

for biodiesel production expanded rapidly in the 2000s (German et al. 2010; German et al. 2011; GRAIN 2013) as a result of large-scale investments in Asia, Africa, and Latin America (Dias et al. 2012). Additionally, some governments implemented various Jatropha cultivation programs for smallholder farmers (GEXSI 2008). In the late 2000s, a variety of challenges came to light such as a low profitability or low productivity of Jatropha projects resulting in discontinuation of several projects (Sanderson 2009), thus leading to shrinking of the Jatropha boom. This occurred mainly because breeding of Jatropha had not progressed much as it was still classified as a wild species (Grass 2009) and yields were lower than had been initially expected (Dias et al. 2012; Osorio et al. 2014). Breeding programs of Jatropha have been tried in Africa and Asia; however, Basha et al. (2009) reported low germplasm variability in these regions. Elevation of Jatropha from a wild species to a profitable biodiesel crop was greatly hindered by the low genetic and phenotypic variability found in different regions around the world. This hampered efficient breeding for high yield and oil content traits (Grass 2009; Dias et al. 2012; Osorio et al. 2014). Therefore, breeding using genetically divergent materials through systematic collection of Jatropha germplasm expands the possibility of obtaining new varieties with desirable agronomic and industrial characteristics (Heller 1996; Basha et al. 2009; Zamarripa-Colmenero et al. 2010). Genetically divergent Jatropha exists in the center of origin, center of domestication (Pecina-Quintero et al. 2014), and other centers of diversity (Dias et al. 2012) where wide genetic variability could permit identification of useful genetic materials for future improvement (Pecina-Quintero et al. 2014). Thus, identification of a wider range of areas with high genetic diversity is essential for efficient or reliable collection of genetically divergent materials.

Jatropha is native to tropical America, but is now found abundantly in many tropical and subtropical regions throughout Africa and Asia (Jongschaap et al. 2007); however, its migratory routes from the center of origin has not yet been established (Pamidimarri and Reddy 2014) nor have the places with high genetic diversity been located yet. Humans have long used and distinguished some plants with beneficial properties by giving them selected names. Indigenous traditional knowledge on medicinal plants (name and usage) has been transmitted orally for centuries from generation to generation and from one ethnic group to another (Dambatta and Aliyu 2011). There are proper names of Jatropha specific to an ethnic group or identical names in different ethnic groups and regions. When the proper name of Jatropha was transmitted to another ethnic group in different location unchanged in its original form, this suggested existence of a transmission route. In this study, the dispersal route of Jatropha from its center of origin will be predicted basing on local names of the plant and probable causes for lower genetic diversity away from the center of origin.

Seed dispersal occurs not only by humans, but also by wind, mechanical explosion, water, and animals, including birds (Howe and Smallwood 1982). Though Negussie et al. (2013) suggested pantropical distribution of Jatropha, today it is mainly explained by intentional anthropogenic dispersal. Jatropha seed dispersion by natural processes such as wind or water is impossible due to their big size (Negussie et al. 2013; United States Department of Agriculture 2015). According to the United States Department of Agriculture (2015), Jatropha seeds can be dispersed by some animals though is uncertain about birds. However, Rivera-Lorca and Ku-Vera (1997) reported that the white-winged dove *Zenaida asiatica* usually eats the Jatropha seeds in Chiapas State of Mexico. When pigeons eat Jatropha seeds, Jatropha seeds potentially get scattered on the ground or spread around. Thus, this study attempts to verify that the seeds of Jatropha are eaten by pigeons in Mexico, a probable center of origin.

15.2 Survey of Local Names of Jatropha

Table 15.1 shows identified local names of Jatropha. The largest number of local names in indigenous languages of Jatropha was found in Mexico. In Central and South America, a large

Table 15.1 Local names of *Jatropha curcas* L. in different locations

Country (State)	Local name[a] [name of indigenous people] (reference)
Mexico	
(Sinaloa)	Sangregado (Arellano-Rodoríguez et al. 2003, UCR 2014)
(San Luis Potosi)	Piloch [Tenek] (UNAM 2015), Tdakpen te' (UNAM 2015)
(Hidalgo)	Achcuauit (UCR 2014), Acheuauit (Arellano-Rodoríguez et al. 2003), Ashcuahuitl [Nahuas] (Gómez-Pompa et al. 2009), Pipián De Árbol (Villavicencio-Nieto and Pérez-Escandón 2005), Ra Demuza (Villavicencio-Nieto and Pérez-Escandón 2005)
(Veracruz)	Ashte (Arellano-Rodoríguez et al. 2003), Ashtél [Nahuas] (Gómez-Pompa et al. 2009), Axte (Arellano-Rodoríguez et al. 2003), Axté (Gómez-Pompa et al. 2009, UCR 2014), Chota [Totonaco] (Escalona-Gutiérrez 2004), Chote [Totonaco] (Escalona-Gutiérrez 2004), Chu.ta [Totonaco] (García-Ramos 2007, UNAM 2015), Chu:tá' [Totonaco] (Escalona-Gutiérrez 2004), Chuahuayohuiaxtli (UCR 2014), Cuahaychuachili (UCR 2014), Cuahuayohuiaxtli (Arellano-Rodoríguez et al. 2003), Cuauyohuatli (Arellano-Rodoríguez et al. 2003), Piñoncillo [Totonaco] (García-Ramos 2007), Scu-lu'u [Totonaco] (Arellano-Rodoríguez et al. 2003), Scu-luú [Totonaco] (Gómez-Pompa et al. 2009), Scu-lúu [Totonaco] (UCR 2014), Skulu [Totonaco] (García-Ramos 2007; UNAM 2015), Skuulu [Totonaco] (García-Ramos 2007), Tempatl [Nahuas] (Gómez-Pompa et al. 2009),
(Mexico)	Avellana Purgante (Arellano-Rodoríguez et al. 2003; Gómez-Pompa et al. 2009)
(Puebla)	Axti (Schmook and Serralta-Peraza 1997; UNAM 2015), Axti [Nahuas] (Martínez-Alfaro 2001), Cahuax [Nahuas] (UNAM 2015), Chu.ta[Totonaco] (UNAM 2015), Chu-ta [Totonaco] (Martínez-Alfaro 2001), Cuauhayohuachtli [Nahuas] (Gómez-Pompa et al. 2009), Cuta (Ross 2003; UNAM 2015), Pitana (UNAM 2015), Taktinau [Tepehua] (Martínez-Alfaro 2001, UNAM 2015), Taxiinau (UNAM 2015), Temuza [Otomí] (Martínez-Alfaro 2001; UNAM 2015)
(Guerrero)	Ts'oom Taxua Cachi [Amuzgos] (Robinson and López 1999)
(Morelos)	Chuahuayohuiaxtli (UCR 2014), Cuahaychuachili (UCR 2014), Cuauixtli (Schmook and Serralta-Peraza 1997), Cuauixtli [Nahuas] (UNAM 2015), Cuauyohuatli (Arellano-Rodoríguez et al. 2003)
(Oaxaca)	Cak Siil [Huave] (Gómez-Pompa et al. 2009), Kusekeey (Schmook and Serralta-Peraza 1997), Kutsekeey (UNAM 2015), Que-ca [Chontal] (Arellano-Rodoríguez et al. 2003, Gómez-Pompa et al. 2009, UCR 2014), Sikil-te (Arellano-Rodoríguez et al. 2003), Tzon Texoa Kichi (UNAM 2015), Vico (Arellano-Rodoríguez et al. 2003, UCR 2014), Yaga-be-lape (UCR 2014), Yaga-be-pale [Zapoteca] (Gómez-Pompa et al. 2009)
(Chiapas)	Cuipi [Tzozil] (Gómez-Pompa et al. 2009), Cuipú (UCR 2014), Nacuala (Arellano-Rodoríguez et al. 2003, UCR 2014), Najuala (Arellano-Rodoríguez et al. 2003, UCR 2014), Sakilté [Tzeltal] (CONABIO 2013)
(Yucatan)	Kxakal-che [Maya] (Gómez-Pompa et al. 2009), Niin (Arellano-Rodoríguez et al. 2003, UNAM 2015), Piij (Arellano-Rodoríguez et al. 2003; UNAM 2015), Piij [Maya] (UCR 2014), Sicilte [Maya] (UCR 2014), Sikilte' (Arellano-Rodoríguez et al. 2003, UNAM 2015), Sikilte' [Maya] (UCR 2014), Xcacalche [Maya] (UCR 2014), Xkakabche' (Arellano-Rodoríguez et al. 2003; UNAM 2015), Xkakabché [Maya] (UCR 2014), Xkakal-ché [Maya] (UCR 2014), X-kakalché [Maya] (UCR 2014), Xsikilte (UNAM 2015), Xsikilte' (Arellano-Rodoríguez et al. 2003), Xsikilte' [Maya] (UCR 2014)

(continued)

Table 15.1 (continued)

Country (State)	Local name[a] [name of indigenous people] (reference)
(Yucatan Peninsula)	Pomolche (Herbario 2010), Sikilte (Herbario 2010), Sikilte [Maya] (Schmook and Serralta-Peraza 1997), Sikil-té [Maya] (Makkar and Becker 1997), Ska'ak'alche' (Arellano-Rodoríguez et al. 2003), Skakalche' (Arellano-Rodoríguez et al. 2003), X-kakalché (Herbario 2010)
(not specified)	Acu-lu'u (Martinez-Herrera et al. 2012), Ashte (Martinez-Herrera et al. 2012), Avellana Purgante (Duke 2009), Axti (Ocampo and Balick 2009), Cahuax (Ocampo and Balick 2009), Cak Siil (Martinez-Herrera et al. 2012), Chuahuayohuiaxtli (Martinez-Herrera et al. 2012), Chuta (Martinez-Herrera et al. 2012), Cuauayohuachtli [Nahuas] (UCR 2014), Cuauyohuatli (Martinez-Herrera et al. 2012), Cuipi (Martinez-Herrera et al. 2012), Cuipu (Duke 2009), habb-el-meluk (Ross 2003), Palo De Piñón (UNAM 2015), Piñón Bronco (UNAM 2015), Piñón Purgante (Duke 2009), Piñón (Makkar and Becker 1997, Martínez-Alfaro 2001; Ocampo and Balick 2009), Pitana (Hanelt 2001; Martínez-Alfaro 2001; Ocampo and Balick 2009, Globinmed 2015), Piñoncillo (Little et al. 1974; Heller 1996; Makkar and Becker 1997; Hanelt 2001; Rodolfo 2001; Grandtner 2005, UNAM 2015), Que-ca (Martinez-Herrera et al. 2012), Sangregado (Little et al. 1974; Rodolfo 2001; Duke 2009), Sangre-gado (Grandtner 2005; Duke 2009), Sangregrado (Duke 2009), Sangre-grado (Duke 2009), Scu-lu'u (Martinez-Herrera et al. 2012), Sicilte (Rodolfo 2001), Sikil'te (Martinez-Herrera et al. 2012), Tempatl (Martinez-Herrera et al. 2012), Xkakal-che (Martinez-Herrera et al. 2012)
Belize	Hazelnut (Grandtner 2005), Physic Nut (Grandtner 2005; Duke 2009; Ocampo and Balick 2009), Sakilte [Maya] (Grandtner 2005), Xkakalche [Maya] (Grandtner 2005)
Guatemala	Coquito (Ocampo and Balick 2009), Pinon (Ross 2003), Piñon (Anzueto and MacVean 2000), Piñón (Heller 1996; Rodolfo 2001; Duke 2009; Ocampo and Balick 2009), Piñoncillo (Ocampo and Balick 2009), Sakilté (Anzueto and MacVean 2000; Ocampo and Balick 2009), Sakilté [Quiche] (UCR 2014), Tempacte (UCR 2014), Tempate (UCR 2014), Tempocte (Duke 2009), Tepate (Anzueto and MacVean 2000), Xcacalché (Anzueto and MacVean 2000), Yupur (Anzueto and MacVean 2000; Grandtner 2005, UCR 2014)
Honduras	Arbol Santo (Grandtner 2005), Cotoncillo (Little et al. 1974; Rodolfo 2001; Grandtner 2005; Duke 2009), Piñón (Grandtner 2005; Duke 2009, Ocampo and Balick 2009), Yupur (Ocampo and Balick 2009)
El Salvador	Coquillo (Rodolfo 2001), Coquito (Ocampo and Balick 2009), Piñón (Ocampo and Balick 2009), Tártago (Ocampo and Balick 2009), Tempate (Rodolfo 2001, Ocampo and Balick 2009, IICA 2010), Tempote (Duke 2009)
Nicaragua	Piñón (Duke 2009, Ocampo and Balick 2009), Purging Physic (Ross 2003), Tempate (Ocampo and Balick 2009; IICA 2010), Tempote (Duke 2009)
Costa Rica	Coquillo (Heller 1996; Hanelt 2001; Rodolfo 2001; Grandtner 2005; Duke 2009; Ocampo and Balick 2009), Coquito (Little et al. 1974; Nuñez-Melendez 1975; Grandtner 2005; Duke 2009; Ocampo and Balick 2009), Kuubín-ua [Boruca] (Grandtner 2005), Piñón (Nuñez-Melendez 1975; Ocampo and Balick 2009), Tapate (Rodolfo 2001; Duke 2009), Tempate (Nuñez-Melendez 1975; Heller 1996; Rodolfo 2001; Duke 2009; Ocampo and Balick 2009; IICA 2010), Tempote (Duke 2009)
Panama	Arbol Santo (Grandtner 2005), Árbol Santo (Rodolfo 2001; Duke 2009), Coquillo (Little et al. 1974; Escobar 1987; Rodolfo 2001; Grandtner 2005; Duke 2009; Ocampo and Balick 2009), Kwiwala (Duke 2009), Kwiwala [Bayano] (Escobar 1987), Piñón (Grandtner 2005), Tempate (Rodolfo 2001)
Venezuela	Tua Tua (Rengifo et al. 1997), Piñón (Duke 2009)
Guyana	Physic Nut (Ross 2003), Pigon D'Inde (Duke 2009)
Surinam	Chijnoot (Grandtner 2005), Schijtnoot (Little et al. 1974, Duke 2009), Sket'noto (Rengifo et al. 1997)

(continued)

Table 15.1 (continued)

Country (State)	Local name[a] [name of indigenous people] (reference)
Colombia	Frailecillo (Little et al. 1974; Rodolfo 2001; Grandtner 2005; Duke 2009), Frailejón (Little et al. 1974; Rodolfo 2001; Grandtner 2005; Duke 2009), Jaquillo (Duke 2009; Ocampo and Balick 2009), Piñón De Purga (Little et al. 1974; Rodolfo 2001; Grandtner 2005; Duke 2009; Ocampo and Balick 2009), Purga (Ocampo and Balick 2009), Purga De Fraile (Rengifo et al. 1997), Túa Túa (Rodolfo 2001), Tuatua (Grandtner 2005), Tuatúa (Little, et al. 1974; Duke 2009)
French Guyana	Médicinier (Duke 2009)
Ecuador	Piñón (Duke 2009)
Peru	Barbasco (Rengifo et al. 1997; Duke 2009), Higo Del Duende (Rengifo et al. 1997; Duke 2009), Huiso Pionis (Duke 2009), Huiso Pionis [Shipibo-conibo] (Rengifo et al. 1997), Josho Pionis [Shipibo-conibo] (Rengifo et al. 1997), Periyanasi [Piro] (Rengifo et al. 1997; Duke 2009), Piñol (Heller 1996; Rengifo et al. 1997; Hanelt 2001; Duke 2009; Ocampo and Balick 2009), Piñon (Ocampo and Balick 2009), Piñón (Rengifo et al. 1997; Grandtner 2005; IICA 2010), Piñon Blanco (Ocampo and Balick 2009), Piñón Blanco (Duke 2009), Piñón Joshó [Amahuaca] (Rengifo et al. 1997; Duke 2009), Piñón Purgativo (Rengifo et al. 1997; Duke 2009), Piñoncito (Rengifo et al. 1997; Ocampo and Balick 2009), Piñóncito (Duke 2009), Pinyanasi [Piro] (Duke 2009), Wapa-wapa Oshe [Ese eja] (Rengifo et al. 1997)
Brazil	Figo Do Inferno (Duke 2009), Mandabi Guacu (Duke 2009), Mundubi-assu (Heller 1996, Hanelt 2001), Pião (Little et al. 1974; Grandtner 2005), Piao Branco (Rodolfo 2001, Ross 2003), Pião Branco (Little et al. 1974; Grandtner 2005; Madaleno 2007), Pinhão (Agra et al. 2008), Pinhão Bravo (Little et al. 1974, Grandtner 2005), Pinhao Manso (Duke 2009), Pinhão Manso (Arruda et al. 2004), Pinheiro Do Inferno (Duke 2009), Pinheo Bravo (Rodolfo 2001), Pinhno (Duke 2009), Pinhno Bravo (Duke 2009), Pinhno Do Inferno (Duke 2009), Pinhno Do Manso (Duke 2009), Pinhno Do Paraguai (Duke 2009), Pinnao De Purga (Ross 2003), Pino (Duke 2009), Pino Branco (Duke 2009), Piñón Manso (IICA 2010), Purgueira (Infopédia 2016)
Bolivia	Chacsis (Grandtner 2005), Higo De Infierno (Rengifo et al. 1997; Rodolfo 2001), Krigre [Trinitario] (Thomas and Vandebroek 2006), Kuchire [Trinitario] (Thomas and Vandebroek 2006), Piñón (Grandtner 2005; Duke 2009; Ocampo and Balick 2009), Vocudyes (Grandtner 2005), Wasicano [Chacobo] (Duke 2009), Wassa Supay (Rodolfo 2001), Yita Ti Rebetibe [Yurakaré] (Thomas and Vandebroek 2006)
Paraguay	Chisasquil (Grandtner 2005), Josho Pionis (Grandtner 2005), Kuri Y Vai [Guaraní] (Grandtner 2005), Kuri'y Vai (Grandtner 2005), Piñón (Grandtner 2005), Piñón Manso (IICA 2010)
Bahamas	Physic Nut (Grandtner 2005)
Cuba	Palo Santo (Duke 2009), Piñón (Grandtner 2005), Piñón Botija (Roig y Mesa 1928; Little et al. 1974; Blanco 1993; Rodolfo 2001; Cano 2004; Grandtner 2005; Duke 2009), Piñón Criollo (Roig y Mesa 1928; Cano 2004; Grandtner 2005; Duke 2009), Piñón De Botija (Duke 2009), Piñón De Cerca (Little et al. 1974; Rodolfo 2001; Grandtner 2005; Duke 2009), Piñón Lechero (Grandtner 2005; Duke 2009), Piñón Lotija (Duke 2009), Piñón Purgante (Blanco 1993; Grandtner 2005; Duke 2009), Piñón Voci (Grandtner 2005), Piñón Vomico (Duke 2009), Purga De Fraile (Duke 2009), Tártago (Grandtner 2005; Duke 2009)
Haiti	Feuilles Médicin (Grandtner 2005), Feuilles Médicinier (Duke 2009), Fey Medsen [Creole] (Grandtner 2005), Gran Medsinye (Grandtner 2005), Grand Médicinier (Grandtner 2005; Duke 2009), Gros Ricin (Duke 2009), Gwo Ricen (Duke 2009), Médicinier (Little et al. 1974; Grandtner 2005; Duke 2009), Médicinier À Grandes Feuilles (Grandtner 2005), Médicinier Beni (Duke 2009), Médicinier Béni (Little et al. 1974, Grandtner 2005), Médicinier Cathartique (Grandtner 2005; Duke 2009), Médicinier Purgantif (Grandtner 2005), Médicinier-barrière (Grandtner 2005), Médicinier-blanc (Grandtner 2005), Medsinye Beni (Grandtner 2005), Medsinye Gran Fey [Creole] (Grandtner 2005) Medsiyen (Duke 2009, Ocampo and Balick 2009), Medsiyen Béni (Duke 2009), Medsiyen Beni [Creole] (Grandtner 2005), Wedsiyen (Ross 2003)

(continued)

Table 15.1 (continued)

Country (State)	Local name[a] [name of indigenous people] (reference)
Dominican Republic	Avellana Purgante (Ruiz et al. 1999), Botija (Ruiz et al. 1999), Physic Nut (Ruiz et al. 1999), Piñón (Grandtner 2005; Duke 2009, Ocampo and Balick 2009; IICA 2010), Piñón Botija (Ocampo and Balick 2009), Piñón Criollo (Ruiz et al. 1999), Piñón De Barbados (Ruiz et al. 1999), Piñón Lechero (Ruiz et al. 1999), Piñón Purgante (Ruiz et al. 1999), Tapate (Ruiz et al. 1999), Tártago (Ruiz et al. 1999), Wild Oil Nut (Ruiz et al. 1999)
Virgin Islands	Physic-nut (Little et al. 1974; Grandtner 2005)
Jamaica	Common Physic Nut (Duke 2009), Physic-nut (Grandtner 2005), Pulza (Grandtner 2005, Mabberley 2008), Purge-nut (Grandtner 2005), Purging Nut (Grandtner 2005), Wild Oil Nut (Grandtner 2005; Duke 2009)
Puerto Rico	Barbados-nuts (Little et al. 1974), Dinon (Ross 2003), Physic-nut (Little et al. 1974), Pignons D'Inde (Little et al. 1974), Piñón (Little et al. 1974, Grandtner 2005), Piñón Purgante (Little et al. 1974, Grandtner 2005), Purgante (Rodolfo 2001), Purging-nuts (Little et al. 1974), Semen Ricini Majoris (Little et al. 1974), Tártago (Little et al. 1974; Heller 1996; Hanelt 2001; Rodolfo 2001; Ross 2003; Grandtner 2005; Duke 2009)
French West Indies	Médicinier Beni (Duke 2009), Médicinier Cathartique (Duke 2009), Noix Des Barbades (Duke 2009)
Guadalupe	Médicinier Barriére (Duke 2009), Médicinier Beni (Duke 2009), Médicinier Blanc (Duke 2009), Médicinier Grand Bénit (Duke 2009), Médicinier Purgatif (Duke 2009), Pignon Des Barbados (Duke 2009), Pignon D'Inde (Duke 2009)
Dominica	Medicinier Blanc (Little et al. 1974; Grandtner 2005; Duke 2009)
Netherlands Antilles	Barbados-Nut (Grandtner 2005), Barbadosnut Nettlespurge (Grandtner 2005), Belly-ache Bush (Grandtner 2005, Chijnoot (Grandtner 2005), Curcas-bean (Grandtner 2005), Grave Physic-nut (Grandtner 2005)
Dutch Antilles	Grave Physic-nut (Little et al. 1974), Schijnoot (Little et al. 1974)
Cape Verde	Pinon Botija (Ross 2003), Pulga (Freitas 1906), Pulgueira (Freitas 1906), Purga (Freitas 1906), Purgueira (Freitas 1906), Tubaang-bakod (Ross 2003)

Local names of *J. curcas* were picked up through books, papers, or data bases from Mexico, Latin America, and Caribbean and the Cape Verde Islands, and those which appear in more than one country were identified

[a]The first letter of local names is 'capitalized' in accordance with Plant Names: A Guide to Botanical Nomenclature (Spencer et al. 2007)

number of local names in indigenous languages were also found. In the Caribbean, many Spanish names were found in Cuba and the Dominican Republic, French names in Haiti, while English names were found in Jamaica. On the other hand, "Tártago" in Cuba and Puerto Rico, as well as "Tapate" in Dominican Republic, was considered to be derived from indigenous languages there. Thus, local names of Jatropha in the Caribbean are a mixture of Euro-American and indigenous languages.

Table 15.2 shows countries where same local names exist and are marked with closed circles. In Mexico and in Guatemala, "Piñoncillo" and "Sakilté" exist in common. In Mexico and Dominican Republic, "Avenilla Purgante" exists in common, while again in Mexico, Cuba, Dominican Republic, and Puerto Rico, "Piñón Purgante" exists in common. "Purgante" means purgative in English, thus Jatropha is deemed to have medicinal properties as purgative in the Caribbean. In Cuba, Dominican Republic, Puerto Rico, and in El Salvador, "Tártago" exists in common. "Tapate" exists in Dominican Republic and Costa Rica. Thus, some relationship is suggested between the Caribbean and Central America. In the Caribbean islands, "Piñón Botija" and "Piñón De Botija" exist in Cuba while "Botija" and "Piñón Botija" are used in Dominican Republic (Table 15.1). On the other hand, "Pinon Botija" is used in Cape Verde. These examples suggest a partial commonality of Jatropha names between the Caribbean and Cape Verde.

Table 15.2 Countries and common local names of *Jatropha curcas*

	Central America								South America									
	Mexico	Belize	Guatemala	Honduras	El Salvador	Nicaragua	Costa Rica	Panama	Venezuela	Guyana	Surinam	Colombia	French Guyana	Ecuador	Peru	Brazil	Bolivia	Paraguay
Piñoncillo	●		●															
Avellana Purgante	●																	
Sakilté	●		●															
Piñón Purgante	●																	
Piñón	●		●	●	●	●	●	●	●					●	●		●	●
Physic Nut		●								●								
Coquito			●		●		●											
Piñon			●												●			
Tempate			●		●	●	●	●										
Yupur			●	●														
Arbol Santo				●				●										
Coquillo					●		●	●										
Tártago					●													
Tempote					●	●	●											
Tapate							●											
Chijnoot											●							
Purga												●						
Médicinier													●					
Josho Pionis															●			●
Piñón Manso																●		●
Purgueira																●		
Piñón Botija																		
Piñón Criollo																		
Piñón Lechero																		
Médicinier Beni																		
Médicinier Cathartique																		

(continued)

Table 15.2 (continued)

	Central America								South America									
	Mexico	Belize	Guatemala	Honduras	El Salvador	Nicaragua	Costa Rica	Panama	Venezuela	Guyana	Surinam	Colombia	French Guyana	Ecuador	Peru	Brazil	Bolivia	Paraguay
Wild Oil Nut																		
Physic-nut																		
Grave Physic-nut																		

	The Caribbean													
	Bahamas	Cuba	Haiti	Dominican Republic	Virgin Islands	Jamaica	Puerto Rico	French West Indies	Guadalupe	Dominica	Netherlands Antilles	Dutch Antilles	Cape Verde
Piñoncillo													
Avellana Purgante				●									
Sakilté													
Piñón Purgante		●		●			●						
Piñón		●		●			●						
Physic Nut	●			●									
Coquito													
Piñon													
Tempate													
Yupur													
Arbol Santo													
Coquillo													
Tártago		●		●			●						
Tempote													
Tapate				●									
Chijnoot											●	●	
Purga													
Médicinier			●										
Josho Pionis													
Piñón Manso													
Purgueira													
Piñón Botija		●		●									

(continued)

Table 15.2 (continued)

	The Caribbean												
	Bahamas	Cuba	Haiti	Dominican Republic	Virgin Islands	Jamaica	Puerto Rico	French West Indies	Guadalupe	Dominica	Netherlands Antilles	Dutch Antilles	Cape Verde
Piñón Criollo		●		●									
Piñón Lechero		●		●									
Médicinier Beni			●					●	●				
Médicinier Cathartique			●					●					
Wild Oil Nut				●		●							
Physic-nut					●	●	●						
Grave Physic-nut											●	●	

Presence of the local names was marked with closed circles based on Table 15.1

Table 15.3 Results of the interview about feeding damage of Jatropha by pigeon in the State of Chiapas, Mexico

Municipality	Village	Date of interview	No. of interviwed	No. of feeding damage by pigeon	%
La Concordia	Niños Héroes	Sept. 17, 2012	8	4	50
Villa Corzo	Tierra Santa	Oct. 2, 2012	10	6	60
Villa Corzo	Ignacio Zaragoza	Sept. 29, 2012	7	4	57
Villa Corzo	Doctor Belisario Domínguez	Sept. 20, 2012	11	11	100
Villa Corzo	Los Amates	Sept. 19, 2012	24	11	46
Villaflores	California	Sept. 13, 2012	23	0	0
Villaflores	Los Angeles	Sept. 13, 2012	20	10	50
Villaflores	Javier Lopez Moreno	Sept. 27, 2012	7	7	100
Total			110	53	48

Tha interview survey was conducted on 110 farmers in 8 villages of Chiapas State, Mexico from September to October, 2012

15.3 Pigeon Damage to Jatropha Seeds

Table 15.3 shows the response from a field survey on pigeon damage to Jatropha seeds in Chiapas State, Mexico. Results show that 48% of farmers who cultivate Jatropha suffer yield losses from seed damage due to pigeons.

15.4 Dispersal Route of Jatropha

15.4.1 Mexico–Central American Route

"Sakilté" is the local Jatropha name commonly used by indigenous Mexicans Tseltal (or Tzeltal) (CONABIO 2013) and Guatemalan people Quiché (UCR 2014). Additionally, "Sakilte," which has no accent however, is used by the Mayan indigenous people of Belize (Grandtner 2005). There are several local names of Jatropha in Central American countries that are thought to have originated from indigenous people such as "Tempate" in Guatemala (UCR 2014), El Salvador (Rodolfo 2001; Ocampo and Balick 2009; IICA 2010), Nicaragua (Ocampo and Balick 2009; IICA 2010), Costa Rica (Nuñez-Melendez 1975; Heller 1996; Rodolfo 2001; Duke 2009; Ocampo and Balick 2009; IICA 2010), "Yupur" in Guatemala (Anzueto and MacVean 2000; Grandtner 2005; UCR 2014), Honduras (Ocampo and Balick 2009), "Tempote" in El Salvador (Duke 2009).

According to Wolters (2001), in the beginning of cultivation of some economic plants, seeds, fruits, roots, or tubers were likely passed on overland from one village to the next, and from one person to another, or taken along on migration. He also reported that dissemination of economic plants on sea routes in the Pacific Ocean between Peru and Mexico as well as the Caribbean and in the Gulf of Mexico can be suspected via trade or immigration. Based on the above, Jatropha could have spread from Mexico, the center of origin to Central America and South America overland, and further to the Caribbean Islands by indigenous people via sea. Detailed discussion is presented in the sections below.

15.4.2 Mexico–Caribbean Route to Cape Verde

With regard to Mexico and the Caribbean islands, the name "Avellana Purgante" exists in Mexico (Arellano-Rodoríguez et al. 2003; Gómez-Pompa et al. 2009) and Dominican Republic (Ruiz et al. 1999), while "Piñón Purgante" is commonly used in Mexico (Duke 2009), Cuba (Blanco 1993; Grandtner 2005; Duke 2009), Dominican Republic (Ruiz et al. 1999), and Puerto Rico (Grandtner 2005). This suggests presence of a dispersal route from Mexico to the Caribbean. Additionally, these local names are Spanish, thus involvement of Spanish people in Jatropha propagation is also suggested.

There exists the name "Botija" as a part of names in common between the Caribbean and Cape Verde. For example, "Piñón Botija" (Roig y Mesa 1928; Little et al. 1974; Blanco 1993; Rodolfo 2001; Cano 2004; Grandtner 2005; Duke 2009) and "Piñón De Botija" (Duke 2009) in Cuba, "Botija" (Ruiz et al. 1999) and "Piñón Botija" (Ocampo and Balick 2009) in Dominican Republic, and "Pinon Botija" in Cape Verde (Ross 2003). Moreover, there is a place called "Botija" in the Porto Novo region in Cape Verde (http://trip-suggest.com/cape-verde/porto-novo/botija/). Although the linkage between the name of the place "Botija" and "Botija" in the name of Jatropha is unknown, some sort of connection is suggested between the Caribbean and Cape Verde.

Heller (1996) suggested the dispersal route of Jatropha as by Portuguese seafarers from the Caribbean via Cape Verde and through the current Guinea-Bissau to Africa and Asia. Freitas (1906) suggested that some slave ships brought Jatropha seeds back to Cape Verde from Brazil or the Caribbean in his book as "Algum navio negreiro, que no seu retorno aportou em Cabo Verde, trouxe do Brazil ou das Antillas a semente do modesto arbusto [Some slave ships which landed on return to Cape Verde from Brazil or Antillas brought some seeds of small shrub (*J. curcas*)]." On the other hand, according to Thomas (1997), anchorage sites of slave trade ships in Latin America and the Caribbean were Veracruz of Mexico, Havana of Cuba, Santo Domingo of Dominican Republic, Portobelo of Panama and Cartagena of Colombia. Results of genetic analysis such as random amplified polymorphic DNA (RAPD), amplified fragment length polymorphism (AFLP), Nuclear ribosomal DNA (nrDNA)—internal transcribed spacer (ITS) (nrDNA-ITS), Pamidimarri and Reddy (2014) showed strong genetic relationship between the Mexican and Cape Verde germplasm. Therefore, Jatropha seeds were likely carried from the center of origin, Mexico through Veracruz to Cape Verde by slave ships.

According to Wolters (2001), there existed sea routes that connected Mexico, Peru, and the Caribbean, and an exchange of crop plants can be anticipated to have occurred on these sea routes. He also cited evidence of spread of economic plants on sea routes by Amerindians since the beginning of the first millennium AD was already found in the Caribbean area. According to Little et al. (1974), Jatropha is not native in Puerto Rico and the Virgin Islands but was introduced long ago, and now, cultivation is widespread and naturalized locally. These suggest that Jatropha was introduced from somewhere to the Caribbean Islands via sea routes by Amerindians in pre-Columbian times. The presence of identical names such as "Tártago" in El Salvador (Ocampo and Balick 2009) and Cuba (Grandtner 2005; Duke 2009), Dominican Republic (Ruiz et al. 1999) and Puerto Rico (Little et al. 1974; Heller 1996; Hanelt 2001; Rodolfo 2001; Ross 2003; Duke 2009; Grandtner 2005), "Tapate" in Costa Rica (Rodolfo 2001; Duke 2009) and Dominican Republic (Ruiz et al. 1999) suggest some kind of connection between Central America and the Caribbean. Therefore, Jatropha was likely brought from Mexico, through Veracruz to Cape Verde by slave ship voyages or from the Caribbean Islands where the plant had been brought by Amerindians via sea routes in pre-Columbian times (Fig. 15.1).

15.5 Reasons for Low Genetic Diversity of Jatropha in Places Other Than the Center of Origin

15.5.1 Medicinal Use

Jatropha has spread widely in the tropics, in particular because of its medicinal characteristics

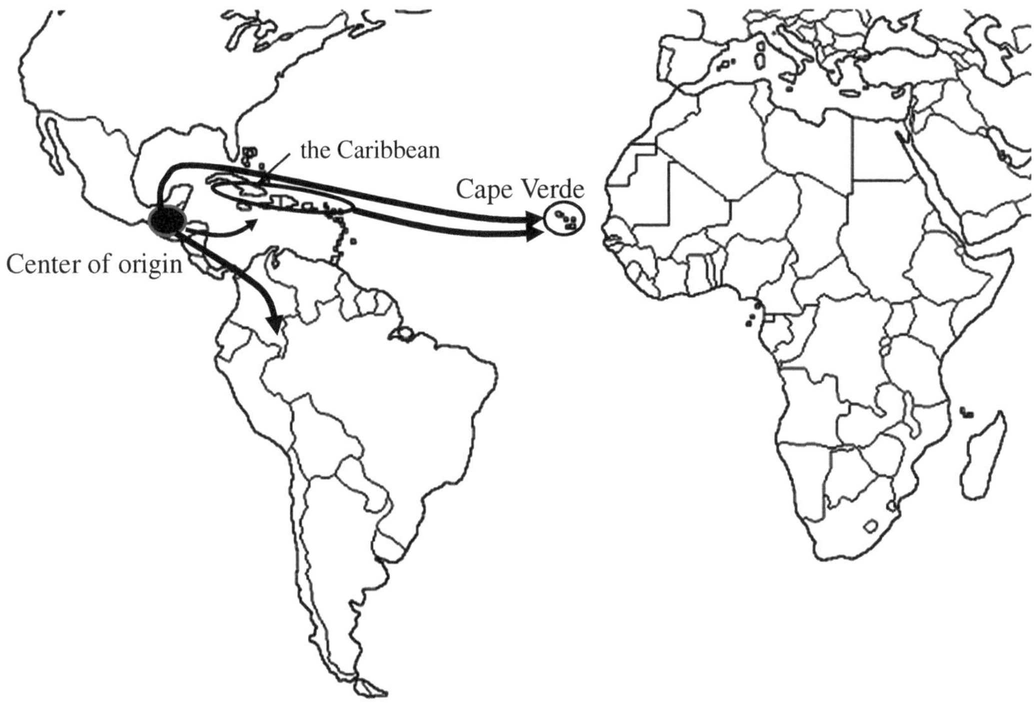

Fig. 15.1 Proposed dispersal route of *J. curcas* L. from the center of origin. *Arrows* show the presumed dispersal routes. One of the dispersal route from the center of origin, Chiapas State of Mexico to the Cape Verde Islands, Jatropha was likely brought by slave ships through Veracruz after the Exploration Age. The other route was from the Caribbean to Cape Verde by Portuguese seafarers as suggested by Heller (1996); however, Jatropha was likely introduced from somewhere to the Caribbean Islands via sea routes by Amerindians in pre-Columbian times. Jatropha might have spread from Mexico to Central America then South America not only overland but also via sea

(FAO 1986). Jatropha has been used for its leaves, bark, roots, seeds, and latex as a whole as an important medicine for animal and humans (Dias et al. 2012). Thus, particular Jatropha traits have not been selected such as high seed yield or high oil content.

15.5.2 Live Fence

Jatropha is widely cultivated in the tropics as a live fence in fields and settlements. When the first European explores arrived, Jatropha was planted around houses in the West Indies and Mexico (Little et al. 1974) and Guatemala (Anzueto and MacVean 2000). This is mainly because it can easily be propagated by cuttings and is not browsed by cattle (Heller 1996). The reason for the genetic similarity of the non-Mexican accessions of Jatropha is likely because a limited amount of germplasm was dispersed to other areas as medicinal plants and propagated vegetatively to build live hedges through cuttings (Basha et al. 2009), which makes limits genetic diversification.

15.5.3 Breeding from a Narrow Germplasm Base

According to Heller (1996), Jatropha was probably spread by Portuguese seafarers from the Caribbean via the Cape Verde Islands and former Portuguese Guinea (now Guinea Bissau) to other countries in Africa and Asia. This suggests some limited amount of Jatropha germplasm was brought to Cape Verde or Guinea Bissau, then some taken to Africa and Asia, thus genetic diversity was presumed to have further reduced.

15.6 Collection of Genetically Divergent Materials

The main barrier to elevating Jatropha from a wild species to a profitable biodiesel crop is low genetic and phenotypic variation found in different regions of the world, thus hampering efficient breeding for productivity traits (Osorio et al. 2014). In other words, the success of breeding programs lies in the identification of genetically divergent materials and development of genetically superior stocks (Basha and Sujatha 2007). Pecina-Quintero et al. (2014) concluded that Chiapas State of Mexico is the most likely center of origin of Jatropha, whereas the states which are closer to the Gulf of Mexico watershed, such as Veracruz, Puebla, Hidalgo, and Yucatan, are the likely sites of domestication. Thus, there is high genetic diversity of Jatropha in Mexico as compared to the genetic resources from outside Mexico (Pamidimarri and Reddy 2014).

On the other hand, Dias et al. (2012) proposed the Northern region of Minas Gerais State in Brazil as a secondary center of diversity of Jatropha, where it presents with a high oil content. This example suggests that high genetic diversity of Jatropha would exist even outside the center of origin. Genetic diversity in crops is the result of an interaction between genetic makeup of the plants and not just environmental (e.g., climate and soil) and biotic (e.g., relatives and pests), but also human factors (Guarino 1995). In other words, in order to increase genetic diversity of Jatropha, it is essential to replicate seed propagation under different environment conditions over a long period of time.

According to the United States Department of Agriculture (2015), Jatropha seeds seem unlikely dispersed by wind, water, birds, or animals directly; however, the report suggested the possibility that Jatropha seeds were dispersed by adhering to bodies of animals. Whereas Rivera-Lorca and Ku-Vera (1997) reported that the white-winged dove (*Zenaida asiatica*) usually feed on Jatropha seeds in Chiapas State of Mexico. Further the United States Department of Agriculture (2015) reported that doves *Turtur leucopterus* and *Chaemaepelia passerine* reportedly eat Jatropha seeds in Jamaica citing earlier research by Gosse in 1847 and Ridley in 1930. On the other hand, the same report citing Ridley expressed doubt whether seeds passed through the bird's digestive tract could maintain germination viability. However, when pigeons eat Jatropha seeds, Jatropha seeds potentially get scattered on the ground or spread around. Therefore, in this study, field survey on Jatropha growers was conducted in the Chiapas State of Mexico and it was confirmed that pigeons feed on Jatropha seeds. This suggests that seed dispersal by pigeons could be possible over a long period of time.

Taken together, in addition to the possibility of Jatropha dispersal from person to person as a medicinal plant, that of gradual spread by animals and pigeons was implied. This may affect to make genetic diversity of Jatropha high in some places such as Minas Gerais State of Brazil to be the secondary center of diversity. Collection of breeding materials for Jatropha is expected not only from center of origin and center of domestication in Mexico, but also from other places with high genetic diversity such as secondary centers of diversity under different environmental conditions.

References

Agra MF, Silva KN, Basílio IJLD, Freitas PF, Barbosa-Filho JM (2008) Survey of medicinal plants used in the region Northeast of Brazil. Revista Brasileira de Farmacognosia 18(3):472–508

Anzueto VAA, de MacVean ALE (2000) Los cercos vivos en Guatemala. Rvista, Universidad del Valle de Guatemala 9:12–18

Arellano-Rodoríguez JA, Flores-Guide JS, Tun-Garrido J, Cruz-Bojórquez MM (2003) Nomenclatura, forma de vida, uso, manejo y distribución de las especies vegetales de la Península de Yucatán. Universidad Autónoma de Yucatán, Facultad de Medicina y Zootecnia, Mérida, México

Arruda FP de, Beltrão NEM de, Andrade AP de, Pereira WE, Severino LS (2004) Cultivo de pinhão manso (*Jatropha curcas* L.) como alternativa para o semi-árido nordestino. Revista Brasileira de Oleaginosas e Fibrosas 8(1):789–799

Basha SD, Sujatha M (2007) Inter and intra-population variability of *Jatropha curcas* (L.) characterized by RAPD and ISSR markers and development of population-specific SCAR markers. Euphytica 156:375–386

Basha SD, Francis G, Makkar HPS, Becker K, Sujatha M (2009) A comprative study of biochemical traits and molecular markers for assessment of genetic relationships between *Jatropha curcas* L. germplasm from different countries. Plant Sci 176(6):812–823

Blanco P, Morales R, Oviedo R, Puig-Samper MÁ (1993) Plantas cubanas y documentos de la Ossa en el real Jardín Botánico de Madrid. Fontqueria 36:117–146

Cano JH, Volpato G (2004) Herbal mixtures in the traditional medicine of Eastern Cuba. J Ethnopharmacol 90:293–316

Comisión Nacional para el Conocimiento y Uso de la Biodiversidad (CONABIO) (2013) La biodiversidad en Chiapas: Estudio de Estado. Comisión Nacional para el Conocimiento y Uso de la Biodiversidad/ Gobierno del Estado de Chiapas. México

Dambatta SH, Aliyu BS (2011) A survey of major ethno medicinal plants of Kano North, Nigeria, their knoledge and uses by traditional healers. Bayero J Pure Appl Sci 4(2):28–34

Dias LAS, Missio RF, Dias DCFS (2012) Antiquity, botany, origin and domestication of *Jatropha curcas* (Euphorbiaceae), a plant species with potential for biodiesel production. Genet Mol Res 11(3):2719–2728

Duke JA (2009) Duke's handbook of medicinal plants of Latin America. CRC Press, Florida

Escalona-Gutiérrez E (2004) Lengua y Cultura: La clasificación botánica totonaca, Tesis de Licenciatura en Lingüística, Escuela Nacional de Antropología e Historia. Instituto Nacional de Antropología e Historia. México, D.F

Escobar N (1987) El desarrollo de las ciencias naturales y la medicina en Panamá: panorama histórico y antología. Universidad de Panamá, Panamá

FAO Foresty Department (1986) Some medicinal forest plants of Africa and Latin America. FAO, Rome

Freitas ASB de (1906) A purgueira e o seu oleo. Instituto de Agronomia e Veterinaria, Lisboa

García-Ramos C (2007) Diccionario: Totonaco—Español Español—Totonaco. Secretaría de Educación de Veracruz, Xalapa, Veracruz, México

German L, Schoneveld G, Skutch M, Andriani R, Obidzinski K, Pacheco P (2010) The local social and environmental impacts of biofuel feedstock expansion: A synthesis of case studies from Asia. Africa and Latin America, CIFOR, Indonesia

German L, Schoneveld GC, Pacheco P (2011) Local social and environmental impacts of biofuels global comparative assessment and implications for governance. Ecol Soc 16(4):29

Gómez-Pompa et al (2009) La Xuta se come: Xuta. Piñón o Aishte, Universidad Veracruzana, Xalapa, Veracruz, Jatropha

GEXSI (2008) global market study on Jatropha—final report. GEXSI LLP, Berlin (http://tinyurl.com/cnyn44)

Global information hub on integrated medicine (Globinmed) (2015) http://www.globinmed.com/index.php?option=com_content&view=article&id=80849&Itemid=101. Accessed Dec 29, 2015

GRAIN (2013) Land grabbing for biofuels must stop: EU biofuel policies are displacing communities and starving the planet. (http://www.grain.org/article/entries/4653-land-grabbing-for-biofuels-must-stop)

Grandtner MM (2005) Elsevier's dictionary of trees, vol 1. Elsevier, Amsterdam

Grass M (2009) *Jatropha curcas* L visions and realities. J Agric Rural Develop Tropics Subtropics 110(1): 29–38

Guarino L (1995) Secondary sources on cultures and indigenous knowledge systems. In: Guarino L, Ramanatha Rao V, Reid R (eds) Collecting plant genetic diversity: technical guidelines. CAB International, Wallingford, pp 195–228

Hanelt P (2001) Mansfeld's encyclopedia of agricultural and horticultural crops. Springer, Berlin

Heller J (1996) Physic nut. *Jatropha curcas* L. Promoting the conservation and use of underutilized and neglected crops. 1. Institute of Plant Genetics and Crop Plant Research, Gatersleben/International Plant Genetic Resources Institute, Rome

Herbario CICY (2010 en adelante). Flora de la Península de Yucatán. http://www.cicy.mx/sitios/flora%20digital/. Accessed Dec 28, 2015

Howe HF, Smallwood J (1982) Ecology of seed dispersal. Annu Rev Ecol Syst 13:201–228

IICA (2010) Atlas de la agroenergía y los biocombustibles en las Américas: II Biodiésel/IICA, Programa Hemisférico enagroenergía y Biocombustibles. IICA, San José

Infopédia (2016) Dicionários porto editora, http://www.infopedia.pt/dicionarios/lingua-portuguesa/purgueira. Accessed March 27, 2016

Jongschaap REE, Corre WJ, Bindraban PS, Brandenburg WA (2007) Claims and Facts on *Jatropha curcas* L.: Global Jatropha curcas evaluation, breeding and propagation programme. Plant Research International B.V., Report, Wageningen

Little EL Jr, Woodbury RO, Wadsworth FH (1974) Trees of Puerto Rico and the Virgin Islands, vol 2. Department of Agriculture, Washington

Mabberley DJ (2008) Mabberley's Plant-book: a portable dictionary of plants, their classifications, and uses. Cambridge University Press

Madaleno IM (2007) Etno-farmacología en Iberoamérica, una alternativa a la globalización de las prácticas de cura. Cuadernos Geográficos 41:61–95

Makkar HPS, Becker K (1997) Potential of *Jatropha* seed meal as a protein supplement to livestock feed and constraint to its utilization. In: Proceedings of Jatropha 97 international symposium biofuel and industrial products from *Jatropha curcas* and other tropical oil seed plants. Nicaragua, 23–27 February 1997

Martínez-Alfaro MA (2001) Catalago de plantas utiles de la sierra norte de puebla, mexico, Cuadernos 27. Universidad Nacional Autónoma de México (UNAM), México, D. F, Instituto de Biología

Martinez-Herrera J, Jimenez-Martinez C, Guemes-Vera N (2012) Use of *Jatropha curcas* L. (Non-Toxic Variety) as traditional food and generation of new products in Mexico. In: Carels N, Sujatha M, Bajadur B (eds) Jatropha, challenges for a new energy crop. Springer, Berlin, pp 333–341

Negussie A et al (2013) Invasiveness risk of the tropical biofuel crop Jatropha curcas L. into adjacent land use systems: from the rumors to the experimental facts. GCB Bioenerg 5:419–430

Nuñez-Melendez E (1975) Plantas medicinales de Costa Rica y su folclore. Universidad de Costa Rica, San José

Ocampo R, Balick MJ (2009) Plants of semillas sagradas: an ethnomedicinal garden in Costa Rica. Finca Luna Nueva Extractos de Costa Rica, S.A

Osorio LRM et al (2014) High level of molecular and phenotypic biodiversity in *Jatropha curcas* from Central America compared to Africa, Asia and South America. BMC Plant Biol 14:77

Pamidimarri DVNS, Reddy MP (2014) Phylogeography and molecular diversity analysis of *Jatropha curcas* L. and the dispersal route revealed by RAPD, AFLP and nrDNA-ITS analysis. Mol Biol Rep 41:3225–3234

Pecina-Quintero V et al (2014) Genetic structure of *Jatropha curcas* L. in Mexico and probable centre of origin. Biomass Bioenerg 60:147–155

Rengifo E, Pinedo M, Cerruti T (1997) Plantas Medicinales de la Amazonía Peruana: Estudio de su uso y cultivo. Instituto de Investigaciones de la Amazonía Peruana (IIAP)

Rivera-Lorca JA, Ku-Vera JC (1997) 1.7 chemical composition of three different varieties of *J. curcas* from Mexico. Biofuels and Industrial Products from *Jatropha curcas*. In: Proceedings of the symposium "Jatropha 97". Managua, Nicaragua. February. pp. 23–27

Robbins M (2011) Fuelling politics. Nature 474(23):22–24

Rodolfo S (2001) Manejo de semillas de 75 especies forestales de América Latina. CATIE, Turrialba

Roig y Mesa JT (1928) Diccionario botanico de nombres vulgares cubanos, Santiago de las Vegas, Habana

Ross IA (2003) Medicinal plants of the world volume 1 chemical constitutents traditional and modern medicinal uses. New York: Humana press

Ruiz CAA, Yermenos LYS, Torres MIO, Mañon D (1999) Estudio de seis plantas medicinales Dominicanas. Acta Medicia Dominicana 21(3):86–93

Sanderson K (2009) Wonder weed plans fail to flourish. Nature 461:328–329

Schmook B, Serralta-Peraza L (1997) *J. curcas*: distribution and uses in the yucatan peninsula of Mexico. In: Proceedings of the symposium "Jatropha 97". Managua, Nicaragua, February. p. 23–27, 1997

Spencer R, Cross R, Lumley P (2007) Plant names: a guide to botanical nomenclature. Royal Botanic Gardens, Melbourne

Thomas H (1997) The slave trade: the story of the atlantic slave trade: 1440–1870. Simon & Schuster Paperbacks, New York

Thomas E and Vandebroek I (2006) Guía de Plantas Medicinales de los Yuracarés y Trinitarios del Territorio. Indígena Parque Nacional Isiboro-Sécure, Bolivia. Santa Cruz: Imprenta Sirena

UCR (University of California, Riverside) (2014) Arboles Tropicales Comunes del Area Maya database. http://www.herbarium.ucr.edu/Arboles.html. Accessed December 30, 2014

UNAM (2015) Biblioteca Digital de la Medicina Tradicional Mexicana. http://www.medicinatradicionalmexicana.unam.mx/index.php. Accessed December 27, 2015

United States Department of Agriculture (2015) Weed risk assessment for *Jatropha curcas* L. (Euphorbiaceae)—Physic nut. Department of Agriculture: Animal and Plant Health Inspection Service, United States

Wolters B (2001) Dissemination of American economic plants on Precolumbian sea routes by Amerindians. Migr Diffus 1(7):40–58

Zamarripa-Colmenero CA, Pecina QV, Avendaño ACH, Solís BJL, Martínez VBB (2010) Genetic diversity of Mexican germplasm collection of *Jatropha curcas* L. In: Proceedings 18th European Biomass Conference and Exhibition 2010. Lyon. pp. 542–543

16 New Clonal Varieties of Jatropha

Zamarripa Colmenero Alfredo
and Víctor Pecina Quintero

Abstract

The objective of this chapter was to present the agronomic and industrial advantages of three new varieties of Jatropha that were developed in Mexico in order to meet the demand of the industry and the producers. It is discussed the importance of genetic resources and particularly of genetic diversity to develop new improved varieties. Based on agronomic and industrial characteristics, promissory genotypes were selected and variety trials were established in 4 tropical environments of Mexico to evaluate agronomic behavior. The main selection criteria used were grain yield, oil content, growth habit, and high presence of female flowers. It was found that the number of inflorescences in the plant and the number of female flowers are good predictor of yield because it was found positively correlated with the yield (0.83 and 0.78, respectively). In a population of 312 plants of various origins, plants with the presence of male and female flowers predominate with 75.5%, followed by flowering plants completely pistillate whose presence reaches 22.75% of the population, and plants with male flowers and hermaphrodites are 1.75%. Jatropha exhibits great variation in yield over the years presenting various types of behavior. It can be observed that there are genotypes with early and sustained production as México-JC 59 and genotypes with late and sustained production as México-JC 41. Also can be noted the genotype México-JC 88 produced late and sustained production but with very low yield. The best genotypes of this clonal trials were two clones with 100% female flowers and one clone with predominant of male flowers, but also with the

Z.C. Alfredo (✉)
Breeder of Tropical Crops, Calle Mina no. 455,
Tuxtla Chico, Tuxtla Gutiérrez, Chiapas, Mexico
e-mail: zamarripaco.alfre@yahoo.com.mx

V.P. Quintero
INIFAP- Campo Experimental Bajío, Km. 6.5
Carretera Celaya, San Miguel de Allende,
Guanajuato, Mexico

S. Tsuchimoto (ed.), *The Jatropha Genome*, Compendium of Plant Genomes,
DOI 10.1007/978-3-319-49653-5_16

presence of female flowers. The females were registered with the names of "Gran Victoria" and "Doña Aurelia," and the clone with the biggest percentage of male flowers was denominated as "Don Rafael." Jatropha is a plant with a tendency to cross-pollination, that is the reason the new varieties should not be spread by seed, but it ought to be done by methods of vegetative reproduction such as cuttings or in vitro propagation.

16.1 Introduction

Earlier programs involving large-scale plantations of Jatropha, *Jatropha curcas,* in Mexico and other American countries such as Brazil and Nicaragua indicated that the crop productivity was very low and could not be profitable. It was considered that one of the causes was the lack of improved varieties (Zamarripa and Solís 2013a).

Plant breeding is an efficient and sustainable economic strategy for the solution of agronomic problems, such as low production, and biotic or abiotic limitations, such as diseases or drought. The genetic improvement is the art and science which allows to change and improve the inheritance of plants (Milton 1981). According to Demarly (1977), it is the search for better ways to elaborate a genetic structure adapted to the criteria and need of mankind from an imperfect construction.

The breeding allows increasing the unit performance and the quality of the product to reach the competitiveness in the market. Nevertheless, the genetic improvement would be successful only if we begin the selection process with high genetic variability that makes it possible to identify genotypes and genes regulating the characters of agronomic and industrial interest. The programs of genetic improvement will progress when the available plant material presents genetic variability.

In this context, this chapter will talk about the new varieties of Jatropha formed from the high diversity existing in the genetic resources of the state of Chiapas, Mexico.

16.2 Genetic Resources

According to the International Plant Genetic Resources Institute (IPGRI), the Plant Genetic Resources (PGR) "represents the living matter that can procreate sexually or asexually. It has an actual or potential value for the nourishment and agriculture, and could be primitive, obsolete or modern cultivars; populations in process of plant breeding, wild populations and related species."

The germplasm collections are containers of PGR that provide the primary matter for crop improvement. This resource accomplishes the vital function in the sustainable and competitive development of the agriculture, as long as it helps to increase the production. The PGR kept and preserved in the germplasm collections is a vital and irreplaceable resource. Also, it is an inheritance that ought to be preserved to have options to face the actual and future problems of agriculture, nourishment, and environment in a globalized world against the climate change and many others challenges.

The value of the genetic resources of germplasm collections of Jatropha depends on the utilization of themselves to domesticate the species and to procure new species. The characterization and evaluation of the genetic resources have the purposes to find the differentiation of the accessions and to know their agronomic and industrial value. Also, it has the purpose to provide genetic variants or genotypes to the plant breeding method that gives the chance to fulfill the global demand of the oil production in the industry.

16.2.1 Characterization

The variation is the appearance of the differences between individuals, caused by the genetic composition or the environment where the differences between the progenies and progenitors were developed. The perceived variation of the plants depends on the interaction between the inherence and the environment. The genetic conformation determines an intrinsic variation of each individual that depends on its origin. The variation owing to the environment is independent of the origin of the individual and is not heritable (Demarly 1977; Hervé 1989).

The variation that is shown in a plant population could be genetic or could be environmental. The appreciation of the relative importance of the different genetic effects presents a big importance in the selection, especially in the estimation of the part about additive effects. In this matter, the particular genetic association called genotype that constitutes every individual will be destroyed at the moment of its reproduction and every relation of interaction will be disordered. Each individual transmits to each one of its descendants only half of its genes. In that way, the transmissible genetic effects with certainty are the additive effects; for this reason, the additivity helps to define the heritability, that is to say, the probability of transmitting a character or even the portion of superiority of an individual that could be transmitted to their descendants. The characterization, defined as the description of the variation that exists in a germplasm collection in terms of morphological, biochemical, molecular, and agronomic characteristics, allows the differentiation of the established accessions in germplasm collections and the knowledge of the existent genetic diversity (Avendaño and Zamarripa 2012). This first step of knowledge hugely determines the success of future commercial crops, as Jatropha.

In Mexico, the "National Research Institute of Forestry, Agricultural and Livestock" (INIFAP) has 422 toxic and non-toxic accessions of Mexican Jatropha collected in the states of Chiapas, Veracruz, Oaxaca, Michoacán, Guerrero, Morelos, Jalisco, Guanajuato, Puebla, Tamaulipas, Colima, and Yucatán, which have formed the base to study the genetic diversity and for the procurement of elite materials that has been evaluated in different regions of the country. This collection has been established in a field in Tuxtla Chico, Chiapas, with an average temperature of 26 °C, altitude of 420 m.a.s.l., and an annual precipitation of 4700 mm (Zamarripa et al. 2012a).

The morphological characterization, that is to say, the characterization of the visually detectable variability, was executed with 90 accessions of the INIFAP's collection for the following characters: plant, leaf, flower, fruit, and seed. In accordance with the analysis of the main components, a morphological variation in the collection of Jatropha was found to be principally in the shape and weight of the fruit and the seeds (Zamarripa and Solís 2013b).

The existence of genetic variability in the germplasm founded in Chiapas has already been reported (Zamarripa et al. 2010; Pecina et al. 2011; Ovando et al. 2011). Zamarripa et al. (2012a) reported wide variation in oil content, protein content, and fatty acid composition in the INIFAP's germplasm collection and also pointed the presence of 12% of pistillate flowering plants in a large studied population. Pecina et al. (2011) showed that the analysis of the genetic relations, molecular variance, and diversity index (ID = 60%) confirmed a large genetic base in the germplasm of Jatropha from Chiapas, Mexico. In a recent study, a representative set of 175 accessions of Jatropha from nine central and southeastern Mexican states was used for diversity analysis by amplified fragment length polymorphism (AFLP) markers. This study suggested the possibility that this area is the center of origin for this species and that domestication took place in the states bordering the Gulf of Mexico (Pecina et al. 2014).

In a recent research of agronomic evaluation in over 300 plants (our unpublished results), the results in the fifth year of harvest indicated that the variation range was quite large, and significant differences were found in all the evaluated characteristics. In the research, we found that the maximum value recorded in number of female flowers was 1604 per plant, while that of male flowers reached 24,012 per plant. The variation in the length and width of the seed was found as reported by Zamarripa et al. (2010). The average weight of the seed oscillated from 0.3 to 1.2 g. The oil content was from 24 to 59%. For the toxicity variability, plants without phorbol ester, the toxic material, from the state of Puebla, and those with the high level of that (3.56 mg/g) were found.

16.3 Jatropha Selection and Breeding

16.3.1 Breeding Methods

Genetic improvement of Jatropha has been reviewed, relatively recently, in articles or books written by Divakara et al. (2010), Surwenshi et al. (2011), and Chikara et al. (2013). Among the most used methods for genetic improvement are the mass selection, recurrent selection, hybridization, and clonal selection.

Mass selection is the simplest and oldest method, without much technicality. It is often used to preserve plant populations. Its effectiveness depends on the rate of selection, the selection environment, and the heritage of the selected characters (Hervé 1989). Mass selection consists of the selection of reproductive individuals based on their individual characteristics, and bulk seeds are used to produce the next generation for genetic improvement. It can be effective to improve characters of the single-gene inheritance and especially those with high heritability. Mass selection has major drawbacks: lack of the pollen source, confusing the environmental effect, and reduced population size leading to inbreeding depression (Divakara et al. 2010).

Yi et al. (2014) reported a new Jatropha variety (JO S2) with improved seed productivity. JO S2 was originated from an open-pollinated seed population (GN) harvested from plus trees of accessions from Malaysia and Thailand. Systematic mass selections were conducted from the initial GN population in Singapore. Plus plants with vigorous growth, early flowering, natural branching, better productivity, and a better oil profile were selected for seed harvest. Open-pollinated seeds from multiple plus trees were harvested and mixed together to form the original breeder seeds of JO S2. This Jatropha variety produced up to 2.95 ton/ha of dry seeds in the first year and up to 4.25 ton/ha of dry seeds in the second year, much better than the local variety used as control.

Recurrent selection is a breeding procedure for increasing the frequency of desirable genes within a population while maintaining sufficient variability for continued selection. The methods generally rest on the appreciation of hereditary capacities of selected individuals based on observation of their offspring.

In Jatropha, 15 open-pollinated varieties have been developed using mass selection and recurrent selection methods in India (Divakara et al. 2010). Jatropha as a facultative cross-pollinated crop shows heterosis, particularly when inbred lines are used as parent (Divakara et al. 2010). This method has great advantages such as agronomic superiority due to the phenomenon of heterosis or hybrid vigor, homogeneity, facilitative cultivation operations, and the possibility of associating within a hybrid genotype, and characters of interest, such as disease or drought resistance, are most often separated into different populations. Currently, there are several breeding programs principally in México, Brazil, and Guatemala to carry out work for obtaining F_1 hybrids of Jatropha.

Clonal varieties are a set of plants that are derived from the same mother plant by vegetative propagation. The first step of the clonal selection is the constitution of a large germplasm collection. These collections allow us a first evaluation of a clone for features of interest such as yield, vigor, growth habits, resistance to diseases, and oil quality. The most interesting plants are multiplied as rooted cuttings and evaluated in clone trials together with standard varieties. The best clones selected in one country are not necessarily suitable for other countries due to the interaction with the environment. Clonal varieties have the property to be homogeneous in the genetic plan although they are generally heterozygotes, and homozygosity has no practical advantage. The clones have a strong potential variability. Heterozygosity of clones can be exploited by simple selfing or outcrossing. The variation is quickly revealed in the first generation after the crossing.

After the selection of the best clones, they can be used to produce new hybrid combinations, which can then used to select new promising clones. Given the ease of asexual propagation of Jatropha, this method is very interesting to be considered in programs of genetic improvement.

16.3.2 Plant Characteristics and Relationship with Yield

In order to know the association grade between some variables of the plant and the yield, it was necessary to study the correlation coefficient (coefficient's correlation) of the characteristics of the plant in inflorescences, flowers, fruit, and seed (Table 16.1). It was found that the number of inflorescences in the plant and the number of female flowers are good predictor of yield because it was found positively correlated with the yield (0.83 and 0.78, respectively). The total number of fruits, number of harvested fruits, weight of the fruit, total number of seeds, and weight of the seeds showed a high positive correlation with the yield of the grain (0.85, 0.98, 0.98, 0.98). It is important to mention that the total number of flowers of the plants and the total number of male flowers presented a very low correlation with the yield (0.19 and 0.15, respectively). Also, it is important to point out that the number of fruits per cluster did not show any significant correlation with the yield (0.08).

16.3.3 Jatropha Flowering

Given that the number of female flowers was positively correlated with grain yield, but the total number of flowers was not, further studies on flowering of Jatropha were made. In a population of 312 plants of various origins, plants with the presence of male and female flowers predominate with 75.5%, followed by flowering plants completely pistillate whose presence reaches 22.75% of the population, and plants with male flowers and hermaphrodites are 1.75% (Table 16.2). In the population of plants with male and female flowers, the ratio was 1:22. The number of plants with female flowers increased from 18% in the first year of study to 25% in the fourth year, while plants with male and female

Table 16.1 Correlation coefficients between 10 plant characteristics and grain yield of *J. curcas* in Chiapas, México

	NI	TNF	NMF	NFF	NFPC	TFN	NFH	FW	SN	SW	YIELD
NI	1										
TNF	0.3603	1									
NMF	0.3124	0.9986	1								
NFF	0.8089	−0.0097	−0.0573	1							
NFPC	−0.1059	−0.2882	−0.3015	0.259	1						
TFN	0.9115	0.1679	0.1158	0.9239	0.2594	1					
NFH	0.8492	0.2117	0.1658	0.7818	0.0932	0.873	1				
FW	0.8328	0.1822	0.1371	0.7967	0.0957	0.8618	0.9838	1			
SN	0.8219	0.1843	0.1392	0.7604	0.1042	0.8561	0.9923	0.9765	1		
SW	0.8336	0.1956	0.1506	0.7877	0.0892	0.8592	0.9826	0.9887	0.9818	1	
YIELD	0.8336	0.1956	0.1506	0.7877	0.0892	0.8592	0.9826	0.9887	0.9818	1.0000	1

NI inflorescences number, *TNF* total number of flowers, *NMF* number of male flowers, *NFF* number of female flowers, *NFPC* number of fruits per cluster, *TFN* total fruit number, *NFH* number of fruits harvested, *FW* fruit weight, *SN* seeds number, *SW* seed weight, *YIELD* dry grain yield

Table 16.2 Flower type in a population of *J. curcas* through four cycles

Type of flowers according to sex (%)			
Year	Female	Male − female	Male − hermaphrodite
1	18	81	1
2	22	76	2
3	26	72	2
4	25	73	2
Average	22.75	75.5	1.75

Table 16.3 Male and female flowers numbers in 14 Mexican genotypes of *J. curcas* in four production cycles

Genotype	Number of male flowers per plant				Number of female flowers per plant			
	Year 1	Year 2	Year 3	Year 4	Year 1	Year 2	Year 3	Year 4
MÉXICO-JC 1	0	0	0	0	405	218	199	551
MÉXICO-JC 2	0	0	0	0	619	963	591	1166
MÉXICO-JC 3	0	0	0	0	217	661	583	958
MÉXICO-JC 4	0	0	0	0	737	1082	855	1176
MÉXICO-JC 5	0	0	0	0	573	226	870	973
MÉXICO-JC 6	0	0	0	0	408	1064	936	996
MÉXICO-JC 7	701	325	0	0	1050	573	220	182
MÉXICO-JC 8	939	178	0	0	746	158	229	143
MÉXICO-JC 9	880	409	0	0	321	453	520	403
MÉXICO-JC 10	11,077	16,864	16,799	24,012	485	779	774	1301
MÉXICO-JC 11	9299	17,883	11,112	7476	459	890	490	396
MÉXICO-JC 12	6058	9960	10,684	12,326	364	398	553	472
MÉXICO-JC 13	5949	12,718	7223	10,698	429	525	399	577
MÉXICO-JC 14	4624	7208	9717	7348	307	309	423	410

flowers decreased from 81% to 73%. In this population, only two plants were found to have male, and hermaphrodite flowers.

Table 16.3 gives number of male and female flowers of 14 genotypes during 4 years of the production. We found that six genotypes had completely female flowers from the first year of the production and remained stable with this feature during 4 years.

Genotypes such as México-JC 7, México-JC 8, and México-JC 9 showed changes in the type of flower production, presenting male and female flowers in the first 2 years, but from the third year the plant produced only female flowers as shown in Table 16.3. This behavior has continued into the seventh year of the production (data not shown). Note that despite the change of sex in the flowers, the number of female flowers did not increase, and therefore, grain yield did not increase.

The genotypes such as Mexico-JC 10, México-JC 11, México-JC 12, México-JC 13, and México-JC 14 showed the presence of male and female flowers during the 4 years. The genotype Mexico-JC 10 was highlighted for a lot of male and female flowers. In the fourth year, its total production was more than 25 thousand flowers.

Table 16.4 provides the ratio between male and female flowers of 14 Mexican genotypes. It may be noted that the first six genotypes with only female flowers are unchanged throughout the 4 years. Likewise, genotypes such as

Table 16.4 Ratio between female and male flowers in 14 Mexican genotypes of *J. curcas* in four production cycles

Genotype	Ratio between female and male flowers			
	Year 1	Year 2	Year 3	Year 4
MÉXICO-JC 1	100:0	100:0	100:0	100:0
MÉXICO-JC 2	100:0	100:0	100:0	100:0
MÉXICO-JC 3	100:0	100:0	100:0	100:0
MÉXICO-JC 4	100:0	100:0	100:0	100:0
MÉXICO-JC 5	100:0	100:0	100:0	100:0
MÉXICO-JC 6	100:0	100:0	100:0	100:0
MÉXICO-JC 7	1:0.7	1:0.6	100:0	100:0
MÉXICO-JC 8	1:1.2	1:1.1	100:0	100:0
MÉXICO-JC 9	1:2.7	1:0.9	100:0	100:0
MÉXICO-JC 10	1:23	1:22	1:22	1.18
MÉXICO-JC 11	1:20	1:20	1:23	1.19
MÉXICO-JC 12	1:17	1:25	1:19	1:26
MÉXICO-JC 13	1:14	1:24	1:18	1:19
MÉXICO-JC 14	1:15	1:23	1:23	1.18

Table 16.5 Flower numbers in four production cycles of two genotypes with male, female and hermaphrodite flowers

Genotype	Total number of flowers	Number of male flowers	Number of female flowers	Number of hermaphrodite flowers
MÉXICO-JC 100	12,001 (±4764)	11,442 (±4543)	21 (±3)	538 (±224)
MÉXICO-JC 200	6947 (±1498)	6586 (±1422)	283 (±97)	78 (±49)

México-JC 10, México-JC 11, México-JC 12, México-JC 13, and México-JC 14 remain without major changes during the four cycles. Only in genotypes such as México-JC 7, México-JC 8, and México-JC 9, plant sex changes to become only female flowers. It is worth noting that in these genotypes, the ratio of male and female flowers was close, between 1:0.6 and 1:2.7.

Table 16.5 shows the average of flower numbers in four production cycles of two genotypes with male, female, and hermaphrodite flowers. México-JC 100 genotype presented a total of 12,001 flowers per plant, 95.3% being male flowers, 4.5% hermaphrodite, and 0.2% female.

16.3.4 Yield Behavior Over Time

In order to understand the behavior in performance over time, the grain yield of genotypes was recorded for 5 years as shown in Fig. 16.1. These genotypes were grown with the same agronomic management and under the same environmental conditions. It can be observed that there are genotypes with early and sustained production as México-JC 59 and genotypes with late and sustained production as México-JC 41. Also can be noted the genotype México-JC 88 produced late and sustained production but with very low yield.

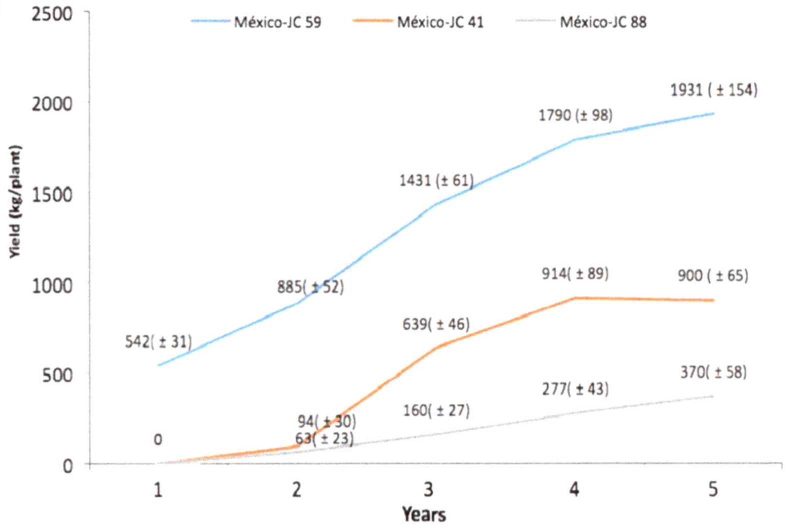

Fig. 16.1 Yield behavior of 3 genotypes of *J. curcas* during 5 years

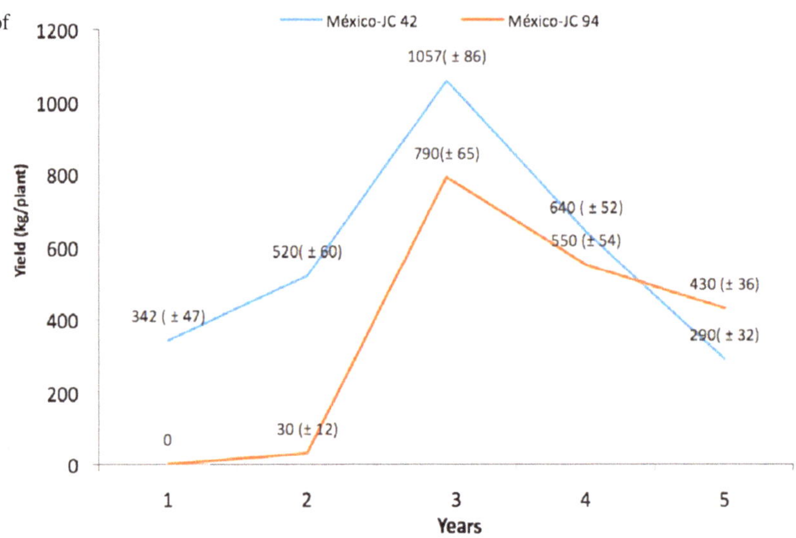

Fig. 16.2 Yield behavior of 2 genotypes of *J. curcas* during 5 years

Furthermore genotypes were found with increasing yield until the third year but with a discontinuous production. In these genotypes, the yields fell from fourth year as shown in Fig. 16.2. The genotype México-JC 42 produced a yield of 1.05 kg per plant at the third year; thereafter, the yield fell to 0.64 kg per plant in the fourth year and continued to decrease until 0.29 kg per plant in the fifth year. Generally, the performance of these genotypes remained low in the subsequent years.

Jatropha exhibits great variation in yield over the years presenting various types of behavior. However, genotypes were found with early and sustained production, and these genotypes were good candidate for multilocation trials.

16.4 New Varieties of Jatropha

The researches made in the south of Mexico by the INIFAP's bioenergy program led to the selection of three varieties of Jatropha with excellent adaptation to the dry tropic climate, high grain yield, high oil content, and good oil quality. The denominated varieties "Doña

Aurelia," "Don Rafael," and "Gran Victoria" were officially registered for its commercialization in September 2014 in the "National Catalog of Plant Varieties" (CNVV) of the National Service Seed Inspection and Certification (SNICS) belonging to Secretariat of Agriculture, Livestock, Rural Development, Fisheries and Food. The making of the varietal description of the new varieties was based on the "Technical guide for varietal description of *Jatropha*" published by SNICS (2014).

The breeding method used to obtain the varieties was clonal selection. This method has been successfully used in many species as coffee, cocoa, sugarcane, mandioca (Montagnon 2000; Resende and Barbosa 2005; Cueto et al. 2007). The varietal selection began with the morphological, biochemical, and genetic characterization of over 1600 plants of Jatropha established in the national germplasm collection from nine states of the southern Mexico (Zamarripa et al. 2012b).

Based on agronomic and industrial characteristics, promissory genotypes were selected and variety trials were established in four tropical environments of Mexico, to evaluate the agronomic behavior. The main selection criteria used were grain yield, oil content, growth habit, and high presence of female flowers.

The best genotypes of these clonal trials were two clones with 100% female flowers and one clone with predominant of male flowers, but also with the presence of female flowers. This is why the females were registered with the names of "Gran Victoria" and "Doña Aurelia," and the clone with the biggest percentage of male flowers was denominated as "Don Rafael." In the following are described the main characteristics of each variety.

16.4.1 Doña Aurelia Variety

It is an early variety of medium vigor, with an abundant ramification, bearing an average of 61 branches per plant. An important characteristic is inflorescences with completely pistillate flowers that remain throughout the year. Doña Aurelia produces semi-compact bunches with medium-sized elliptic fruits. The seed is medium-sized, with an average weight of 0.83 g. Table 16.6 shows some of the Doña Aurelia variety's characteristics that made based on the "Technical guide for the varietal description of *Jatropha*" elaborated by the SNICS of Mexico (SNICS 2014).

Table 16.6 Description of the Doña Aurelia variety according to descriptors of the technical guide for the varietal description of *Jatropha*

Characteristics	Description
Plant: habit	Open
Leaf blade: length/width ratio	Medium
Leaf blade: shape of base	Cordate
Leaf blade: shape of apex	Acuminate
Leaf blade: lobes	Medium
Petiole: length	Long
Inflorescence: male/female flowers ratio	Completely female
Female flowers: length/width of petal ratio	Big
Fruit: thickness of pericarp	Medium
Fruit: length/width ratio	Medium
Fruit: shape	Elliptic
Fruit: length of peduncle	Medium
Seed: shape	Elliptic
Seed: length/width ratio	Medium
Endocarp: intensity brown	Dark
Endocarp: rugosity	Medium

This variety has been proved to be stable with regard to its characteristics through years of study, and to produce an average yield in good environments of 0.49 tons/ha in the first year of the production. It reached 3 tons of dry grain per hectare in the third year of the production. The average oil content in the variety's seeds is 53.4%. Table 16.7 provides information about the physicochemical composition in seeds and oil of the Doña Aurelia variety.

As for the composition of the fatty acids, this variety was found to have a bigger portion of polyunsaturated fatty acids (39.75%) than

Table 16.7 Physicochemical composition in seeds and oil of the Doña Aurelia variety

Characteristics	Values
Seed weight (g)	0.83
Almond weight (g)	0.512
Shell weight (g)	0.302
Humidity (%)	3.844
Almond (%)	62.842
Head seed (%)	37.158
Oil content (%)	53.44
Saturated fatty acids (%)	35.36
Monounsaturated fatty acids (%)	24.89
Polyunsaturated fatty acids (%)	39.75
Oleic acid (%)	30
Linoleic acid (%)	32
Iodine ($gI_2/100$)	98.571
Density 15 °C (g/cm^3)	0.921
Viscosity 40 °C (mm^2/s)	34.894

saturated ones (35.36%). The smaller portion of fatty acids was the monounsaturated (24.89%). The main fatty acids in the seed of the variety were the oleic acid with content of 30% and linoleic acid with 32% content.

16.4.2 Gran Victoria Variety

It is an early variety, with a large vegetative vigor and a very abundant ramification. It presents an average of 95 branches per plant. An important characteristic of this variety is that it produces inflorescences with completely pistillate flowers and it keeps them throughout the years. Grand Victoria produces very compact bunches with fruits of medium size. The seed is medium in size, with an average weight of 0.77 g. Table 16.8 shows some of the characteristics of the Grand Victoria variety realized based on the technical guide for the varietal description of Jatropha (SNICS 2014).

Table 16.8 Description of the Gran Victoria variety according to descriptors of the technical guide for the varietal description of *Jatropha*

Characteristics	Description
Plant: habit	Medium upright
Leaf blade: ratio length/width	Big
Leaf blade: shape of base	Cordate
Leaf blade: shape of apex	Apiculate
Leaf blade: lobes	Medium
Petiole: length	Long
Inflorescence: ratio male/female flowers	Completely female
Female flowers: ratio length/width of petal	Medium
Fruit: thickness of pericarp	Thick
Fruit: ratio length/width	Medium
Fruit: shape	Oblate
Fruit: length of stalk	Medium
Seed: shape	Obovate
Seed: ratio length/width	Medium
Endocarp: intensity brown	Medium
Endocarp: rugosity	Medium

The Grand Victoria variety has demonstrated stability in its characteristics through the years of evaluation and gives an average production in good environment of 0.60 tons/ha in the first year of field establishment. By the third year, it reaches a production of 3.6 tons of dry grain per hectare. The average oil content is 53.4%. Table 16.9 gives information about the physicochemical composition in seeds and oil of the variety.

As for the fatty acid composition, it was found to have a bigger portion of polyunsaturated fatty acids (43.94%) rather than monounsaturated ones (34.95%), and a low portion of saturated ones (20.92%). The main fatty acids in the seeds of the Grand Victoria variety were the oleic acid with a content of 33% and the linoleic acid with a content of 40%.

Table 16.9 Physicochemical composition in seeds and oil of the Gran Victoria variety

Characteristics	Values
Seed weight (g)	0.77
Almond weight (g)	0.49
Shell weight (g)	0.258
Humidity (%)	4.496
Almond (%)	65.530
Head seed (%)	34.470
Oil content (%)	53.43
Saturated fatty acids (%)	20.92
Monounsaturated fatty acids (%)	34.95
Polyunsaturated fatty acids (%)	43.94
Oleic acid (%)	33
Linoleic acid (%)	40
Iodine ($gI_2/100$)	109.16
Density 15 °C (g/cm^3)	0.921
Viscosity 40 °C (mm^2/s)	32.793

16.4.3 Don Rafael Variety

It is an early medium variety, with medium vegetative vigor, and a scarce ramification, producing an average of 45 branches per plant. It bears male and female flowers with dominating male flowers above the female on a 20:1 ratio and does not have a significant variation throughout the years. It produces a high quantity of male flowers, an average of 9700 flowers per plant. According to Bhattacharya (2005) who reported a production of 1600 grains of pollen per flower, and having the Don Rafael information, it is estimated to have a very high pollen production of 15.5 millions grains of pollen that puts this variety as a good pollinator progenitor for female plants. Don Rafael variety presents very compact clusters with fruits of an elliptic shape and big size. The seed is big in size and has an average weight of 0.82 g. The average content of oil is 51%. Table 16.10 shows characteristics of the Don Rafael variety based on the technical guide for the varietal description of Jatropha (SNICS 2014).

The Don Rafael variety has demonstrated stability in its characteristics through years of evaluation and produces an average production of 0.4 tons of dry grain per hectare. By the third year, it reaches a production of 1.7 tons of dry grain per hectare. Table 16.11 gives information about the physicochemical composition of the Don Rafael variety.

Table 16.10 Description of the Don Rafael variety according to descriptors of the technical guide for the varietal description of *Jatropha*

Characteristics	Description
Plant: habit	Upright
Leaf blade: length/width ratio	Big
Leaf blade: shape of base	Cordate
Leaf blade: shape of apex	Apiculate
Leaf blade: lobes	Medium
Petiole: length	Long
Inflorescence: male/female flowers ratio	Predominantly male
Female flowers: length/width of petal ratio	Big
Fruit: thickness of pericarp	Thick
Fruit: length/width ratio	Big
Fruit: shape	Elliptic
Fruit: length of peduncle	Long
Seed: shape	Elliptic
Seed: length/width ratio	Big
Endocarp: intensity brown	Dark
Endocarp: rugosity	Medium

According to the fatty acid composition, it was found to have a bigger portion of saturated ones (35.94%) rather than monounsaturated (29.10%) or polyunsaturated ones (34.95%). The main fatty acids in the seeds of Don Rafael variety are the oleic acid with a content of 22% and the linoleic acid with a content of 38%.

Figure 16.3 shows photographs of the varieties and also drawings of the structures in their adulthood. The three varieties show contrasting architectures. The most upright and tall variety is Don Rafael. Doña Aurelia presents an open and ungainly portage, with prostrate branches. The variety with the best architecture and bigger ramification is Grand Victoria, which also presents a bigger grain production.

Table 16.11 Physicochemical composition in seeds and oil of the Don Rafael variety

Characteristics	Values
Seed weight (g)	0.82
Almond weight (g)	0.537
Shell weight (g)	0.285
Humidity (%)	4.943
Almond (%)	65.379
Head seed (%)	34.621
Oil content (%)	51
Saturated fatty acids (%)	35.94
Monounsaturated fatty acids (%)	29.10
Polyunsaturated fatty acids (%)	34.95
Oleic acid (%)	22
Linoleic acid (%)	38
Iodine ($gI_2/100$)	105.390
Density 15 °C (g/cm^3)	0.920
Viscosity 40 °C (mm^2/s	33.922

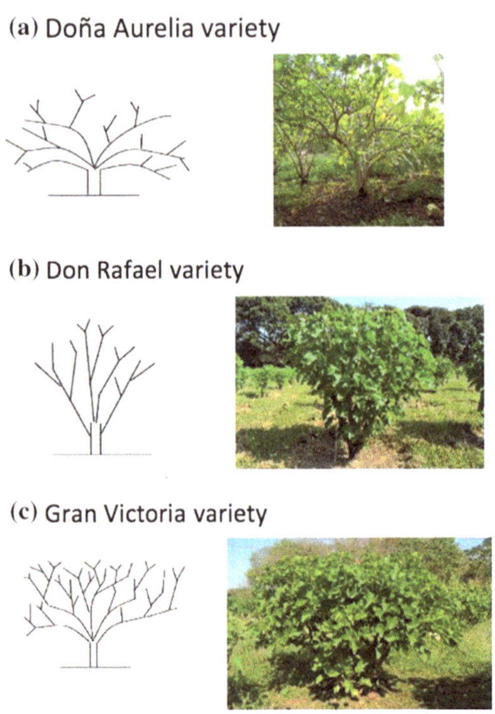

Fig. 16.3 Images show the structures of the new Mexican varieties of *Jatropha curcas* L.

The varieties must be established in the field as polyclonal varieties, and the following arrangement is suggested: 3:1:3, is, to say, 3 rows of Gran Victoria, a row of the pollinator Don Rafael, and 3 rows of Doña Aurelia.

16.5 Propagation of the New Varieties

The new clonal varieties of Jatropha with high fruit yield and a high oil production must be propagated and distributed to the producers by vegetative reproduction. Jatropha is a plant that can produce fruits through selfing and cross-pollination, but it has the tendency to cross-pollinate with rates higher than 76% (Qing et al. 2007), which is the reason why the propagation through seeds does not allow to keep the genetic information of the varieties and loses its characteristics of agronomic and industrial value (Zamarripa et al. 2009; Iracheta et al. 2015). The two ways of vegetative reproduction with more potential for the plant's production are propagation by cuttings from clonal gardens and the propagation by somatic embryogenesis in liquid medium.

To form clonal gardens, we obtain cuttings of branches from the mother plants of the improved varieties. The cuttings could be from 0.40 to 0.50 cm long and with a diameter higher than 3 cm in order to grow vigorously, and let the new buds emerge faster. Besides the factors such as the cutting date, size, and thickness of the rods, success of the cutting depends on the genotype. A research in Mexico demonstrated the effect of the variety in the rate to obtain new plantlets by finding values from 7 to 98% (Zamarripa and Solís 2013b). The varieties Doña Aurelia, Don Rafael, and Gran Victoria successfully produce the rates of 93, 98, and 95%, respectively, 5 weeks after the cutting.

Using this method, and starting from a clonal garden of 1 ha with 2000 plants, in 3 years we would obtain 5.6 millions of cuttings, and in 5 years, we would reach the production of 341 millions of cuttings. This would allow to establish over 200 thousand hectares of Jatropha and also to reach an oil production of approximately 250 thousand tons annually.

In order to establish new plantations from the new varieties of Jatropha, the most potential way of propagation that has the advantage of producing plants in big numbers with a low cost in a short span is the in vitro propagation. The plant tissue culture focused on the massive propagation of selected genotypes is a desirable goal (Foidl and Eder 1997; Sujatha et al. 2008). This will allow providing genetic material quickly without diseases in large numbers during any season of the year. In Mexico, INIFAP began the first research for the establishment and induction of in vitro morphogenesis of some elite genotypes in 2009 (Pérez et al. 2011; Martínez et al. 2011; González et al. 2011; Cruz et al. 2012). In 2015, multiplication of buds and embryos of the new Jatropha varieties was accomplished with high rates (Iracheta et al. 2015).

According to Iracheta et al. (2015), starting from 300 mg of embryogenic callus with the actual reached rate (127 vegetative structures per explant), the in vitro potential is higher than 90 million in each 6-month cycle. If we consider the 80% survival rate in the acclimatization phase of in vitro plants, the potential production would be higher than 73 million each 6 months. Using this propagation method along 2.5 years, it could produce 365 millions of seedlings to establish over 200 thousand hectares utilizing the new varieties.

If we compare both production methods, it is noticeable that we can reach an equal production of plantlets. Nevertheless, the in vitro propagation is faster than the propagation in clonal gardens, which takes half of the time (30 vs. 60 months). Another great advantage of the in vitro method is that the plant production is neither exposed to biotic factors such as pests and diseases, nor abiotic factors such as hurricanes and floods, which could put the massive production in danger.

References

Avendaño Arrazate CH, Zamarripa Colmenero A (2012) Guía gráfica de descriptores varietales de piñón mexicano (*Jatropha curcas* L.). Publicación especial No.1. INIFAP. Campo experimental Rosario Izapa, p 162

Bhattacharya A, Datta K, Datta SK (2005) Floral biology, floral resource constraints and pollination limitation in *Jatropha curcas* L. Pak J Biological Sci 8(3):456–460

Chikara J, Prakash A, Mastan SG, Ghosh A (2013) Genetic improvement in *Jatropha curcas* through selection and breeding. In: Bahadur B, Sujatha M, Carels N (eds) Jatropha, challenges for a new energy crop Volume 2: Genetic improvement and biotechnology, pp. 119–133. Springer, New York

Cruz T SA, L Iracheta D, CG González J, A Zamarripa C, P López G, Juárez MC (2012) Efecto de auxinas y citocininas en la inducción de callo embriogénico de *Jatropha curcas*. In: Memorias del VII Congreso Estatal de biotecnología. Tapachula, Chiapas, p 3

Cueto-Moreno J, Aguirre-Medina JF, Iracheta-Donjuan L, Zamarripa-Colmenero A, Olivera- De los Santos A, Grajales-Solís M (2007) El mejoramiento del cultivo de cacao (*Theobroma cacao* L.) en México. Libro Técnico No. 4. Instituto Nacional de Investigaciones Forestales, Agrícolas y Pecuarias. Campo Experimental Rosario Izapa. Tuxtla Chico, p 176. ISBN: 978-970-43-0207-8

Demarly Y (1977) Génetique et amelioration des plantes. París, Collection Sciencies Agronomiques, p 287

Divakara BN, Upadhyaya HD, Wani SP, Gowda CLL (2010) Biology and genetic improvement of *Jatropha curcas* L.: a review. Appl Energy 87:732–742

Foidl N, Eder P (1997) Agro-industrial Exploitation of *Jatropha curcas,* Linn. In: Gübitz GM, Mittelbach M, Trabi M (eds) Biofuels and industrial products from *Jatropha curcas*. Austria Dbv-Verlag, Graz

González J, Iracheta Donjuan L, A Zamarripa C, I Méndez L, BR Pérez P, JC Godínez A, SA Cruz T, D Keller M (2011) Cultivo in vitro de meristemos apicales de *Jatropha curcas* L. In: Memorias del XIV Congreso Nacional de la Sociedad Mexicana de Ciencias Hortícolas. Culiacán Sinaloa, p 272

Hervé Y (1989) Introduction a l' amelioration des plantes. Ecole Nationale Superieure Agronomique de Rennes. Ministére de l' Agriculture, p 79

Iracheta Donjuan L, Díaz Fuentes VH, Basulto Graniel J, González Jiménez A, Rico Ponce HR, López Gómez P (2015) Propagación in vitro del piñón mexicano (*Jatropha curcas* L.). INIFAP. Campo Experimental Rosario Izapa. Folleto técnico No. 33. Tuxtla Chico, p 40

Martínez PMA, Iracheta Donjuan L, Méndez L, Zamarripa C, Castellanos J, Cruz T, Gonzalez J, Mikery G

(2011) Influencia de citocininas en la brotación de estacas de piñon (*Jatropha curcas* L.) para la obtención de cultivos in vitro. In: Memorias del XIV Congreso Nacional de la Sociedad Mexicana de Ciencias Hortícolas, Culiacán Sinaloa, p 270

Milton PJ (1981) Mejoramiento genético de las cosechas. Limusa, p 453 Kindly check the inserted year of the Reference (Milton 1981) is appropriate.

Montagnon C (2000) Optimisation de gains genétiques dans le schema de sélection récurrente reicproque de *Coffea canephora* Pierre. Ph.D. Thesis. Ecole National Superieure Agronomique de Montpellier, France

Ovando-Medina FJ, Espinosa-García JS, Núñez-Farfan, Salvador-Figueroa M (2011) State of the art of genetic diversity research in *Jatropha curcas*. Sci Res Essays 6(8):1709–1719

Pecina-Quintero V, Anaya-López JL, Zamarripa-Colmenero A, Montes-García N, Núñez-Colín CA, Solís-Bonilla JL, Rangel MRA, Langarica HRG, Bustamante DJM (2011) Molecular characterisation of *Jatropha curcas* L. genetic resources from Chiapas. México through AFLP markers. Biomass Bioenergy 35:1897–1905

Pecina-Quintero V, Anaya-López JL, Zamarripa-Colmenero A, Núñez-Colín CA, Montes-García N, Solís-Bonilla JL, Jiménez-Becerril MF (2014) Genetic structure of *Jatropha curcas* L. in México and probable centre of origin. Biomass Bioenergy 60:147–155

Pérez SJR, Iracheta Donjuan L, Méndez L, Zamarripa C, Cruz T, González J, Mikery G (2011) Influencia de la fertilización de piñón (*Jatropha curcas*) en la contaminación y callogénesis de explantes foliares. In: Memorias del 6° Congreso Estatal de Biotecnología Chiapas, Ocosingo Chiapas, p 2

Qing Y, Ping PD, Biao DZ, Liang WZ, Xiang SQ (2007) Study on pollination biology of *Jatropha curcas* (Euphorbiaceae). J S China Agricult Univ 28(3): 62–66

Resende MDV, Barbosa MHP (2005) Melhoramento genético de plantas de propacao assexuada. Empresa Brasileira de Pesquisa Agropecuaria Embrapa Florestas. Ministerio da Agricultura, Pecuaria e do Abastecimento, Colombo, p 130

SNICS (2014) Guía técnica para la descripción varietal de jatropha (*Jatropha curcas* L.). Servicio Nacional de Inspección y Certificación de Semillas. Secretaría de Agricultura, Ganadería, Desarrollo Rural, Pesca y Alimentación, p 19

Sujatha M, Reddy TP, Mahasi MJ (2008) Role of biotechnological interventions in the improvement of castor (*Ricinus communis* L.) and *Jatropha curcas* L. Biotechnol Adv 26:424–435

Surwenshi A, Kumar V, Shanwad UK, Jalageri BR (2011) Critical review of diversity in *Jatropha curcas* for crop improvement: a candidate biodiesel crop. Res J Agric Sci 2(2):193–198

Yi C, Reddy C, Varghese K, Bui TNH, Zhang S, Kallath M, Kunjachen B, Ramachandran S, Hong Y (2014) A new *Jatropha curcas* variety (JO S2) with improved seed productivity. Sustainability 6:4355–4368

Zamarripa Colmenero A, Solís Bonilla JL (2013a) Estado del arte y novedades de la bioenergía en México. La bioenergía en América Latina y El Caribe. El estado de arte en países seleccionados. Oficina regional para América Latina y el Caribe. RLC. ONU. Santiago de Chile, Chile, pp 277–311

Zamarripa Colmenero A, Solís Bonilla JL (2013b) *Jatropha curcas* L. Alternativa bioenergética en México. Libro Científico No. 1. INIFAP. Campo experimental Rosario Izapa, México p 162

Zamarripa Colmenero A, Solís Bonilla JL, Martínez Valencia B, Pecina Quintero V (2012a) Agronomic and biochemical study of *Jatropha curcas* L. in México. In: BIT'S 2nd Annual world congress of bioenergy. Xi'an, p 115

Zamarripa Colmenero A, Solís Bonilla JL, Iracheta Donjuan L, Martinez Valencia B, Maldonado Méndez JJ (2012b) Mejoramiento de insumos agropecuarios para la producción de biocombustibles. Informe final. Instituto Nacional de Investigaciones forestales, Agrícolas y Pecuarias. Campo Experimental Rosario Izapa, México p. 28

Zamarripa Colmenero A, Pecina Quintero V, Avendaño Arrazate CH, Solís Bonilla JL, Martínez Valencia B (2010) Genetic diversity of mexican germplasm collection of *Jatropha curcas* L. In: 18 th European biomasa and exhibition, 3–7 May 2010 Lyon

Zamarripa CA, Ruiz CPA, Solís BJL, Martínez HJ, A. de los Santos AO, Martínez B (2009) Biocombustibles: Perspectiva de producción de biodiesel a partir de *Jatropha curcas* L., en el trópico de México. INIFAP. Folleto Técnico No.12, p 46

MIX
Papier aus verantwortungsvollen Quellen
Paper from responsible sources
FSC® C105338

If you have any concerns about our products,
you can contact us on
ProductSafety@springernature.com

In case Publisher is established outside the EU,
the EU authorized representative is:
Springer Nature Customer Service Center GmbH
Europaplatz 3, 69115 Heidelberg, Germany

Printed by Libri Plureos GmbH
in Hamburg, Germany